Science Confronts
the Paranormal

Science Confronts the Paranormal

BEST OF
SKEPTICAL INQUIRER
VOLUME 2

EDITED BY

KENDRICK FRAZIER

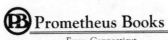
Prometheus Books
Essex, Connecticut

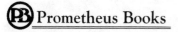Prometheus Books

An imprint of Globe Pequot, the trade division of
The Rowman & Littlefield Publishing Group, Inc.
4501 Forbes Blvd., Ste. 200
Lanham, MD 20706
www.rowman.com

Distributed by NATIONAL BOOK NETWORK

British Library Cataloguing in Publication Information Available

Library of Congress Cataloging-in-Publication Data Available

ISBN 9781633889644 (paper : alk. paper) | ISBN 9781633889651 (ebook)

♾™ The paper used in this publication meets the minimum requirements of American
National Standard for Information Sciences—Permanence of Paper for Printed Library
Materials, ANSI/NISO Z39.48-1992

Table of Contents

Part II: Evaluating Fringe Science

Introduction

Why examine the paranormal? Why should intelligent scholars, scientists, teachers, students, and laymen pay any attention at all to claims of psychic powers, ESP, clairvoyance, psychokinesis, or poltergeists—let alone the fringe-science subjects of astrology, UFOs, speculative archaeology, creationism, and cryptozoology? Even cursory consideration shows that the subjects are replete with exaggeration, deceit, fraud, misperception, self-delusion, and other prominent foibles of the human race. Shouldn't they perhaps just be ignored and automatically discounted?

Certainly many scientists and scholars do take this attitude. It is hard to blame them. In the narrowest sense, the scrutiny of paranormal claims offers little professional gain. Examination of the latest psychic fad that infatuates the public is unlikely to lead to scientific advancement or prestigious publication. Doing science is time-consuming and difficult, and there's little enough time to attend to one's own specialized research without worrying about investigating exotic claims that have little likelihood of validity and devoting time to disabusing the public about them. That view is understandable. But it is also an example of living in an ivory tower or, to mix metaphors and altitudes, of sticking one's head in the sand.

I will suggest at least four reasons responsible scholars, and their students, should devote some attention to examining the paranormal. The first I will consider at some length; the others, more briefly.

1. *The public is interested.* Indeed, large proportions of the public believe in paranormal powers. More than likely you have a close friend, relative, or next-door neighbor who has a strong interest and a more than mild belief in one or more aspects of the paranormal. Almost everyone has had some extraordinary and mystifying experience that may seem unexplainable except by resorting to the paranormal. Paranormal claims abound in the media. This is so not just in the notorious (if amusing) supermarket tabloids with their latest headlines about the woman who has been impregnated by a (choose one) UFO alien, Bigfoot monster, or roving horny spirit. Your own daily newspaper is very likely to carry at least one national or local story a week on some claim that comes under the loose rubric of paranormal or fringe-science beliefs. It almost certainly prints a

daily astrology column. (I doubt very much if it has a daily astronomy column.)

A widely syndicated Washington political columnist who won a Pulitzer Prize for his investigative acumen has devoted not one but five columns to the assertion that people in the Pentagon and the intelligence services are taking seriously claims of psychic warfare and spying by mental "remote-viewing." His own attitude toward these assertions seems to range from fun-poking to uncritical acceptance.

Magazines, books, and radio and TV talk-shows all have a full complement of psychics offering advice and making predictions. Serious magazines like *Forbes, Computerworld,* and the *Journal of Defense and Diplomacy* have recently published lengthy articles repeating claims of psychic powers without even a hint of critical scientific perspective. Reader's Digest Books and more recently Time-Life Books have heavily promoted and distributed books or book series on "mysteries of the unknown." The paranormal sells. It wouldn't sell if people weren't broadly interested and, to a large degree, accepting.

A 1984 Gallup poll of 506 American teenagers, aged 13 to 18, found these answers to the question, "Which of the following do you believe in?": angels, 69 percent; ESP, 59 percent; astrology, 55 percent; clairvoyance, 28 percent; Bigfoot, 24 percent; witchcraft, 22 percent; ghosts, 20 percent; Loch Ness monster, 18 percent. Such polls are decidedly unsatisfying. I always want to know the depth of the "belief" in question and on what assumptions it is based. But they do give a general idea of levels of paranormal belief, which despite fluctuations are always fairly high.

I divide the public's interest in the paranormal into two broad types: The first stems from the intrinsic appeal of the subject. It appeals to our curiosity, our sense of wonder, and the human need for fantasy and diversion. It can be fun. It is fascinating to consider whether some people have the power to read minds, see distant events, cause objects to move, or know the future. It is intriguing to ponder amazing-sounding coincidences. Who can deny that it is interesting to wonder if somewhere there are large unknown creatures roaming the earth's land or lakes or if we have ever been visited by extraterrestrial beings?

The second is the understandable human need for comforting beliefs. To some, the daily challenges of life may be a bit more bearable if they feel their destiny is shaped by cosmic forces (the positions of the planets) or other factors beyond their control (their date of birth). It can be an impersonal world, and the thought that the universe has some personal interest in and influence over one's destiny has broad appeal. The same can be said if there are people one can consult who seemingly have special insight into one's life and personality and special powers to know what's in store.

Note that one doesn't have to believe wholeheartedly in such matters to find some such comfort. It's enough to feel that, just perhaps, there's something to it—just a slight influencing of the odds in one's favor. What can it hurt? When a chronically or fatally ill relative has exhausted all medical help, who can blame the human impulse to consult unconventional practitioners and untested remedies that promise a cure? And who, really, relishes the thought of a loved one dying and for evermore being cut off from one's life? Spiritualism, the forerunner of much of modern parapsychology, was motivated by such universal concerns. Many other people find their fundamentalist religious views threatened by science's discoveries, e.g., that the earth is 4.6 billion years old, that all life slowly evolved over time, and that we ourselves are a result of the long process of evolution. For people whose belief structure would crumble if they had to accept that the story of Genesis is a beautiful parable rather than a literal historical account, adamant resistance to all competing notions is understandable, if regrettable.

Now we begin to see why the hold that paranormal systems of thought have had on human culture is so pervasive; why, even in a scientific and technological world, paranormal beliefs are no less evident than in earlier historical times; and why they will undoubtedly always be with us.

We can also begin to see why it is important to understand the effect of the broad public interest and belief in the paranormal on human thought and culture. Where interest is so high and the need to believe so great, the seeds for misinterpretation are sown. Psychologists have produced an abundant body of knowledge on how our own beliefs, biases, and preconceptions influence our perception. In a very real sense, we believe what we want to believe, and we see what we want to see. Our minds are wondrous mechanisms, capable of extraordinary feats. But, in filtering out all the available perceptual data except that most important to the task at hand, our minds, with our own subconscious help, can deceive us. We are always selecting what to focus attention on and, except in those infrequent times when we force ourselves to operate in the most analytical and objective mode, we have a way of searching for information that reinforces our beliefs and ignoring all that doesn't.

All this makes us very human. It also make us susceptible to error. Such distortions are generally unimportant, because in our daily life our perceptions are continually being compared with the real world and undergoing correction. We couldn't cross the street or drive a car if they weren't, let alone function in our jobs. But beliefs in the paranormal, because they operate at the level of deep-seated psychological needs, seem to be remarkably immune to the error-correcting processes of daily life. And thus many people go on believing in things that more objective inquiries may indicate either are probably not so or are far less likely to be so than is generally believed.

All this is fertile ground for scholarly inquiry. It leads us into discovering more about how our minds work. It allows us to understand better how the natural world does, and does not, operate. Hardly trivial questions! It also makes us struggle with the problems of the gap between what scientists (who *are* involved in a continual, organized, error-correcting process) know about the natural world and what the public thinks it knows. That gap has always existed and always will, but if it becomes too large science and society become excessively decoupled, to the eventual detriment of both. Scholars and others who cannot at least understand why people are attracted to the paranormal and don't know the arguments being made are not in a good position to counter the many misconceptions that arise. Nonscientific people whose aspirations, motives, and psychological needs fail to gain at least some degree of sympathetic hearing from scientists and scholars are going to be pushed toward pro-mystical positions antagonistic to science.

2. *Public education.* If an informed and rational citizenry is indeed important to a democracy (as I believe it is), then scientists have an obligation to help the public understand the difference between sense and nonsense, good science and bad science, scientific speculation and outright fantasy. They must help their students and the public sort out the many competing claims for truth in controversies of interest, including the paranormal. Scientists who wish to counter paranormal claims or to correct misinformation about them must become well informed. It is not enough to state positions. One must be armed with facts. And one must be prepared for attacks from unexpected directions. Sometimes these attacks are carried out with surprising cleverness and effectiveness. The scientist who expects his scientific knowledge is enough to carry the day in a public debate with a committed believer in a paranormal belief system will soon learn otherwise. The early years of the debates with creationists showed that one has to know what arguments to expect and what specific counter-arguments to marshal.

3. *Intellectual honesty.* Scientists and scholars are supposed to be engaged in discovering the truth, wherever it may be. To the scientific mind many of the claims of paranormal powers may seem highly unlikely, even outrageous, in light of well-accepted scientific principles. But scientists must never rule out a claim on that basis alone. All claims should be put to the test of evidence. Many paranormal claims have been tested and found wanting, but that doesn't rule them out; it only decreases further the probability of their being valid. The proper approach to an *un*tested claim is open-minded skepticism. Trained psychologists and natural scientists must be willing to devote some time to examining and testing claims on the fringes of science for several reasons. First, if they don't, persons with far fewer credentials for credibility will, and that allows more potential misinformation into the system. Second, there's always a chance they will find something important or, at least, scientifically interesting. More than likely it will not be what the proponents of paranormality expected or

wanted, but there is still a chance some novel properties of nature or of human capacity can be identified. If not that, at least we start to learn more about how the mind and perceptions work to make people *think* that paranormal powers are at play or, to take a typical fringe-science claim, that alien spacecraft have landed. Third, just plain intellectual honesty requires investigation before reaching conclusions. The investigation doesn't have to be first-person; no one can investigate everything. It can be part of the cumulative scientific effort by all manner of qualified specialists, who then publish the results for examination by others. But, if a number of responsible, well-trained scholars don't devote some time to examining claims, then science stands vulnerable to accusations of being closed-minded. That is not a good position to be in.

4. *Opportunity to teach real science.* The appeal of the paranormal provides a wonderful opportunity to teach real science. The natural fascination people have with the paranormal and with astrology, UFOs, and the like can be converted into a curious audience willing to hear about the science involved. Astronomers and physicists can use questions about astrology to pull students into discussions of the cultural basis of constellations, the problem of precession, the scale of the solar system and the universe, the principles of gravitation and electromagnetism, and virtually any other topic in astronomy and physics. They can do the same thing with UFOs. Psychologists can use case studies about psychic claims to make any number of points about human perception, the limitations of observation, the flexibility of the mind, and so on. The interest in fringe medicine can be used as a springboard to discussions of the mind/body problem, the placebo effect, and human physiology. Questions raised about creationism can lead into the most central topics of biology, geology, and astronomy.

Carl Sagan has said that the wonders of real science far surpass the supposed and imagined mysteries of fringe science, and I agree. Scientists have an opportunity to show that science deals with awesome mysteries and concepts, from the mind-boggling information content of human DNA to the physics of the first 10^{-30} second of the existence of the universe. It's a fascinating world, and science is always on the frontier looking out into the unknown. People want to share in that adventure and experience. And they will, with just a bit of encouragement and guidance. Then they are with you, learning and exploring together, in a partnership of science.

In 1976, the Committee for the Scientific Investigation of Claims of the Paranormal (CSICOP) was founded in Buffalo, New York, to consider all of these matters. This international group of prominent philosophers, psychologists, physical and biological scientists, science writers, and several magicians (experts in deception and in the investigation of "psychic" powers) were concerned that the public fascination with the paranormal was beginning to have some negative consequences for education and science and society at large. They wanted to encourage the investigation of paranormal and fringe-science claims in a responsible, scientific way. One important goal was to obtain and disseminate to the public accurate, scientifically reliable information about the paranormal, in contrast to the mish-mash of mostly unevaluated claims and misinformation that was then available. They founded a quarterly publication, the *Skeptical Inquirer,* to carry the results of these inquiries and serve as a forum for discussion of the wide range of issues involved. The *Skeptical Inquirer* has achieved some prominence in this role. Hundreds of scientists, scholars, and writers from around the world have contributed their evaluative skills and talents. It has published several thousand pages of critiques, essays, and research reports on virtually every topic in the realm of the paranormal and fringe science. These represent a considerable body of evaluative material that never before existed.

The chapters in this volume are taken from articles published during the past five years in the *Skeptical Inquirer.* (A companion volume, *Paranormal Borderlands of Science,* consists of articles published in the first five years, 1976-1981.) The selection process was difficult. Some excellent articles could not be included. I have tried to emphasize articles especially substantial in content while also general in their approach and, I hope, useful to the reader. I have organized them into two somewhat arbitrary sections: those that address claims involving, in the broadest sense, alleged unknown mental powers and other matters that might fall into the fields of psychology or parapsychology ("Assessing Claims of Paranormal Phenomena") and those that address exotic claims from outside science about the rest of the natural world ("Evaluating Fringe Science").

I hope you find them interesting, useful, and enjoyable. I also hope that, in sifting through the world of things that may or may not be so, you retain your sense of wonder and mystery, which is the motivating force for all science and exploration.

KENDRICK FRAZIER

Part I

Assessing Claims
of Paranormal Phenomena

Parapsychology and Belief

1

Debunking, Neutrality, and Skepticism in Science

Paul Kurtz

The term *paranormal* was not invented by the Committee for the Scientific Investigation of Claims of the Paranormal but has been widely used, first by parapsychologists and later by others to refer to anomalous phenomena that allegedly could not be explained in terms of the existing categories of science. "Paranormal" refers to that which is "beside" or even "beyond" the range of normal experience and explanation. It is used to depict phenomena like clairvoyance, precognition, telepathy, psychokinesis, levitation, poltergeists, astral projection, automatic writing, communication with discarnate spirits, and so on.

Most skeptics deny that the term *paranormal* has any clearly identifiable meaning. Like the "noumenal," "occult," or even "supernatural," its precise referents are vague and ambiguous. The boundaries of human knowledge are constantly expanding and being refined, and what was unknowable yesterday may become scientifically explicable the next day; thus the DNA code, the concept of black holes, and newly postulated subatomic particles surely cannot be said to have been "paranormal" when they were initially proposed. Is the paranormal simply equivalent to that which is "unfamiliar" or "strange" at one state in the development of human knowledge? If so, that would not make it unusual. The term *paranormal* has also been stretched far beyond parapsychology to other, so-called mysterious powers within the universe not contained within the parameters of our existing conceptual framework. It has been used to refer to such disparate phenomena as reincarnation, life after life, biorhythms, astrology, UFOs, Chariots of the Gods, the Bermuda Triangle, monsters of the deep—whether Nessie, Chessie, or Champie—Bigfoot, cattle mutilations, human spontaneous combustion, psychic archaeology, and faith healing; in short, almost anything that comes within the range of human imagination and is thought to be "incredible."

On the current world-scene, belief in the paranormal is fed and reinforced by a vast media industry that profits from it; and it has been transformed into a folk religion, perhaps the dominant one today. Curiously, it is often presented as "scientifically warranted" and as a new, if bizarre, conception of reality that is breaking down our naturalistic-materialistic view of the universe.

Contemporary science is rapidly expanding in many directions: On the macrolevel, astronomy reports exciting new discoveries. The quest for extraterrestrial life is one of the most dramatic adventures of our time. This is grist for science fiction and the poetic imagination, outstripping that which has been verified or is technologically feasible today. On the microlevel, physicists postulate new particles in an attempt to unravel the nature of physical reality. And in the life sciences, biologists are decoding the genetic basis of life and are on the threshold of creating new forms. At the same time, the information revolution unfolds stunning new applications.

Men and women have always been fascinated by the depths of the unknown. As far back as we can trace there has been an interest in the occult and the magical. The persistence and growth of ancient paranormal beliefs in our highly educated scientific-technological civilization is a puzzling phenomenon to many of us. There are many reasons for this, not the least of which are the fast pace of scientific progress, the role of science fiction in stimulating the imagination, and the breakthrough into space beyond our planet. And so people ask, for example, why is it not possible for the mind to engage in remote viewing of far distant scenes and events, precognate or retrocognate, or to exist in some form separate from the body. Present-day science for many seems to demonstrate that virtually *anything* is possible, and that what was once thought to be impractical or unreal can later be found to be so. And they think perhaps psi phenomena, biorhythms and horoscopes, faith healing and extraterrestrial UFOs are genuine. There is some confusion in the public mind between the possible and the actual, and for many people the fact that something is possible converts it into the actual.

Some skeptics have dissented, maintaining that since paranormal concepts contradict the basic conceptual categories by which we understand and interpret the world, they may be rejected on a priori grounds. In my view, it is difficult to impose preconceived limits to inquiry or to rule out such claims as logically "impossible." The history of science is littered with such vain attempts. Whether or not paranormal phenomena exist and, if they do, how they may be interpreted can only be determined in the last analysis within the process of scientific verification and validation and not antecedent to it.

What Should Be the Role of Science?

Now the question is often raised: How should science deal with the paranormal?

One familiar response is that science should ignore the paranormal entirely. Many scientists until recently considered it beneath their dignity to become involved in what they viewed as patent nonsense. This has not been the response of those scientists and scholars associated with CSICOP. We believe that such claims ought to be investigated because of the widespread public interest and also because some paranormalists on the borderlands of science claim to have made significant discoveries.

If one decides to examine such claims, how does one proceed? One way is to debunk nonsensical paranormal beliefs. Martin Gardner quotes H. L. Mencken to the effect that "one horse-laugh is worth a thousand syllogisms." Some people have insisted that debunking is not an appropriate activity, particularly for academic scientists. To "debunk" means "to correct a misapprehension, to disabuse, set right, put straight, open the eyes or clear the mind, disenchant, or dispose of illusion, unfoil, unmask, or tell the truth" (*Roget's Thesaurus*). Some of the claims that are made—even by scientists and scholars—are preposterous and debunking is not an illegitimate activity in dealing with them. Sometimes the best way to refute such a claim is to show how foolish it is, and to do so graphically. Indeed, debunking, in its place, is a perfectly respectable intellectual activity that any number of great writers have engaged in with wit and wisdom: Plato and Socrates, Voltaire, Shaw, and Mencken, to mention only some. Surely it has a place within philosophy, politics, religion and on the borderlands of science and pseudoscience. It should not, however, be abused but should be used with caution; and it should be based upon a careful examination of the facts.

But there are dangers here: Sometimes what appears to be bunkum because it does not accord with the existing level of "common sense" may turn out to be true. Mere prejudice and dogma may supplant inquiry. If one debunks, he had better command an arsenal of facts and marshal evidence to show *why* something is improbable or even downright false. We can ask, Does sleeping under a pyramid increase sexual potency? Do plants have ESP and will talking to them enhance their growth? Do tape-recorders really pick up voices of the dead? All of these claims have been proposed by paranormalists within the past decade. They should not be rejected out of hand. On the other hand, at some point—after inquiry, not before—they may deserve forceful debunking; this is particularly the case when scholarly critiques of inflated claims go unnoticed by the public. Jeane Dixon and Uri Geller, for example, seem as unsinkable as rubber ducks—though some of us have attempted to make duck soup out of them. Thus we are concerned not simply with paranormal beliefs in the laboratory but with their dramatization in the media.

Another response to the paranormalists is to maintain that we should examine each and every claim—however far-fetched—that anyone makes, and to give it equal and impartial hearing. There are literally thousands of claims pouring forth each year. One cannot possibly deal with them all. We receive a goodly number of calls and letters every week at the offices of CSICOP from people who claim that they have prophetic powers, are reincarnated, or have been abducted aboard UFOs. Some of our critics nevertheless have insisted that this is the only appropriate response for science to make: to be neutral about them all. After all, were not Galileo and Semmelweis, and even Velikovsky, suppressed by the scientific and intellectual establishments of their day? And might not we in our day likewise reject an unconventional or heretical point of view simply because it is not in accord with the prevailing intellectual fashion? I repeat: This is a danger that we need especially to avoid. For the history of science is full of radical departures from established principles. Thus we must keep an open mind about unsuspected possibilities still to be discovered.

However, one should make a distinction between the open mind and the open sink. The former uses certain critical standards of inquiry and employs rigorous methodological criteria that enable one to separate the genuine from the patently specious, and yet to give a fair hearing to the serious heretic within the domain of science. Isaac Asimov has made a useful distinction between *endoheresies*, which are deviations made *within* science, and *exoheresies*, which are deviations made *outside* of science by those who do not use objective methods of inquiry and whose theories cannot be submitted to test, replication, validation, or corroboration. Even here one must be extremely cautious, for an exoheretic may be founding a new science. A protoscience may thus be emerging that deserves careful appraisal by the scientific and intellectual community. Or the exoheretic may simply be a crank—even though he or she may have a wide public following and be encouraged by the powerful effects of extensive media coverage. Simple neutrality in the face of this may be a form of self-deception.

Philosopher Paul Feyerabend has maintained that there are virtually no standards of scientific objectivity and that one theory can be as true as the next. But I submit that he is mistaken. If we cannot always easily demarcate antecedent to inquiry pseudo from genuine science, we can after the fact apply critical standards of evaluation. Within these limited confines, then, I submit that some debunking is not only useful but necessary, particularly if we are to deal with the realities of belief in our media-coddled society. Given the level of ready public acceptance of the "incredible" and a tendency toward gullibility, one horse-laugh in its appropriate setting may be worth a dozen scholarly papers, though never at the price of the latter.

There is still another response to bizarre claims. In the last analysis this is the most important posture to assume; namely, if a paranormal claim is seriously proposed and if some effort is made to support it by responsible research methods, then it does warrant serious examination. I am not talking about antiscientific, religious, subjective, or emotive approaches to the paranormal, which abound, but efforts by serious inquirers to present hypotheses or conclusions based upon objective research. This is the case with parapsychology, which today deserves a fair and responsible hearing. Going back at least a century, some of the important thinkers—philosophers, psychologists, and physical scientists—have investigated the psychical: William James, Henry Sidgwick, H. H. Price, Oliver Lodge, William Crookes, and more recently Gardner Murphy and J. B. Rhine. Their work deserves careful analysis, though it is not immune to strong criticism on methodological and evidential grounds. Similarly for some aspects of recent UFO and astrological research. If there are falsifiable claims and conceptually coherent theories, then they need rigorous testing and careful logical analysis by independent scientific investigation. And here neutrality in the process of evaluation is the only legitimate approach; take a hypothesis, examine the experimental data reported, attempt to replicate the experiment, make predictions, and see if the theories are logically consistent and can be verified.

That this same neutrality should apply to fortune telling, horoscopes, tarot cards, palmistry, fortune cookies, and other popular fields is another matter. Take them into the laboratory to see if you can get results. But if you get no results, then the only response often is to debunk them.

What Is Skepticism?

Now it is no secret that CSICOP has been identified with the skeptical position. We have said that we do not find adequate support for many or most of the claims of the paranormal that have been made both within and without science. We have been bitterly attacked by paranormal magazines and newspapers (such as *Fate* magazine) for publishing debunking articles at the same time these publications purvey misinformation to the public and seek to sell everything from crystal balls to Ouija boards. We believe that both debunking and careful scientific examination should be done. In regard to the latter, we often find in the parasciences a lack of replication, inadequate experimental design (as in J. B. Rhine's early experiments), and questionable interpretation of statistical data (as in the remote-viewing experiments of Targ and Puthoff). Sometimes—but only sometimes—there is fraud or deceit (as in the case of S. G. Soal, Walter Levy, and others), but underlying it all there is a strong will to believe (as Project Alpha has shown).

Skepticism is among the oldest intellectual traditions in philosophy, and it can be traced back to ancient philosophers like Carneades, Pyrrho, and Sextus Empiricus, and in modern thought to Descartes, Locke, Berkeley, Hume, and Kant. Today skepticism is essential to the very lifeblood of scientific inquiry.

There are many forms that skepticism can assume. One form it may take is universal doubt, the attitude that the reality of the senses and the validity of rational inference should be mistrusted. For this form of skepticism one must adopt an *epoché* in regard to all things; that is, assume the role of the agnostic and suspend judgment. Since one position is as good as the next, and all positions may be equally false, none can be said to be true. In philosophy, this has led to extreme solipsism, where one doubts not only the reality of the external world but one's own existence. In ethics, it has led to extreme subjectivism, a mistrust of reason, and a denial that there are any objective ethical standards; for values, it is held, are rooted in personal taste and caprice. In science, universal skepticism has led to methodological anarchism, the view that all scientific positions depend upon the mere prejudices of the scientific community and the shifts in paradigms that occur. If this is the case, astrology would be as true as astronomy and psychic phenomena as real as subatomic physics. Such a form of skepticism is easily transformed into the kind of "neutralism" discussed above—since all positions may be equally true or false, we have no way of judging their adequacy.

Universal skepticism is negative, self-defeating, and contradictory. One cannot consistently function as a total skeptic but must assume certain principles of inquiry, some of which turn out to be more reliable than others. We must act upon the best evidence we have, as our beliefs confront the external world independent of our wishes. Moreover, we do have well-tested hypotheses that may be held with varying degrees of probability and incorporated into the body of knowledge. The skeptic's own universal principle that there is no reliable knowledge must apply to itself; and, if so, we are led to doubt its range of applicability. A universal skepticism is limited by its own criteria. If we assume it to be true, then it is false; since if it applies to everything, it applies to itself, and hence universal skepticism cannot be universal. I do not wish to become impaled by the logic of types. The point I want to make is simply that the most meaningful form of skepticism is a *selective* one. This maintains that doubt is limited to the context of inquiry. We cannot at the same time doubt all of our presuppositions, though we may in other contexts examine each in turn. The doubt that properly emerges is within a problematic context of inquiry and thus can be settled only by the relevant evidence—though perhaps not completely.

What I mean is that the scientific community is always faced with new research problems, and it seeks solutions to these problems (*a*) in the theoretical sciences, through explanations of what is happening and why, (*b*) in the technological and applied sciences, by resolving questions of application. There are alternative theories or hypotheses that may be proposed and compete for acceptance. Some of them may fall by the wayside; those that win out seem to accord best with the relevant data and the conceptual framework at hand, though these may in turn be eventually modified.

Clearly, a researcher should suspend judgment until he can confirm his hypothesis and until it is corroborated by other inquirers. However, no one law or theory can be said to be final or absolute, or to have reached its ultimate formulation. Here Charles Peirce's famous "principle of fallibilism" plays a role: for we may be in error, we may uncover new data, or alternative hypotheses may be found to fit the data more adequately. Thus we must be prepared to admit new hypotheses, however novel or unlikely they may at first appear. Science is open to revision of its theories: The self-corrective process is on-going. We must always be willing to entertain and not rule out new ideas. This applies to the established sciences, but also to newly emerging proto- or para-sciences.

Conclusion

I am often asked why belief in the paranormal is so strong in the world today, and especially in highly developed and highly educated scientific-technological societies like our own. There are many explanations that can be and have been given. I wish to conclude by mentioning only two.

First is the fact that we exist in a religious culture of longstanding historic traditions, and dissenting points of view in the area of religion are not given a fair hearing. Since belief in the supernatural and occult remains largely unchallenged, the paranormalist finds a receptive audience. There are at least two cultures existing side by side. On the one hand, the religious, and on the other the rational-philosophic-scientific. Until the religious is submitted to intellectual critique openly and forthrightly, the paranormal will continue to flourish on the fringe of science.

The second reason is that although we are a scientific culture we have not thus far succeeded in our curricula of scientific education in conveying the meaning of science. There is a widespread appreciation for the benefits of scientifc technology, particularly for its economic value, as new industries are being spawned at a breathtaking pace. But at the same time there exists fear of science and its possible implications for other aspects of life. Sadly our elementary and high schools, colleges and universities, turn out specialists who may be extremely competent in their narrow fields of expertise, but who lack an appreciation for the broader scientific outlook. Within their own fields students are able to master their subject matter and apply the methods of science and critical intelligence, but these methods often do not spill over to other areas of belief.

In my view, a major task we face is proper education in science, both in the schools and for the general public. There is a failure to appreciate the importance of skeptical thinking. A truly educated person should come to appreciate the tentative character of much of human knowledge.

The burden of proof always rests upon the claimant to warrant his claim. If all the facts do not support it, then we should suspend judgment.

Science surely is not to be taken as infallible, and some of the defects found in the pseudo- and para-sciences can be found in the established sciences as well, though on a reduced scale. Scientists are fallible, and they are as prone to error as everyone else—though it is hoped that the self-corrective process of scientific inquiry will bring these errors to the light of day. Similarly, it would be presumptuous to maintain that all intelligence and wisdom is on the side of the skeptic; for he may be as liable to error as the next person. Fortunately, we have our critics and they are only too willing to point that out—for which we should be grateful. We have made mistakes and have sought to correct them. We should not trust anyone to have all the truth, and this applies to ourselves as well.

Whether life can be lived truly rationally and whether all of our beliefs can be tested before we accept them is a topic that philosophers have long debated. Suffice it to say that selective skepticism can have a constructive and positive role in life, that some degree of skepticism is important, and that reflective individuals will learn to appreciate its value.

2

The New Philosophy of Science and the 'Paranormal'

Stephen Toulmin

Over the past thirty years, there has been a major shift of focus on the part of many philosophers of science. This has been associated with a new recognition of the depth and importance of historical change as a factor in shaping our scientific beliefs, ideas, and presuppositions, and in determining the contexts of scientific discovery and even the methods of scientific research. The most widely read book in this new vein has, of course, been Thomas Kuhn's *The Structure of Scientific Revolutions*. But this work is only the tip of an iceberg, and it is perhaps too vague, superficial, and lacking in detail to help us in any examination of the claims of the paranormal. It is too easy, for instance, for parapsychologists and others to claim the Kuhnian protection of working according to "a different paradigm." So, here, let me indicate in my own terms—without any resort to the jargon of paradigms—what implications the current shift in philosophy of science has for the work of the Committee for the Scientific Investigation of Claims of the Paranormal.

Our immediate predecessors (as is well known) hoped that the work of defining a proper "method" for the sciences would yield a *unique* method. applicable to the subject matter of any science and notably to a given science at any stage in its historical development. Thus they sought to move beyond the position of (say) Aristotle, who claimed that each different kind of problem and subject matter needed to be analyzed, discussed, and explained in correspondingly different terms. In turn, they hoped that they could find the basis for a sharp, clear, and above all *permanent* criterion—the so-called demarcation criterion—for telling sciences whose students have some genuine claim to be proceeding rationally and methodically from "pseudosciences," which might pretend to deserve scientific status but fail to meet the required tests.

The newer approach to philosophy of science moves away from the more extreme claims of (say) the Viennese philosophers of the 1920s and 1930s. It begins from the observation that any *universal and timeless* account of scientific method—and, with it, any universal and timeless demarcation criterion for telling "real" sciences from pseudosciences—is also, unavoidably, an *abstract* account. We may respectfully tip our hats to the idea that true scientists should pay attention to, for instance, the quality and quantity of the "evidence" available to them, the strength of the "support" it lends to any given "hypothesis," the "observations" that might conceivably "falsify" that hypothesis, and the scope of the "predictions" and "explanations" to which it leads. But all of this is so much apple pie. The universality of the demands in question is purchased at the price of losing touch with the real work of the sciences.

For example, the standards that determine what counts as a scientific "prediction" vary between one science and another, and even from one phase to another in the development of any particular science. So, before we can apply these formal, abstract demands to actual scientific situations, we must pay attention *also* to certain specific, concrete features that are distinctive of any given science at this or that stage in its history; and it turns out that most of these features are highly variable, both among different sciences and among historical epochs. Correspondingly, all the general terms that the Viennese philosophers looked to as providing the universal and timeless indices of a science's rational status—*verify, falsify, predict,* and the like—turn out to be *multivocal:* They have a determinable sense only when understood in the particular ways appropriate to the problems of the science in question at the time in question.

So far, this is only prologue. If we take the new approach to philosophy of science seriously, we shall have to call up for reconsideration a dozen assumptions over which—as it seems now, in retrospect—our Viennese forerunners were too complacent. Two of the items requiring review, as a result, will be: (1) the very idea of a permanent "demarcation criterion" for telling real sciences from pseudosciences, and (2) the ways in which we draw the boundaries between the "normal," the "abnormal" (or "anomalous"), and finally the "paranormal" (or putatively mysterious). If the new approach is at all sound, we may expect to find that these distinctions have themselves to be drawn in historical and contextual terms. So, this article will discuss two topics: namely, the implications of this new approach, on the one hand, for defining any program and method for inquiring into the claims of the paranormal and, on the other hand, for specifying the content of such a program.

Shifts in Demarcation Criteria

As to the first of these two questions, we should begin by reconsidering the meaning of the very term *normal* and its two parallel antonyms, *abnormal* and *paranormal*. Beforehand, we might be inclined to assume that the idea of *normality* must be a permanent item in the inventory of human thought; or that it is at the very least a necessary presupposition of modern science.

Surely human beings (particularly, scientific investigators) have always had such a notion?

The slightest historical investigation reveals that this is not the case. The idea of the "normal," in its modern sense, is curiously recent, certainly far more recent than the rise of modern science. Other earlier notions (to be sure) are clearly cognate or ancestral to our own modern idea of normality. But the first use of the term *normal* in this sense recorded in the *Oxford English Dictionary* dates from the year 1840; as such, it coincides with the start of the professionalization of scientific work. Notice that 1840 was also the year when, in the course of his presidential address to the new British Association for the Advancement of Science, William Whewell introduced his freshly coined word to replace all those vague words and phrases hitherto used when referring to scientific students of natural phenomena—namely, "natural philosophers," "savants," "virtuosi," and the like—the brand new word, *scientist*, which Whewell invented on the model of the much older and better established word *artist*. (Sociologists will certainly see it as no accident that the rise of a professional class of "scientists" also coincided with fresh attempts to mark off the "normal" course of natural events from other, more surprising or mysterious phenomena.)

As to the contrary term, "abnormal," this turns out to be even more recent and ill defined. Indeed, this is one of the rare words for which the *O.E.D.* reserves a critical commentary. About "abnormal" it says, in what connoisseurs will recognize as fighting words, "Few words have shown such a series of pseudoetymological perversions"; and the earliest uses of the word in a recognizably modern sense date from around the 1860s and 1870s. (Not surprisingly, the word "paranormal" does not appear at all, either in the *O.E.D.* or in its supplements!)

During the two preceding centuries, scientists had spoken of exceptional or mysterious happenings and phenomena in different terms. For eighteenth-century natural philosophers, the operative contrast was that between the "natural" and the "unnatural," with the "supernatural" standing on one side (so to say) for emergency use only. For their seventeenth-century forerunners, the basic distinction was that between the "natural," on the one hand, and the "divine" or "miraculous," on the other. In framing his account of Nature, for instance, Isaac Newton took it for granted that the observed phenomena embrace both passive, natural processes and also the active interventions of Divine Agency; conversely, when Leibniz and the Cartesians criticized Newton's hypothesis of universal gravitation, they objected to it as being "either miraculous or imaginary."

Why was there this shift, around 1840, away from the ideas of Nature and the "natural" and toward those of the "normal," or "what is normally expected"? Aside from the beginnings of professionalization, at least two other factors need attention. For a start, from the 1830s, the Romantics had acquired something of a monopoly over "Nature," and as a result it was felt to have lost some of the precision it needed to be of use in science. In addition, this change was probably associated with the rise of the new "positivism" of Auguste Comte, which in turn acquired a special status among theoretical physicists and chemists, most of all in France.

In order to find out what is *natural*, all that one need do was to *look and see*. The "natural" is simply what we find to be natural as a matter of experience: before passing judgment about what is "possible" or "impossible," we must await the outcome of experience. (Pigs *don't* fly. Stick around them, and you will never see an airborne porker!) By contrast, we can decide what is "normal" beforehand only if we have built up a theoretical picture of relevant aspects of the natural world that is clear, definitive, and reliable enough to distinguish those things that "are normally to be expected" from those that can be believed in only by credulous and gullible people. The "normal" is thus what is theoretically intelligible: We have theoretical reasons for distinguishing the "conceivable" from the "inconceivable." (Pigs *can't* fly. You needn't waste time sticking around them, as the very idea of a porker taking off unaided is contrary to the laws of physics!)

Once we see how deeply the idea of the "normal," together with the contrary ideas of the "abnormal" and the "paranormal," is implicated with the current state of *theory*, we can recognize a serious problem. The content of scientific theories changes, sometimes drastically; so we may be tempted to rule out, as "inconceivable" and "paranormal," things that a later stage of science acknowledges as quite possible, or even actual. Thus we may unwittingly cross the line between skepticism and dogmatism, and so rule out of court phenomena and hypotheses that historical imagination should warn us not to dismiss. The task of drawing a line between what is and is not "theoretically conceivable," and so what is or is not "normal" (as contrasted with "paranormal" or "scientifically inconceivable"), has a long and complex history of its own; and if we only reflect on the course of that history, it will induce a certain modesty even about our skeptical doubts.

History's Sobering Lessons

When we turn to the second of the topics we set ourselves to consider here, this point is only reinforced. When it comes to compiling a list of supposedly "paranormal" phenomena, which are rightly suspect today, we cannot afford to overlook those natural phenomena that we accept as quite natural, normal, and genuine today but which were dismissed in earlier times merely because, as yet, they had won no place in the accepted theories about the natural world. A short list of these is sobering.

The seventeenth-century mechanical materialists (for a start) could accept as "natural" constituents of the physical world only those material particles or corpuscles that were strictly passive. These were not themselves sources of motion, but they could be set in motion, either when they collided with similarly material particles or else (exceptionally) when they were acted on by "nonmaterial" agents. Interactions of the latter kind were not in themselves suspect. Although they were ruled out from the purely material (or physical) wing of Descartes's ontology, they had to be accepted as the means by which "mental action" alone could bring about physical effects. So, for instance, we find Giovanni Borelli's treatise *On Animal Motion* insisting that all of a living creature's muscles and bones are merely mechanical and that the ultimate source of its motion must be an immaterial but "vital" agency operating through the vehicle of its brain.

Newton, for his part, had no worries about admitting both passive physical phenomena and active divine intervention into his overall picture of the world: Indeed, all the phenomena that were later attributed to "fields" Newton regarded as vehicles for divine intervention, and so attracted Leibniz's mockery of his reliance on "miracles." (In Leibniz's eyes, miracles could be tolerated in the realm of Grace, but they were not to be admitted into the realm of Nature.) As for the notion of a "thinking machine," that would have struck all the leading natural philosophers of the seventeenth-century as a *contradiction in terms.*

Over the 350 years between 1620 and 1970, in point of fact the meanings of the terms *matter* and *machine* have shifted quite drastically, and with them that of the correlative notion of an "immaterial entity" also. In the seventeenth century, "matter" was so defined that *thought* was, by definition, an immaterial thing; but so too were gravitational fields, electric currents, and even (on some accounts) gases. Only their successors in the eighteenth century, e.g., Julien de la Mettrie and Joseph Priestley, had the perception to see, and the courage to declare, that the seventeenth-century definition of "matter" had been arbitrary in the first place and that the resulting dualism was therefore unnecessary. For the time being, however, a seventeenth-century physical theorist would have been quite in order if he had looked at an electronic computer and said, "That's not what I mean by a machine at all!"

In the late eighteenth century, the topic of hypnotism aroused much of the suspicion and hostility that was to surround the topic of unconscious motivation a century later. In fact, Lavoisier undertook on behalf of the French Academie des Sciences an inquiry into the status of "animal magnetism," as hypnotism was then called. During the nineteenth century, likewise, the wave theory of light, cross-generic hybridization, and the molecular basis of genetics and morphogenesis were called into question; but the most suspect topic of all was, of course, Darwin's theory of evolution by variation and natural selection. Indeed, right up to the time of

Darwin's death, there were those who believed that his theory was "inconsistent with physics"; and he himself was conscious enough of the force of their objections that he could meet them only with the hope that one day they would prove needless and "go away."

The basis of these objections was Darwin's estimate of the age of the earth and the length of time needed for the action of natural selection to produce the living creatures we know at the present time. On the best calculations of the time required for a spherical object the size of the earth to cool to a habitable temperature, according to Newton's Law of Cooling, it would be habitable for no more than 25 million years before becoming too cold to continue supporting life: At least, this was evidently the case if all the sources of the earth's heat were its initial high temperature and the solar radiation falling onto its surface. (In any case, as was pointed out by Lord Kelvin, the sun, too, must be cooling down in the same way and "could not" have remained at the required temperature for as long as Darwin needed.) So it was easy enough for Darwin's contemporaries to dismiss his views on the best available scientific grounds; and it took a man with the caution of Charles Lyell to see that the results of all these calculations held good only on the assumption that nineteenth-century physics already knew all the actual sources of the earth's heat. As it turned out, of course, it did not. And, since 1900, the extreme antiquity of the sun and the earth has been securely underpinned only with the discovery of two radioactive processes: nuclear fusion to maintain the sun's heat, and the inner sources of radioactivity within the earth to generate mountain building and provide the warmth needed to support life.

Yet even greater surprises were in store for orthodox physicists before the turn of the century. Max Planck's theory of quantization, introduced in 1899 in hope of saving James Maxwell's theory of electromagnetism from the "scandal" of the black body radiation spectrum, was greeted by many physicists as a major betrayal of established ideas. Worse still, J. J. Thomson's first paper in the *Proceedings* of the Cambridge Philosophical Society, reporting the discovery of minute negatively charged particles ("electrons") having a mass less than one-thousandth of the mass of a single hydrogen atom, was at first suspected of having been published as a hoax! After all, nineteenth-century atomism left no place for material units lighter than a single hydrogen atom; so (it was assumed) one *knew* that no such smaller particles could conceivably exist. It would be hard to illustrate more elegantly the alliance between skepticism and dogmatism.

The Blurred Periphery

To say this is not to argue for any fashionable "relativism to the current paradigm": Appeals to "paradigm switches" can, in any case, too easily be used to snow the gullible. Still less is it to call into question the general program of investigating, in a critical and skeptical spirit, the claims made on behalf of the "paranormal" today. It is to argue only that we need to be discriminating, even in the reach of our own skepticism. As we can now see in retrospect, some of our predecessors, e.g., the founders of the Society for Psychical Research in the nineteenth century, worked with ideas that were naive and dogmatic in just the sorts of ways we most need to avoid. We are right to recoil from the chance of being gulled; but, in doing, so, we must take care not to recoil so far that we put ourselves in the same camp as (say) the critics of J. J. Thomson. We can hope to find out what the proper scope of our doubts should be only in the same way that we discover other scientific truths, that is, *as we go along*.

At any stage in the development of science, there is a point beyond which we cannot know for certain *exactly* what it is that we do and do not understand and *exactly* where a line should be drawn between phenomena that are as yet mysterious and happenings that are frankly incredible. This is not to say that the changes we must look for, in this respect, will be so drastic as to wipe out all our current boundaries and distinctions. Some common core of understanding may be expected to survive all the future changes in scientific theory, just as Newton's ideas survive in our own quite different intellectual context. Still, at the edges, the periphery will always remain blurred; and, in the future as in the past, as our scientific understanding is still further refined, that blurred boundary will sometimes come into sharper focus in quite unexpected ways.

For the moment, the world of the paranormal embraces enough plainly unscrupulous quackery and exploitation to keep us fully occupied. Whether we concentrate on finding alternative ways of accounting for the supposedly *non*-normal features of all these things, or whether we focus rather on the sociological factors and political motives involved in such situations, there is lots to keep us busy. That being so, we shall do well if we try to avoid wasting our fire on marginal and questionable targets.

3

Parapsychology's Past Eight Years: A Lack-of-Progress Report

James E. Alcock

On the weekend of April 30 1976, critics of the paranormal met on the SUNY-Buffalo campus in a symposium entitled "The New Irrationalisms: Antiscience and Pseudoscience." Out of that meeting emerged the Committee for the Scientific Investigation of Claims of the Paranormal (CSICOP). It is now eight years later.

What changes have occurred in parapsychology since that first meeting? Some milestones are obvious: On the human level, the passing of the years has claimed the lives of several of the founders of modern parapsychology: Joseph Banks Rhine, Louisa Rhine, Gardner Murphy, J. Gaither Pratt, Arthur Koestler and Margaret Mead, two of its most important proponents, and the famous Dutch psychic sleuth Gerard Croiset and his mentor, Professor W. H. C. Tenhaeff, all passed away during this period.

The past eight years have also witnessed demises of another sort: Star psychic Uri Geller, once the darling of both the public and parapsychology alike, no longer is considered by most parapsychologists to have genuine psychic powers; this change was brought about in large part by the efforts of James Randi. As for Gerard Croiset's reputation as a psychic sleuth, while it may live on for some years in the public mind, anyone who has read Piet Hein Hoebens's account in the SKEPTICAL INQUIRER about the fraudulent nature of Croiset's claims (vol. 6, nos. 1 and 2) need no longer wonder about his supposed psychic abilities. (Professor C. E. M. Hansel [1966] first put the claims of Croiset and Tenhaeff in doubt eighteen years ago, but Croiset's psychic career continued to flourish.) Lawrence Kusche's book on the Bermuda Triangle, which appeared not long before CSICOP's founding, seems to have taken the wind from the sails of those who disseminated the nonsense about that particular "mystery." Carlos Castaneda's claims about his adventures with a superpsychic Yaqui shaman are now, one would hope, thoroughly discredited in the eyes of any intelligent reader as a result of the investigation by Richard de Mille, which he reported in his book as well as in the pages of the *Skeptical Inquirer* and elsewhere.

CSICOP and the *Skeptical Inquirer* cannot of course take the credit for these "demystifications"; many or all of them would have occurred without CSICOP. However, CSICOP has served to create a sense of community among many critics of the paranormal. By providing an excellent outlet for critical commentary in the form of the *Skeptical Inquirer*, and through its efforts to bring attention to the weakness of claims made by proponents of parapsychology and UFOlogy and the generally unfair bias of the media, it has helped bring about an atmosphere that has encouraged significant numbers of people to contribute constructively and critically to the debate about claims of the paranormal.

But what about *progress* in parapsychology during these past eight years? Has any new evidence been forthcoming that puts paranormal psychology on surer scientific ground? The answer here is surely "No." The past eight years have been no kinder to those seeking compelling evidence about the reality of paranormal phenomena than were the previous eighty: The long-sought reliably demonstrable psychic phenomenon is just as elusive as it has always been.

Yet one can sense a certain excitement in the recent writings of some parapsychologists. It is as though something very big is in the wind. Much is made of the supposed compatibility between parapsychological claims and some interpretations of quantum-mechanical theory. For the first time, parapsychologists—some of them at least—are proudly talking of testable theories. Witness, for instance, the recent words of Gertrude Schmeidler (1982):

> . . . back in the 1940s, and even to the 1960s, critics used to say that parapsychology gave facts without theory, yet here in the 1970s we have almost an embarrassment of theoretical riches. [These theories] . . . are astonishing because another critical argument used to be that physics showed parapsychological data to be false, both because the data showed no diminution of ESP success as spatial distance increased and because precognition contradicted what we know of time relations—and yet here we find, with modern physics, that both these apparent anomalies fit well into solutions of equations in quantum mechanics. What used to seem major arguments against parapsychology are now only historical curiosities. [pp. 140-141]

That there should be some embarrassment associated with these theories is undeniable, but it is hardly because of their richness. Rather, the embarrassment should come from the unabashed efforts to twist certain features of the theory of quantum mechanics, and to presuppose certain solutions to some of the well-known quantum mechanical enigmas, in such a way as to "make possible" the supposed ability of psychic forces to leap the constraints of both time and space. This process of "quantum mechanization" is followed by trumpeting to all who wish to listen that, rather than being a research area of disputed credibility from the point of view of mainstream science, parapsychology is on the scientific forefront, waging battle shoulder to shoulder with quantum physics to force a reluctant nature to give up its secrets.

Although it is true, as Schmeidler pointed out, that parapsychologists have been criticized on the grounds that their claims are compatible with no known theory, the production of rather contrived theories based loosely on quantum mechanics to "explain" data whose very validity is at the heart of the dispute between critics and proponents is hardly likely to put criticism to rest. Indeed, this brings to mind a letter that appeared in the *Journal of the American Society for Psychical Research* in 1971. In this letter, J. T. McMullan of the School of Physical Sciences at the New University of Ulster in Northern Ireland proposed a theoretical explanation, based on thermodynamics, to account for the energy expended during poltergeist activity. He noted that poltergeist reports are consistent in their mention of a lowering of the room temperature by as much as five degrees Celsius during a poltergeist display. He then calculated that enough energy would be given up by a one-degree lowering of room temperature in a room of normal size to raise a 25-kilogram table vertically through a distance of some 200 meters. Thus, he suggested, the energy required for the spectacular movings and jarrings of objects that typify a poltergeist incident can be accounted for by means of what is already known and accepted about the conversion of energy from one form to another. Would it be in consequence appropriate to say that thermodynamical theory predicts, allows for, or is compatible with the occurrence of poltergeist activity? Obviously not, although were one to do so this would no doubt, in many eyes at least, lend scientific credibility to poltergeist claims. (Even parapsychologists have not been responsive to McMullan's notions.) However, McMullan cannot be faulted for his creativity; were one able unambiguously to verify the non-normal nature of putative poltergeists, and were one about to verify temperature change, then McMullan's speculations might be of some interest. Without a clearly demonstrated phenomenon, and without, in consequence, the need to try to account for its occurrence, the proposition is little different from medieval arguments about the number of angels that can dance on the head of a pin. Demonstrate that the accomplishments of poltergeists occur without benefit of natural cause, show that the temperature really does change, or verify the presence of at least some angels, and then one is in a position to seek or propose explanation.

If parapsychologists showed so little interest in McMullan's ideas, why, then, should anyone be excited by the so-called paraphysical theories, which depend on some controversial (to say the least) interpretations of quantum-mechanical theory to explain data whose very validity is in serious question? The answer lies, I believe, in the fact that, while it might be quite difficult for many people to accept that a small decrease in temperature can produce enough energy to toss a chair around, it is much easier to believe that very subtle psychic influences can affect things on a subatomic scale to create subtle influences on large-scale events, influences that can only be determined statistically.

It is hardly contentious to say that the strongest parapsychological data claims are intimately tied to statistical analyses. Apart from such analyses, the only other evidence is either anecdotal or of a case-study nature, and neither of these sources has proved capable of producing evidence of any weight. Indeed, it is safe to say that parapsychology emulates experimental psychology's preoccupation with statistical *significance*, to the point of all but ignoring the size of the actual effects. During the 1930s and the 1940s, experimental psychologists engaged in lengthy debates with parapsychologists about the appropriateness of statistical models and tests used by the latter, in the belief that parapsychological evidence was merely the outcome of misapplied statistical tests. These battles did lead to improvements in the quality of statistical analyses used within parapsychology, and gradually the psychologists broke off their attacks based on that ground. In fact, it can now be argued that, if one looks at the *best* research that modern parapsychology has to offer, the quality of the statistical analysis is often as good as or better than much of what appears in psychological journals. However, there is a very crucial difference between psychology and parapsychology in this regard: Psychologists run *experiments*, by which term they mean that they vary an independent variable and look for concomitant changes in a dependent variable; in other words, they employ control groups. For example, one might randomly assign subjects to one of two groups and then go on to treat the two groups identically except for the varying of the independent variable. If in consequence it is found that statistically significant differences exist between the two groups, one can then infer that the differences are due to the independent variable, since it was the only thing that differed in the treatment of the two groups. Parapsychologists are unable to study directly their putative psi in this way, since they have no way of treating it like an independent variable because they have no way of turning it on or off, or even of knowing whether it is in operation at any specific point in time during the study. Unlike the psychologist who can contrast two groups of scores, the parapsychologist must argue his/her case on the basis of departures from a chance model said to describe the population from which the data arises. This imposes severe limitations on the extent to which one can draw inferences from the statistical conclusions.

At the risk of again being accused of belaboring the obvious, I want to emphasize that statistical conclusions cannot say anything at all about the existence or nonexistence of psi. All one can gain from statistical procedures is an indication that the observed results are unlikely to have been observed by chance, as described by the particular probability model being used. Unfortunately, despite claims to the contrary, parapsychologists routinely interpret statistical departures from "chance expectation" as evidence that psi is involved. Sometimes the attempt is made to argue that "psi" is just a label to describe the departure from chance and that no particular explanation is implied. This claim is demonstrably false, and all one has to do to establish that is to turn to the parapsychology research journals to see how departures from chance are typically interpreted. They are indeed treated as manifestations of psychic forces. An alternative explanation of the same departures might be that one is using an inappropriate model or that some unrecognized "normal" influence may be responsible for the departure of the data from what the model would predict. To decide that this unrecognized influence is a "psychic" influence is no more logically compelling than to decide that invisible creatures from another solar system are hovering in the laboratory and causing the observed departures from chance.

Some of the new directions parapsychologists have been following strengthen my belief that there indeed is nothing "psychic" about the departures from chance. For the sake of argument, let us suppose that *all* departures from chance reported by parapsychologists are due, not to psi, but to something I shall refer to as "empirical/statistical artifact," or simply "artifact" for short. By this term I simply mean to describe statistical departures from chance expectation brought about by not only the vicissitudes of chance (as manifested in such things as subject selection, selective publishing, and statistical Type Two error, or false rejection of the chance hypothesis when it is true) but all subtle influences that can lead to bias in a probability experiment. This may include sensory leakage, recording errors, experimenter expectancy, or whatever. This "artifact hypothesis' would suggest that departures from chance should be expected not only in the classic sender-receiver telepathy paradigms, or the old dice-rolling PK studies, but in some subset of all situations in which subjects are attempting to alter statistical outcomes. Of course the more tightly controlled the research situation, the fewer the sources of artifact, and thus the fewer the observed departures from the chance model. Since empirical as well as statistical artifact is involved, one is not protected from this problem by the statistical analysis itself. (Artifact turns up, of course, from time to time in "normal" science; it is through the use of such measures as control groups, when relevant, and through the demand that findings be replicable by neutral others that makes artifact only a minor problem in that domain.)

Thus, for example, if "sensory leakage" were not involved and if we were to attempt to match up subjects' calls in a card-guessing study in which, unbeknown to the subject, the cards were all turned yesterday, we should observe significant effects with the same frequency that we observe them if the cards are turned over only after the order of the entire deck has been described, or even if the cards are turned over two years hence. It should make no difference whether the cards are turned over in Moscow and the agent is in New York, or whether the sender and receiver are just next door to each other. Yet this is just what seems to show up in studies of psi: The independence of psi effects from considerations of time and space are well known, and only rarely questioned, within parapsychology.

By itself, this artifact hypothesis offers nothing new; it merely represents the viewpoint of all those critics who believe that empirical parapsychological claims are based in error of some kind rather than in psi. What I do wish to offer as new, however, is the proposal that recent trends in parapsychological research, which on the surface would seem to build to even greater heights the wondrous capabilities of psi, actually *weaken* the psi interpretation and give more credence to the notion that so-called psi effects are due to artifact. For example, when Rex Stanford can state, as he did in the *Handbook of Parapsychology*, that:

> In short, PK success does not depend upon knowing the PK target, upon knowing the nature or existence of the REG [random event generator], upon knowing one is in a PK study, upon the complexity or the design of the REG, or upon subjects knowing anything about the mechanics of the REG. These findings are highly consistent and appear in different forms throughout the PK literature. [1977, p. 341]

then, rather than adding to the case for psi, this seems to me to add support to the artifact hypothesis. That is, nothing unusual is going on and the observed departures from chance are due to unmeasured biases in the situation as well as to the vicissitudes of chance itself. Similarly, recent evidence suggesting that people apparently are capable of succeeding in psi tasks even when they are totally unaware that psi is involved in the study (e.g., see Palmer [1978]), and the apparent capability of some subjects to successfully shuffle decks of cards so as to have the final order statistically match a concealed target (e.g., see Morris [1982]) also are in line with the artifact hypothesis.

Even more support for the artifact hypothesis is forthcoming when one considers one of the newest areas of excitement in parapsychology, that of retroactive psychokinesis: The so-called observational theories of Schmidt and Walker predict that random events can be affected by the simple fact that the outcome is being observed by people, even if those people only observe it at some future time (e.g., see Bierman & Weiner [1980]). This has led to studies in which, for example, several sequences of tones, whose presence or absence during successive one-second intervals is determined by a random-event generator, are tape-recorded on separate tapes. At a later time, subjects are given a randomly selected tape to listen to and asked either to try to keep the tone on or to try to keep it off. The tapes are then analyzed to see whether or not the proportion of tones departs from what would be expected by chance. In some studies, tapes given subjects asked to keep the tone on have apparently contained significantly more tones on than those of the subjects asked to keep them off. This has been taken to suggest that the subjects' influence extended backwards in time so as to affect the original recording process. (Morris [1982] and Schmeidler [1982] have independently suggested that it is also possible that the experimenter used psi to select those tapes that varied above the chance expectation and gave them to the subjects in the "tone on" group. This would "explain" the observed results without need of backward influence in time.) In a related vein, there is the so-called checker effect, which describes the apparent possibility that ESP scores might be influenced retroactively by the person who checks or analyzes the data (Palmer [1982]).

One could go on and on in citing parapsychological claims that strain credulity: e.g., Helmut Schmidt's (1970) finding that cockroaches were apparently able to influence a random-event generator in such a way as to cause them to be shocked more often than would be expected by chance led Schmidt to suggest that perhaps it was his own psychic power that, because of his dislike of cockroaches, led to the observed effect. As a psychologist, I would have more readily opted for an explanation based on psychic masochism!

I hope that by now my point is clear: The more that parapsychologists report evidence of new statistical departures from chance that suggest new and *wilder* psychic effects, the *less* credibility their claims should have. If artifact is responsible for the reported deviations from chance in the classic psi experiments, then we should expect that similar statistical procedures carried out on data gathered in similar circumstances (i.e., by comparing two sets of scores, one generated by subjects, another generated by some chance process), by researchers with similar degrees of belief in the putative phenomena being studied, should at least sometimes produce similar significant effects. The finding that psi effects turn up whether one uses cockroaches or college students, whether the effects are to be generated in the present or the future or the past, whether the subjects know that there is a random generator to be affected, whether a sender and receiver are inches or continents apart—this *generalizability* of psi to, it seems, almost any situation in which one matches subjects' scores against a list—weakens rather than strengthens the case for parapsychology.

What I attribute to artifact, parapsychologists in effect attribute to psi. Their term then leads them to interpret their data as supporting psychic beliefs. If parapsychologists wish to argue that "psi" is more than "artifact," then the first thing they should do is to inform us as to what conditions, or what tasks, *never* give rise to extra-chance effects. In other words, when does this so-called psi *not* manifest itself? To respond that this occurs when controls are too rigid, or when the experimenter is too skeptical or too emotionally "cold," is not satisfactory.

Conclusion

Despite the enthusiasm for the new "quantum mechanical" theories, nothing of substance has occurred in parapsychology in the past eight years. The same old reasons for skepticism—the lack of public replicability, the problems of defining just what it is that "paranormal" signifies, the circular reasoning inherent in explaining departures from chance in terms of a "psi effect," the unfalsifiability that enters the picture whenever it is suggested that the experimenter's own characteristics or even his/her own psi or lack thereof may prevent him/her from ever observing psi, the failure of a century of research to improve the evidence—are as strong arguments against the psi position today as they were in the past. A new reason for skepticism is that, no matter how wild the hypothesis may seem, statistical evidence can be adduced that supports the claim; this suggests that artifact rather than "psi" is the most probable explanation for the statistical deviations reported in parapsychological research.

References

Bierman, D. J., and Weiner, D. H. 1980. A preliminary study of the effects of data destruction on the influence of future observers. *Journal of Parapsychology* 44:233-234.

McMullan, J. T. 1971. A possible physical source for the energy needed in poltergeist activity. Correspondence. *Journal of the American Society for Psychical Research* 65:493-494.

Morris, R. L. 1982. Assessing experimental support for true precognition. *Journal of Parapsychology* 46:321-336.

Palmer, J. 1978. Extrasensory perception: Research findings. In *Extrasensory Perception*, vol. 2 of *Advances in Parapsychological Research*, ed. by S. Krippner, 59-244. New York: Plenum.

Schmeidler, G. R. 1982. PK research: Findings and theories. In *Advances in Parapsychological Research*, vol. 3, ed. by S. Krippner. New York: Plenum.

Schmidt, H. 1970. PK experiments with animals as subjects. *Journal of Parapsychology* 34:255-261.

Stanford, R. G. 1977. Experimental psychokinesis: A review from diverse perspectives. In *Handbook of Parapsychology*, ed. by B. B. Wolman, 324-381. New York: Van Nostrand.

4

Sense and Nonsense in Parapsychology

Piet Hein Hoebens

Parapsychology is indistinguishable from pseudo-science, and its ideas are essentially those of magic.

Parapsychology is a farce and a delusion, along with other claims of wonders and powers that assail us every day of our lives.

These somewhat unflattering remarks are taken from the concluding paragraphs of two recent books in which the pretensions of parapsychology are examined from a skeptical point of view. The first is from James Alcock's *Parapsychology: Science or Magic?*[1] and the second from James Randi's *Flim-Flam!*[2]

It was to be expected that such sentences would provoke the indignation of the parapsychologists—who in fact were quick to point out what they perceived as gross unfairness on the part of both authors. The complaint most frequently heard was something to the effect that both Alcock and Randi have overstated their case by generalizing their (often justified) criticisms of a subset of paranormal claims to the entire field of parapsychology—thereby tarring all proponents with the same brush. It is argued that, in their eagerness to exorcise the demons of the New Nonsense, the skeptics have failed to take into account the differences between "serious parapsychology" and the less than serious variety.

In his witty (and by no means unsympathetic) review of *Flim-Flam!* in *Theta*,[3] Douglas M. Stokes writes: "In fact, almost all of the phenomena and claims Randi critiques in the book would be equally quickly dismissed by any competent parapsychologist as well. Only the lunatic fringe is going to be outraged by Randi's exposure of Conan Doyle's pictures of fairies, the underwater pyramid and road near Bimini, the space voyages of Ingo Swann and Harold Sherman, the Sirius 'mystery,' ancient astronauts, 'transcendental levitation,' biorhythms, N-rays, psychic surgery, or the oversexed spirits of Kübler-Ross."

This quote is interesting not only because it reveals what Stokes thinks of several of the best-known practitioners of Future Science but also because it implies the existence of a class of persons deserving the label "competent parapsychologists" and easily distinguishable from the crackpots who believe in the Cottingley Fairies, psychic surgeons, and the cosmic outings of Mr. Swann.

I suspect that both Stokes and the skeptics somewhat oversimplify matters: the skeptics, by underestimating the internal differences within parapsychology; Stokes, by projecting an all-too-neat competent/incompetent dichotomy on the complex and confusing reality of modern psychical research.[4]

It simply will not do to reproach the critics for discussing certain outlandish claims in the context of a critique of "parapsychology" or for attacking "weak" and "unrepresentative" cases, since the "parapsychological community" itself cannot agree on the criteria for "strength" and "representativity." We are faced with a similar problem if we want to decide who does or does not belong in the "community." Several proponents have suggested that membership in the Parapsychological Association (PA) and/or a record of publications in the PA-affiliated journals be regarded as a suitable criterion. However, membership in the PA and a record of publications in the serious journals does not guarantee the absence of the sort of beliefs Stokes thinks characterize the lunatic fringe. In parapsychology, the chaff and the wheat overlap to such an extent that a neutral observer often finds it hard to tell the difference.

This of course does not justify overgeneralizations on the part of the critics. Precisely because parapsychology is an ill-defined field lacking a shared "paradigm," it would be unfair to hold each "parapsychologist" individually co-responsible for everything that is claimed by his or her nominal colleagues.

My purpose in this essay-review is to illustrate the previous points by comparing three recent books written by prominent European parapsychologists. One of these books is a clear refutation of the claim implicit in some critical publications that parapsychology is ipso facto antagonistic to skeptical inquiry. The two other books demonstrate with equal clarity that the sort of parapsychology skeptics rightly find objectionable is *not* confined to the *National Enquirer* and the ad pages of *Fate*.

Martin Johnson

Martin Johnson, the Swedish professor of parapsychology at Utrecht State University, is a somewhat controversial figure in the Netherlands—because local "believers" suspect him of being a closet skeptic. When, around 1973, the university authorities announced their intention to appoint Johnson "professor ordinarius," the Dutch Society for Psychical Research, dominated by the redoubtable Dr. Wilhelm Tenhaeff (who, much to his chagrin, had never been promoted from his second-rate status as a "special professor"), initiated an unprecedented and outrageous press campaign against that "Nordic woodchopper" who, because of his "gross incompetence," would "destroy the life's work of the nestor of Dutch parapsychology." (For more about Tenhaeff, see my two-part article on Gerald Croiset in Chapters 12 and 13 of this volume.) Newspaper offices were flooded with angry letters. Questions were asked in Parliament. With a few notable exceptions, the Dutch media supported the "genius Tenhaeff" against the intruder from the Lapp tundra. The university was forced to accept a compromise. Johnson was appointed ordinarius but in addition Henri van Praag was appointed "special professor" to guard Tenhaeff's heritage.

Since then, the "special professor" has kept Dutch occultists happy with breathtakingly uncritical books, articles, and lectures on Sai Baba, flying saucers, Rosemary Brown, reincarnation, fairies and leprechauns, Uri Geller, Ted Serios, psychic surgery, and the imminent Age of Aquarius, while Johnson quietly established what has now become one of the most prestigious and respected parapsychology laboratories in the world.

Parapsychologie (originally published in Swedish) is Martin Johnson's first book.[5] It is intended as a general introduction to this controversial field. In refreshing contrast to most such introductions, it contains no pompous statements to the effect that the existence of psi has been demonstrated beyond any doubt and that only blind materialist prejudice keeps the scientific community from joining the parapsychological revolution.

To the contrary: Johnson agrees with the skeptics that the evidence for psi is weak and ambiguous and quite unable to support the grandiose cosmological claims others have tried to base on it. On the other hand, he believes some of the evidence is sufficiently suggestive to warrant further research based on the reality of psi as a working hypothesis. While Johnson personally is inclined to predict that future investigations will vindicate the psi hypothesis, he insists that the hoped-for breakthrough can only result from applying more rigorous research methods and from exercising more self-criticism.

Johnson agrees with his colleagues that there is considerable empirical support for the claim that something in the nature of ESP and PK (psychokinesis) exists, but he does not believe that this empirical support amounts to anything like proof positive. In a concise survey of the evidence presented hitherto he notes some promising developments (such as Helmut Schmidt's work with random-event generators and some research into psi and personality) but concludes that even with the most sophisticated experiments potentially fatal problems remain. In this context, he is remarkably candid about the role unconscious manipulation and deliberate fraud may play in his field. In this book, we are spared the ritual complaints about C. E. M. Hansel's supposed pig-headedness. Instead, Hansel's critique is welcomed as an interesting contribution to the debate.

There is an intelligent discussion of the replicability problem that has bedeviled parapsychology ever since its inception. Johnson points out that the concept of a "repeatable experiment" is more complex than is often assumed by critics. In mainstream science, opinions wildly differ as to the level of replicability required for academic respectability, whereas history has shown examples of perfectly repeatable observations based on collective misconceptions. However, Johnson does not invoke these methodological subtleties in order to excuse parapsychology's shortcomings. He is quite firm in stating that replicability in parapsychology is insufficient, especially given the extraordinary nature of the claimed phenomena.

A fairly long chapter deals with the numerous attempts to make theoretical sense of psi. Johnson concludes that almost all such attempts precariously depend on "more or less fantastic auxiliary hypotheses" and usually raise more questions than they answer.

In a hilarious chapter on "Miracle Men" Johnson practices some hard-line debunking at the expense of the Uri Gellers and the Sai Babas—and of the parapsychologists who have uncritically endorsed these psychics. "Personally, I am amazed that an intelligent and honest man such as Erlendur Haraldsson [the Iceland parapsychologist who published some remarkably naive eyewitness-accounts of the Indian saint's feats] seriously considers the possibility that Sai Baba, or the Babas of lesser caliber, could be anything but ordinary frauds," he writes.[6]

The section on Uri Geller and other metal-benders is devastating—and should make some of Johnson's fellow parapsychologists blush with embarrassment. At the occasion of the 1976 Utrecht parapsychology conference, which he hosted, Johnson invited a Swedish amateur magician, Ulf Mörling, to demonstrate "psi" for the benefit of the assembled participants. From the outset, Mörling clearly stated that he did not claim any paranormal ability whatsoever and that all his feats were based on conjuring tricks. Alas, Johnson writes, after the performance was over several prominent parapsychologists became "skeptical" and started to speculate seriously about whether Mörling might be a genuine psychic without being aware of it. The PA member who most staunchly defended this theory was—the reader will have guessed—Ed Cox, former associate of the late Dr. Rhine and a self-proclaimed foolproof expert on magic.

Johnson is appalled by the credulity some of his colleagues exhibited at the height of the Geller psychosis. He believes that this greatly contributed to the skeptical backlash of the second half of the seventies.

Johnson is not overly optimistic about his field's immediate outlook: "I think that parapsychology is presently in a critical stage. More unambiguous and robust findings will have to be presented if we want to justify its continued presence at the universities." And: "Time will tell whether psi research will bring about a conceptual revolution—or will languish in the backyards of the established sciences."

Having read *Parapsychologie* several times I am struck by the remarkable similarities between Martin Johnson's views and those of Ray Hyman, the skeptical psychologist who, among other things, is a member of the Executive Council of CSICOP. The book hardly contains a single statement to which a skeptic could reasonably object—unless he resorts to the a priori argument that the inherent absurdity of the concept of psi renders any serious attempt at investigation a waste of time.

Hans Bender

Professor dr. med. dr. phil. Hans Bender is a big name in international parapsychology. Arguably, he is the most renowned representative of the field in continental Europe. His credentials are impressive. He is a (now retired) professor at the Albert-Ludwig University in Freiburg, a former president of the PA, founder of the serious *Zeitschrift für Parapsychologie und Grenzgebiete der Psychologie*, a onetime host to international conferences, a contributor to John Beloff's state-of-the-art volume, *New Directions in Parapsychology*, and author of numerous papers published in reputable journals. By any definition, he belongs to the core of the international parapsychological community.

Bender too is a controversial figure at home. While thousands of Germans adore him as a prophet of the New Age of post-mechanistic spitituality, the highbrow media in West Germany derisively refer to him as "der Spukprofessor." Bender has frequently and bitterly complained that he has been the victim of unfair criticism. He certainly has a point here: The average postwar German skeptic is hardly noted for polemical subtlety. Bender's enemies have mercilessly exploited an embarrassing incident that took place a few years ago. (The magazine *Der Spiegel* alleged that, for three decades, Bender had falsely sported a "dr. med." degree. The professor was unable to produce evidence to the contrary. Recently, he obtained a genuine medical degree on the strength of an extremely curious thesis on poltergeists.) All too often, they have indulged in ad hominem attacks and in misrepresentation of the claims they had set out to debunk. The anti-parapsychological writings of the Mannheim jurist Dr. Wolf Wimmer in particular contain a number of deplorable examples.

A closer examination of Bender's publications, however, may to a certain extent explain why parapsychology continues to arouse such hostile feelings among German rationalists.[7]

Like Martin Johnson's *Parapsychologie*, Hans Bender's *Unser sechster Sinn* ("Our Sixth Sense"), a revised and enlarged edition of which became available in 1982, is a general introduction intended for a lay public.[x] The authors of such books bear a special responsibility, since they must assume that, for the average reader, *this* book will be the most authoritative source of information on parapsychology he will ever be exposed to. General introductions, especially if written by university professors, decisively influence opinions and beliefs. That is why we may demand that the authors carefully refrain from overstating their case and give a fair presentation of the pros and the cons. Johnson's book adequately meets this criterion, as did a small number of earlier publications, such as those by West and Beloff.[9,10] *Unser sechster Sinn*, I am afraid, does *not* belong in this category. The purpose of this book is propagandistic rather than informative. The reader is urged to accept Bender's beliefs as scientifically established facts and is not alerted to possible rational objections to the author's views. The weaker points of parapsychology are carefully glossed over. Instead, we are regaled with the success story of a triumphant new science with revolutionary implications for our views of God, Man, and the Universe.

I believe that I am not the only reader to gain the impression that Bender basically is not interested in evidence, except when it can be used to illustrate a transcendent Truth that he personally would be happy to embrace without any evidence at all.

From Bender's discussion of so-called spontaneous phenomena, the casual reader will never guess why informed critics (including several prominent parapsychologists) resolutely refuse to accept such anecdotes at face value. Examples of seemingly perfect cases that were later conclusively exposed as due to error or fraud are conspicuously absent, although such examples are essential for understanding the controversial status of psi. The "normal" psychological factors that may lead to an "occult" interpretation of nonparanormal events are hardly mentioned at all. Alternative hypotheses to account for the data are either ignored, dismissed, or presented as applying only to an untypical subset of cases. I have reasons to take Bender's anecdotes with a grain of salt. The fact that he repeats the long-discredited claim that Jeane Dixon "predicted the assassination of John F. Kennedy" should suffice as a warning.

A similar bias is apparent in the sections on the mediumistic phenomena that were the main subject of pre-Rhine psychical research. Eusapia Palladino is discussed without any mention of the numerous occasions she was caught in fraud. The exceptionally important writings by the great German skeptics who flourished in the first decades of this century are ignored, except in one instance where Bender gives the wrong author for the chapter on the Schneider brothers in the classic "Drei-Manner-Buch"[11] and dismisses the critics' arguments without even telling us what these arguments amounted to.

The classic laboratory experiments of the Rhine era are dealt with in a similar spirit. Hansel's criticisms of the celebrated ESP tests with Hubert

Pearce are summarily dismissed as having been conclusively refuted by Honorton and Stevenson. The naive reader gains the impression that there never was any *serious* dispute over the work at Duke University.

Inexcusably, Bender has chosen to leave the section on Soal's experiments with Basil Shackleton virtually unchanged in the 1982 "revised edition." These experiments are presented as having provided extraordinarily strong evidence for ESP. Only in a later chapter—in a totally different context—does Bender casually remark that "tragically, doubts later arose as to the accuracy of some of Soal's protocols." Given the well-nigh incontrovertible evidence that this psychical researcher faked the most sensationally successful experiment in the history of parapsychology, Bender's discussion of Soal's work is—to put it mildly—utterly misleading.

The sections on the Rhine/Soal type of ESP and PK experiments performed in Freiburg present us with problems of a different nature. Bender claims fantastically significant results, but such claims are meaningless unless complete reports are available for skeptical scrutiny. The exact conditions prevailing during these experiments are anybody's guess. Inquiries in Germany revealed that no detailed reports were ever published. We have no means of knowing to what extent possible skeptical counter-hypotheses are consistent with the data. Could the significant results of the Achtert-Zutz experiments have been brought about by a coding system? Could the high-scoring subjects in the Pinno-Czechowsky experiment have filled in their scoring sheets *after* the random-event generator had produced the targets? Bender cannot blame the critic for being suspicious, especially since he himself compares some of these Freiburg experiments to the Soal-Shackleton series (p. 62).

In the (new) section on Gellerism, Bender alludes to attempts on the part of certain anonymous magicians to expose the Israeli metal-bender as a trickster, but he typically fails to provide the sort of details that might persuade the intelligent reader to agree with the prosecution. His conclusion is that Geller may on occasion have resorted to trickery ("as do almost all mediums when they are unsuccessful") but that "in *The Geller Papers* the physicist Charles Panati has published experimental results that *prove* psychokinesis." (Emphasis added.) No mention is made of the devastating criticisms of *The Geller Papers* by, among others, Martin Gardner and Christopher Evans.[12] We are not even allowed to know *which* "experimental results" Bender thinks have proved Geller's PK.

The metal-bending star-subject at the Freiburg institute—the Swiss Silvio M.—is introduced as a genuine psychic who has been able to perform his feats while observed by an acquaintance who is a member of the Berne magic circle. Bender does *not* tell us that Silvio was unable to demonstrate any PK while observed by the prominent German magician Geisler-Werry and by Freiburg's own trick expert Lutz Muller or that he was caught cheating on several occasions. While I do not deny that there may be an as yet unexplained (as opposed to inexplicable) residue in the Silvio evidence, I object to Bender's suppressing facts that might cause his readers to doubt the authenticity of the Silvio phenomena.[13] Similarly, I object to Bender's uncritical endorsement of Ted Serios's "thoughtography." *Unser sechster Sinn* is completely silent about the serious doubts that have been raised by skeptics and critical parapsychologists alike concerning the paranormality of these feats.

The sections on Bender's favorite clairvoyant—the late Gerald Croiset of Holland—are nothing short of disastrous. Bender credits Croiset with having paranormally located the remains of a missing Scottish woman in the early seventies, whereas in fact her body has never been found. He further enthusiastically relates the astonishing results of the 1953 "chair test" in Pirmasens, where Croiset is supposed to have given a highly accurate precognitive description of two persons who, at a specified moment in the future, would happen to be seated in specified chairs. Not only is Bender's "paranormal" interpretation of this case absurd; his account also abounds with factual errors. Bender cannot claim ignorance in this instance, as he himself had been the chief experimenter, and the "raw data" are kept at his own institute. Pirmasens is one of Bender's prize cases, and he has referred to it in numerous books, articles, and lectures. My own investigations into this alleged miracle have raised serious doubts about Bender's credibility as a reporter of unusual events.[14] In his evaluation of the 1969 "transatlantic chair test" U.S. parapsychologist Jule Eisenbud made with Croiset, there is a curious discrepancy with the original 1972 edition. In 1972, Bender called this experiment "successful." In 1982 he calls it "controversial." What has caused Bender to change his mind? We are not allowed to know. Needless to add that Bender refrains from informing his readers of the reasons many of his fellow parapsychologists now regard the late Wilhelm Tenhaeff, Croiset's chief chronicler, as a disgrace to the profession. Instead, Tenhaeff is hailed as one of the pioneers of psychical research.

The section on poltergeist phenomena naturally gives pride of place to the celebrated Rosenheim case of 1967-68, which was investigated by Bender and his team. Rosenheim is generally considered one of the most striking ghost stories of all time, and not without justification. From the available material it seems difficult to think of a nonparanormal scenario to account for the data without leaving an uncomfortable number of "loose ends."

However, the case is certainly not as strong as Bender suggests. No full report of the investigations has ever been published, so we are in no position to check to what extent the parapsychologists have been successful in excluding naturalistic explanations. A case in point is the heavy (about 175 kilograms) cabinet that is said to have been moved 30 centimeters away from the wall by a paranormal agency. It is implied that Annemarie S., the young office-girl who was seen as the focus of the disturbances, could never have achieved this by normal means. However, in Bender's accounts one searches in vain for the answers to such essential questions as: Did anyone witness the actual movement of the cabinet? Did the cabinet weigh 175 kilograms when empty or is the weight of the files that were kept there included? Did the cabinet have handgrips? What experimental evidence has convinced Bender that 19-year-old girls cannot move 175-kg cabinets? (Experiments with my own 230-kg piano suggest that they can.)

Worse is that Bender omits from his account the highly significant fact that Annemarie was caught in fraud by a policeman. Neither does he mention the inconclusive but curious discoveries reported by the Viennese magician Allan after a visit to the Rosenheim office during the poltergeist outbreak.[15] He states that it was possible to capture a "phenomenon" (a painting turning around "120 degrees"—that is 200 degrees less than was claimed in Bender's first report!) on Ampex film. He does not tell us why persons who know something of the background of that incident refuse to be impressed with this piece of evidence.

In none of his publications of which I am aware has Bender ever referred to the suspicious features of the case.[16] Presumably, his silence has misled Eysenck and Sargent, in their militantly pro-psi book *Explaining the Unexplained*,[17] to claim that "despite the fact that many people—highly trained in different disciplines—were looking for evidence of fraud all the time, no hint of it was ever sniffed."

The publication of Bender's most recent book, *Zukunftvisionen, Kriegsprophezeiungen, Sterbeerlebnisse* ("Precognitive visions, war prophecies, death experiences"),[18] has done little to restore my faith in the nestor of German parapsychology. A detailed examination of this incredible work would be beyond the scope of this article. I will restrict myself to exposing what to the uninitiated reader must appear to be a perfect proof of the reality of precognitive ESP. On the first page of the book, under the chapter heading "Visions of the Future from a Scientific Perspective," he tells the story of the American student Lee Fried, who, Bender says, in 1977 dreamt about a recently deceased friend who showed him a newspaper bearing a future dateline. The headlines referred to a collision of two 747's over Tenerife with 583 people dead. Fried informed the president of his university of his premonition. Ten days later, the terrifying dream came true to the letter. According to Bender, "the opponents" will try to explain away such miracles by questioning the accuracy of the facts but in the case of the Tenerife prediction they stand no chance, for the documentation of the facts cannot be faulted. (In the chapter on parapsychology that Bender and his apprentice Herr Elmar Gruber contributed to *Kindlers Handbuch Psychologie*, it is stated that the Fried prophecy is "reliably documented.") All the stubborn skeptics could possibly do, Bender says, would be to resort to the preposterous hypothesis that the perfect match between premonition and actual disaster could have been brought about by chance.

Alas, the paranormal warning-system does not seem to have worked for Hans Bender when he wrote down those paragraphs. His version of the facts would indeed seem to preclude a naturalistic explanation. However, he managed to get all the crucial facts wrong. The Fried "prophecy" is a well-known, much-publicized, well-documented, and confessed hoax. In Bender's account, the facts have been distorted almost beyond recognition.

Fried of course never *told* the president of the university of the impending Tenerife disaster.[19] What he did do was to put an envelope, said to contain an unspecified "prediction," in a locked drawer in the president's office. When the catastrophe had taken place, the envelope was opened and a piece of paper with the words "583 Die in Collision of 747's in Worst Disaster in Aviation History" was produced. Soon after, Lee Fried frankly admitted that he had planned the prediction as a stunt. The slip of paper containing the "prophecy" was inserted only *after* the disaster, by sleight of hand.

According to *The Second Book of the Strange*,[20] the Fried episode has shown that "the credulousness of at least a proportion of the news-consuming public is almost unlimited."

We cannot really blame the public for occasionally failing to distinguish between fact and fraud. However, we are entitled to expect better from the most prestigious representative of scientific parapsychology in Germany.

Discussion and Conclusion

Hans Bender has stated that his conviction that the paranormal exists is "unshakable." Furthermore, he has repeatedly affirmed his belief that a wider acceptance of psi will be highly beneficial to mankind. Parapsychology is the supreme weapon against the "mechanistic world-view" of the intellectual establishment—a world-view that Bender holds responsible for many of modern society's most serious problems. From the vantage point of the moralist, such considerations would justify a certain nonchalance vis-à-vis the scientific facts. As soon as one's convictions become unshakable, evidence ceases to be relevant—except as a means to convert the unbelievers—and factual inaccuracies in the parapsychological propaganda are excusable in the light of the Higher Truth. I do not wish to impugn Bender's integrity. I am satisfied that his public statements and actions are consistent with his personal values. These values, however, are clearly incompatible with the spirit of scientific inquiry.

It is typical of the pre-paradigmatic status of parapsychology that Bender continues to be regarded as one of the foremost representatives of the field. Alcock has posed the question: Is parapsychology science or magic? No unambiguous answer can as yet be given. While Martin Johnson has shown that at least *some* parapsychologists are engaged in activities virtually indistinguishable from what critics mean by "skeptical inquiry," the case of Hans Bender demonstrates that the demarcation line separating scientific parapsychology and fringe occultism is by no means as sharply drawn as some proponents have optimistically claimed.

Notes

1. J. Alcock, *Parapsychology: Science or Magic?* Pergamon Press, Oxford, 1981.

2. J. Randi, *Flim-Flam! Psychics, ESP, Unicorns, and Other Delusions,* Prometheus, Buffalo, N.Y., 1982.

3. D. Stokes, review in *Theta,* vol. 9, no. 1, Winter 1981.

4. Elsewhere in his review, however, Stokes writes of "a wide continuum of parapsychologists, ranging from the skeptical to the credulous, with no clear line of demarcation separating the two groups."

5. M. Johnson, *Parapsychologie, Onderzoek in de Grensgebieden van ervaring en wetenschap,* De Kern, Baarn, Holland, 1982.

6. Dr. Haraldsson kindly sent me his comment on Dr. Johnson's criticisms. He argues that Johnson implicitly adopts a criterion that would justify summary rejection of *any* anomalous claim.

7. To a certain extent only. Too many German critics have ignored the fact that, after the emergence of the "new conservatism" in German parapsychology, the field is no longer monopolized by Bender. Ideologically, these new conservatives (e.g., Eberhard Bauer, Gerd Hövelmann, Klaus Kornwachs, and Walter von Lucadou) are close to Martin Johnson. For an excellent survey of recent developments in Germany see E. Bauer and W. von Lucadou (eds.), *Spektrum der Parapsychologie,* Aurum Verlag, Freiburg, 1983.

8. H. Bender, *Unser sechster Sinn,* 1st ed., 1972; revised and enlarged edition, Wilhelm Goldmann Verlag, Stuttgart, 1982.

9. D. West, *Psychical Research Today,* Duckworth, London, 1954.

10. J. Beloff, *Psychological Sciences,* Crosby Lockwood Staples, London, 1973. Remarkably well-balanced chapter on parapsychology.

11. W. Von Gulat-Wellenburg, C. Von Klinckowstroem, and H. Rosenbusch, *Der Physikalische Mediumismus,* Ullstein, Berlin, 1925. Classic critique of the so-called "physical phenomena." Soon acquired the sobriquet "Drei-Männer-Buch" (Three men's book).

12. M. Gardner, *Science: Good, Bad and Bogus,* Prometheus Books, Buffalo, N.Y. 1981. C. Evans, review of *The Geller Papers* in the *Humanist,* May/June 1977.

13. Silvio's cheating is documented in *Zeitschrift für Parapsychologie und*

Grenzgebiete der Psychologie, vol. 23, no. 2, 1981. In recent years, this journal has adopted an open policy. Skeptical contributions are welcome. Here, the growing influence of the "new conservatives" is felt.

14. My exhaustive critical analysis of this alleged·miracle is scheduled for publication in the *Zeitschrift*.

15. Allan, Schiff, and Kramer: *Falsche Geister Echte Schwindler*, Paul Zsolnay, Vienna, 1969.

16. It is only fair to mention at this point that inadequate reporting on poltergeist cases is not Bender's monopoly. In 1978 the criminologist Dr. Herbert Schäfer told the press that Heiner Scholz, focus person of the celebrated Bremen case of 1965-66, had made a complete confession. Bender and his colleague Johannes Mischo have pointed out serious flaws in the fragmentary press accounts of this exposé. Schäfer never published a complete report of his findings, nor has he publicly replied to the parapsychologists' counter arguments.

17. H. Eysenck, and C. Sargent, *Explaining the Unexplained*, Book Club Associates, London 1982.

18. H. Bender, *Zukunftvisionen, Kriegsprophezeiungen, Sterbeerlebnisse*, Piper, Munich, 1983.

19. Personal communication from Terry Sanford, president of Duke University, July 1983.

20. *The Second Book of the Strange*, by the Editors of *The World Almanac*, World Almanac Publications, N.Y., 1981

Expectation and Misperception

5

On Coincidences

Ruma Falk

Flammarion recalls (in *The Unknown*) that when he was writing a chapter on the force of the wind for his *L'Atmosphere*, a gust of wind carried his papers off "in a miniature whirlwind beyond hope of recovery." A few days later, he received the proof of that chapter from his publisher, complete with the missing pages. They had been deposited in a street frequented by the publisher's porter, who often carried Flammarion's material, and who gathered them up under the impression that he must have dropped them.

From the chapter "Coincidences" in *Phenomena: A Book of Wonders* (1977) by Michell and Rickard (p. 90)

We are all intrigued by coincidences. Extraordinary coincidences that were perceived as miracles have attracted people's attention since early days. It is easy to get people to talk about this subject because everyone has some incredible anecdotes stored away. Coincidences puzzle us in daily life as well as in various fields of scholarship and thought. Thus, the quotation: "A coincidence? Perhaps. Still, one can't help wondering . . ." might have been drawn out of a criminal novel, a legal debate, or a gossipy chat. Actually, it is extracted from a scientific article entitled "Emotional Stress and Sudden Death" by G. Engel in *Psychology Today* (November 1977, p. 114).

The subject finds lyric and symbolic expression in literature. Coincidences play a key role in the plots of some Greek tragedies. Incredible coincidences sometimes assume a special emotional significance also in contemporary literature. Thus, for example, much of the power and the startle of Boris Pasternak's *Doctor Zhivago* (1958) derives from the tragedy of fate that coincidentally brought the heroes together at the most crucial moments of their lives.

A Unique Event or One Case Out of Many?

Nathan Shaham tries explicitly to impress his readers in his book *Aller Retour* (1972, in Hebrew). He tells about two Israeli friends who had not seen each other for a long time and did not know of each other's whereabouts who one day coincidentally sat in adjacent seats in a London cinema:

> He repeated again and again: "What a remarkable coincidence!" and for a long time after that, he was still dwelling on an analogy that occurred to him. The chance that two people, who have not arranged a meeting and who live far away from each other, would meet at exactly the same time in neighboring seats in a movie theater is smaller than the chance of a domesticated monkey to type the British anthem on a typewriter.

However, when a similar accidental meeting occurred again, he wrote:

> His hair stood on end when he saw them, following the usher, heading toward the two empty seats next to his. If they are seated here, he told himself, it would be such an amazing occurrence from the statistical point of view that it could excite even a cold-blooded computer, since that legendary monkey ... could have typed out an entire Shakespeare play before such an occurrence would have happened twice in succession.

Strange coincidences always seem less surprising when they happen to others. My critical faculties were aroused reading Shaham's story. Somehow, I did not wish him to be the one selected by "divine forces." I remembered, though, incredible events that had happened to me. My attempts to understand the secret of their surprise led to the following analysis: It is conceivable that the element of surprise in such a small-world meeting stems from the fact that we attend to all the detailed components of the event as it actually occurred. The probability that an event would happen *precisely that way* is indeed minute.

When I happened to meet, while in New York, my old friend Dan from Jerusalem, on New Year's Eve and precisely at the intersection where I was staying, the amazement was overwhelming. The first question we asked each other was: "What is the probability that this would happen?" However, we did not stop to analyze what we meant by "this." What precisely was the event the probability of which we wished to ascertain?

I might have asked about the probability, while spending a whole year in New York, of meeting, at any time, in any part of the city, anyone from my large circle of friends and acquaintances. The probability of this event, the *union* of a large number of elementary events, is undoubtedly large. But instead, I tended to think of the *intersection* of all the components that converged in that meeting (the specific friend involved, the specified location, the precise time, etc.) and ended up with an event of minuscule probability. I would probably be just as surprised had some other combination of components from that large union taken place. The number of such combinations is immeasurably large; therefore, the probability that at least one of them will actually occur is close to certainty. Shaham's hero did not ask himself in advance whether he would encounter that very friend as his immediate neighbor on that particular visit to the cinema. On the contrary, he had nearly forgotten about that friend. Everybody has an enormous number of that sort of "good friend." Furthermore, there are plenty of occasions on which we are on the streets, at a movie, or in other public places. Is it really surprising that at least on some of those occasions we should meet some acquaintance?

Analysis of Surprises

How do people who hear stories about strange coincidences interpret these stories? Are they impressed by the intersection of all the factors that combined in the incident, or do they implicitly take into account that they are considering one out of a large set of events that could possibly have taken place? A questionnaire containing four sections was devised to try to answer this question.

Four stories described different coincidences (an unexpected meeting, a fortunate hitchhike, a convergence of birthdays, and a peculiar numerical combination). Each was based on a true story and each appeared in four versions. Seventy-nine adult subjects (in Israel) rated each story according to the perceived likelihood of the described event.

The past intersection (PI) version described the event as it happened in the *past*, emphasizing the *intersection* of all the amazing details. In one of the stories, for example, a soldier I had recently met told me that the previous day, her nineteenth birthday, she tried to hitchhike from the outskirts of Jerusalem. The first driver who stopped was going precisely to her destination, a little-known settlement in the Western Negev, Mivtahim, where she worked as a teacher.

The past union (PU) version told the same story, again in *past* tense. However, it also gave the reader several hints about the *union*, that is, about the universe of possibilities of which this story represented just one. In the hitchhiking story, for example, the soldier mentioned that she often hitchhiked and that the day before had been her first lucky day in the year. The story also mentioned that Mivtahim is situated on a major highway, suggesting to the reader that many people might have been driving on that road to places farther away and could have offered her a ride to her destination.

In the future union (FU) version, the same *union* of events was described but the question of likelihood referred to a *future* event. The soldier is about to start serving as a teacher. She wants to know her chances of receiving a ride directly to Mivtahim (which lies on a major highway) sometime during the year.

The future intersection (FI) version asked about a *future* event, but the question was very specific and concerned the *intersection* of all the above-mentioned factors. This soldier now wants to know what the chances are that precisely on her nineteenth birthday she will, on her first attempt, get a ride from Jerusalem directly to Mivtahim, the settlement where she works.

The other three stories were similarly improbable when one considered the intersection of all the components but very probable when the question related to the union of all the alluded possibilities. The questionnaire, given to each subject, included the four different stories, each in a different version (PI, PU, FI, and FU).

Since the temporal setting of an event should make no difference in its likelihood, and since unions should be more likely than intersections, one could expect both past and future intersection stories to be judged less probable than the two union stories.

Of all versions, FI is the simplest to judge. One asks about a future event, and all the components constituting that event are unambiguously enumerated. The comparison between the responses to version FI and version PI (the usual form of a coincidence) seems crucial. If the subjects perceive the event in version PI as one of many that could have happened, they may rate version PI as more likely than FI, possibly as high as the union stories PU and FU.

However, it is not inconceivable that the time setting as well as the scope of the event might affect the rating. The knowledge of the fact that an event took place may render it, in hindsight, more credible relative to the same event described hypothetically. Fischhoff (1975) got results to that effect based on subjects' historical judgments. The presentation of coincidences in both tenses may test for the existence of a hindsight effect on the perceived likelihood of such stories. If the temporal setting will affect the judgments to lend more credibility to past events, one should expect *both* intersection and union stories to be perceived more likely in the past than in the future.

This experiment was replicated at Decision Research in Oregon in collaboration with my colleague Don MacGregor. The subjects were 91 adults from Eugene, most of them students at the University of Oregon. Similar stories were presented in the same four versions. Names, places, and other minor details were changed to give the stories an American setting. Subjects rated the stories for surprisingness rather than for likelihood.

The pattern of the results, both from Israel and from Oregon, showed that a past-intersection story was perceived (at least partly) as a union event, and the time factor did *not* affect the judgments. The overall pattern best matched the order depicted schematically in Figure 1; i.e., the past-intersection and past-union stories were judged about equally likely and both were perceived more likely than the future-intersection and less likely than the future-union. In symbols: $P(FI) < P(PI) \approx P(PU) < P(FU)$, where P(version) should be read as: the perceived probability of that version.

The "union" story in the past (PU) was probably not judged as likely as the future union because of the difficulty in describing a union event in the past.[1]

Version FI was evaluated significantly less probable than each of the other three versions and, most important, less probable than version PI. This may answer the initial question of how the usual coincidence story, involving an intersection of events in the past, is perceived. Such a PI story is judged more probable than the hypothetical occurrence of a comparable intersection in the future. This cannot be due merely to the temporal difference between past and future, since a reverse ordering was obtained between the judgments of versions PU and FU.

The most plausible interpretation of the results suggests that, upon hearing a story about a past coincidence, one is aware of a wide range of possibilities and considers the event as one of many that could have happened. One is probably not encoding the story with all its specific details as told, but rather as a more general event "of that kind."

What Strikes People?

The results showed that the additional hints about the other possibilities,

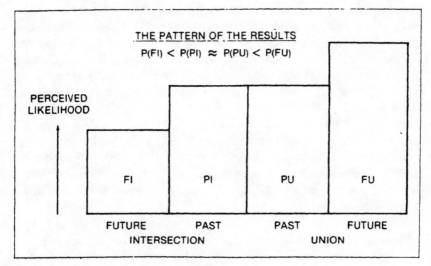

FIGURE 1

in version PU, did not increase the judged probability of PU stories over that of PI (see Figure 1). People do not seem to need those hints. They do not perceive the one event that did happen as more surprising than the occurrence of at least one of a set of events. What, then, is the source of the well-known amazement that nevertheless strikes us when such events occur?

One possible answer would be that we are amazed when a coincidence happens to us personally and are not particularly surprised to hear about such events happening to others. In the course of the Israeli experiment, many of the subjects, who (according to the results) had not viewed version PI stories as particularly surprising, told me spontaneously about incidents that had happened to them. However, these stories did not appear *to me* any more surprising than the four stories drawn from my personal experience that comprised set PI of the experiment.

Our Oregon subjects were specifically asked to write and rate (for surprisingness) a coincidence that had happened to them. These self coincidences were later given, along with our four PI stories, to a new group of 146 subjects to read and rate. The results showed that the subjects who wrote their own stories judged those stories to be more surprising than our PI stories, whereas the new group of subjects found the subject-authored stories less surprising than the PI stories. The self coincidences were also judged to be significantly less surprising to other subjects than they were to their authors.

Subjects reading stories written by others probably perceived the writer of a coincidence as one of many people and every event as a realization of one of many possibilities. They understood that "strange events permit themselves the luxury of occurring" (Gardner 1957, p. 307); hence the stories did not seem particularly improbable to them. However, when such an event happened to them personally, they found it difficult to perceive it as one of a set. One's uniqueness in one's own eyes makes all the components of an event that happened to oneself seem like a singular combination. It is difficult to perceive one's own adventure as just one element in a sample space of people, meetings, places, and times. The hardest would probably be to perceive oneself as one replaceable element among many others. When I analyzed, above, the story of my meeting Dan, I succeeded in extending that specific friend to many other acquaintances, the particular place to the whole area of New York City, and the time to an entire year. I kept only myself constant throughout all these extensions. Martin Gardner (1972, 112) expressed some of these ideas in a succinct fashion:

It is easy to understand how anyone personally involved in a remarkable coincidence will believe that occult forces are at work. You can hardly blame the winner of the Irish Sweepstakes for thinking that Providence has smiled on him, even though he knows it is absolutely certain that someone will win.

Maybe the same line of subjective reasoning applies to the other end of the "luck continuum." Perhaps that inability to perceive oneself as one element in a sample space hinders the realistic assessment of certain risks. It may contribute to the prevailing belief that "this can't happen to me," since population statistics do not apply to me (see, also, Snyder 1980; Slovic, Fischhoff, and Lichtenstein 1980).

A Fallacy of Selection

One characteristic of coincidences is that we do not set out to seek them in a predetermined time and place; they simply *happen* to us. Since they stand out in some strange combination, we single them out and observe them under a magnifying glass. A *selection fallacy* is here in action.

When we design an experiment based on a random sample, carry it out, and show the probability of its results to be extremely low under the hypothesis of mere chance (the null hypothesis), we are entitled to reject that hypothesis. Such a statistically significant result is justly viewed as supporting an interpretation other than chance. On the other hand, when we single out an extraordinary coincidence and claim that it is significant, we commit a logical error. Computing the probability that a surprising event of that kind would take place is not permissible, since the design of the experiment was not determined in advance and the sample space was not clearly outlined. Our "experiment," in the course of which we came across our anecdote, was not preconfined by time, space, or population. It attracted our attention because of its rarity; otherwise we would never have stopped to consider it. Having arrived at it this way, we are not permitted to claim that it is highly unlikely. Instead of starting by drawing a random sample and then testing for the occurrence of a rare event, we select rare events that happened and find ourselves marveling at their nonrandomness. This is like the archer who first shoots an arrow and then draws the target circle around it.

Arthur Koestler devoted several of his books to a detailed description and discussion of coincidences and parapsychological phenomena. His main conclusion (Koestler 1972) was that these phenomena are the result of beyond-causal lawfulness, as yet unexplainable within the framework of accepted physical theory. However, on close examination, one sees that those conclusions were based mostly on anecdotal events, occurrences not obtained in a systematic predesigned study but, rather, contributed by the people involved. In *The Case of the Midwife Toad*, Koestler (1971) tells the life story of the Austrian biologist Kammerer, who was the first to attempt to deal with coincidences "systematically." Kammerer wrote a book in which he described one hundred coincidences—dealing with people, dreams, words, and numbers—that he had collected in the course of twenty years. Some of these stories are indeed bewildering. However, on second thought, it appears that Koestler did not give due weight to the fact that these anecdotes were collected in a process of deliberate selection. Is it so surprising that a scientist who got a reputation as a collector of anecdotes would succeed in accumulating one hundred such stories in the course of twenty years?

In a similar vein, Hintzman, Asher, and Stern (1978) showed that a selection process is also operating on our *memory*. They claim that coincidences *seem* to occur too often to attribute to chance just because of selective remembering of meaningfully related events. Events that do not enter into coincidences are more likely to be forgotten than events that do.

How Small Is Our World?

The Nobel laureate physicist Luis Alvarez (1965) related, in a letter to *Science*, a strange event that had happened to him: A casual phrase in a newspaper triggered a chain of personal associations that reminded him of a long-forgotten figure from his youth. Less than five minutes later, having turned the pages of the paper, he came across an obituary for that same person. Alvarez noted that these two closely spaced recollections of a person forgotten for 30 years, with the second event involving a death notice, follow the classical pattern of many popular parapsychology stories. Such coincidences often make people feel that there must be a causal relation between the two events, as, for example, by thought transference. However, considering the details of this case, it was obvious that no causal relationship could have existed between the two events.

Alvarez felt that scientists should bring such stories to public attention to show that such apparently improbable coincidences do, in fact, occur by chance. He also endeavored to compute its probability by estimating the number of people that an average person knows and the average frequency with which one remembers one's acquaintances. Using rather conservative assumptions, he arrived at the conclusion that the probability of a coincidental recollection of a known person in a 5-minute period just before learning of that person's death is 3×10^{-5} per year. Given the population of the United States, we could expect 3,000 experiences of the sort to occur every year, or about 10 per day. Considering coincidences of other kinds as well increases this number greatly. "With such a large sample to draw from, it is not surprising that some exceedingly astonishing coincidences are reported in the parapsychological literature as proof of extrasensory perception in one form or another" (Alvarez 1965). Paradoxically, the world is so big that quite a few of these "small-world phenomena" are bound to occur.

Underestimating the total number of surprising possibilities could be one of the roots of our amazement in the face of the realization of one of them. Thus one is universally astonished to learn that the probability of at least 2 out of 23 randomly selected people having the same birthday is a little over 1/2. This may be due to our limited ability to appreciate the number of pairs one could form out of 23 objects.

In *Chariots of the Gods?* (1969), Erich von Däniken showers his readers with coincidences designed to convince them of the validity of his theory that ancient intelligent creatures visited earth long ago. For example:

> Is it really a coincidence that the height of the pyramid of Cheops multiplied by a thousand million ... corresponds approximately to the distance between the earth and sun? ... Is it a coincidence that the area of the base of the pyramid divided by twice its height gives the celebrated figure $\pi=3.14159$?*
> ... one might reasonably ask whether the "gods" did not have their say here, too ... If the facts noted here are not coincidences—and it seems extremely difficult to believe that they are—then the building site was chosen by beings who knew all about the spherical shape of the earth and the distribution of continents and seas. [p. 96]

A weak point in this supposedly scientific proof is the fact that the author had not predicted a priori that the ratio between the base and the height of that specified pyramid would equal 2π. Such a prediction would have made the conclusion unavoidable that the result is not coincidental. Why was the connection between the height of the pyramid and the distance from the earth to the sun selected for examination? Other magnitudes, or ratios between magnitudes, drawn from any pyramid's dimensions could have matched some mathematical, geographical, or astronomical parameter. The number of possible "surprises" is enormous; this fact reduces the surprising effect of each individual one once it has been discovered.

In an ingenious experiment reported in Hardy, Harvie, and Koestler's *The Challenge of Chance* (1973, pp. 72-109), Harvie showed that one can produce a considerable number of coincidences just by *randomly* combining a *large* number of elements. Harvie performed a random simulation of a large-scale experiment that had been conducted by the authors to test for telepathy. They employed many subjects who had to guess an unseen target stimulus many times. Thus a huge number of responses were collected. Indeed, remarkable results were obtained: The responses of different subjects on the same trial were at times incredibly similar to each other. The rate of such "coincidental thoughts" seemed surprisingly high.

---

*Von Däniken's facts and mathematics are wrong. These particular "coincidences" not only lack all meaning but are not even "true." See E. C. Krupp's *In Search of Ancient Astronomies* (Doubleday, 1978). p. 275—ED.

Moreover, these matches were qualitatively so astounding (pp. 43–46) that they could have converted even the extreme skeptic. However, Harvie's "mock experiment" showed that a completely random and noncausal rearrangement of the same responses yielded equally surprising outcomes.

Explaining Away Everything

Gardner (1957, pp. 303-05) has described how a selection mechanism often operates in the interpretation of ESP experiments. When a subject fails to guess with great success the target card presented in a sequence of cards, one sometimes finds that his or her guesses significantly match a card which is about to be presented (forward displacement). Sometimes the responses turn out to match the target card of the previous trial (backward displacement). Displacement of the target can eventually extend to two or three steps forward or backward. When the number of hits is particularly low, the researchers claim that this is a manifestation of negative telepathy (avoidance of the target). Considering the union of all the different ways in which ESP can manifest itself, there is little wonder that at least one of them does appear "nonrandomly."

Diaconis (1978) makes a similar point with respect to a series of supposedly extraordinary card tricks demonstrated by a young man called B.D. who was tested in the Psychology Department of Harvard University: "A major key to B.D.'s success was that he did not specify in advance the result to be considered surprising. The odds against a coincidence *of some sort* are dramatically less than those against any prespecified *particular one* of them" (p. 132). Even B.D.'s frequently missed guesses either were considered "close" or were regarded with sympathy rather than doubt and seemed only to confirm the reality of his unusual powers. This ability to explain whatever comes up is described by Fischhoff (1980) in terms of capitalization on chance: "The data are fixed and undeniable, while the set of possible explanations is relatively unbounded; one hunts until one finds an explanation that fits" (p. 23).

The selection fallacy can operate unconsciously even on a very conscientious investigator (Gardner 1957). Suppose 100 investigators decide to test for telepathy, each one using a single subject. Fifty of them are going to obtain, quite by chance, above-average results. They will go on and run a second experiment, whereas the other fifty will drop the research. In the next stage, we can expect another half of the investigators to drop out, and so forth in the following stages. Finally, we are left with one investigator whose subject achieved particularly high scores in six or seven successive sessions. "Neither experimenter nor subject is aware of the other ninety-nine projects, and so both have a strong delusion that ESP is operating. The odds are, in fact, much against the run. But in the total (and unknown) context, the run is quite probable" (p. 303). The reasonable question to ask at this point is whether the result will not be disconfirmed if the experimenter continued testing that subject. Sure enough, the phenomenon does tend to recede. Now the investigator will resort to those post hoc explanations like fatigue and the "typical decrease" in experiments of that kind, which is, of course, also the typical decrease in random sequences, better known as "regression to the mean." The main lesson to be learned from awareness of these pitfalls is that one should subject the nonrandom finding to a retest.

Miracles Do Not Recur

The very essence of a scientific rule is a claim that certain relationships are replicable. A coincidence, however, is not expected to recur in repeated trials.

A (statistically) significant result is an outcome obtained through a predesigned experiment with a random sample. The probability of such a result (or a more extreme one) being obtained at random is so small that we dare claim it is improbable that it occurred only by chance. Therefore, we can expect it to reappear under a similar design. A significant result is not a miracle; it is believed to be replicable (Brown 1957). When we pray for a miracle, we are hoping for a rare event that we cannot expect to repeat itself consistently. Those who are drowning in the ocean and yearn for a miracle to save them are aware of the small probability that a helicopter or a boat will materialize at the right moment. The point of resemblance between coincidences and significant statistical results is the low probability of the event in question. However, valid scientific results recur in a reproduction of the conditions, whereas coincidences do not. The probability of obtaining a significant result is very small *under the chance hypothesis*; but when we obtain one, we *reject* the chance hypothesis and replace it with another hypothesis. The validity of that alternative hypothesis guarantees the replication of the result under similar circumstances.

A Matter of Perception

Our perception of randomness is afflicted by typical fallacies and distortions (Falk 1975). The main bias in this intuitive perception is known as the "gambler's fallacy." When a tossed coin turns up with "heads" several times in succession, gamblers bet a huge sum on "tails," believing that, since the two faces are equiprobable, it is time for balance to be restored. The idea that the probability of tails at this point is higher is incompatible with the concept of sequential independence. While it is true that the probability of heads appearing n times in succession is very small a priori when n is large, the conditional probability of heads, following $n-1$

continuous heads, is still 1/2. This is because, unlike the gambler, the coin has neither memory nor conscience.

That fallacy often assumes another form when we observe a random cluster of identical events—like accidents or misfortunes—and conclude that such a run could not be the result of chance. We reject the hypothesis of randomness in light of a sequence of identical outcomes (a "run") and replace it by some "causal" mechanism (like the devil's interference). Some people start believing that "when it rains, it pours," as did the king in Shakespeare's *Hamlet*:

> When sorrows come, they come not single spies,
> but in battalions!
>
> [Act IV, Scene V]

It seems that we expect random events always to appear in their appropriate proportions even in short subsequences (Kahneman and Tversky 1972). The random runs we encounter often evoke suspicion of violation of randomness. Koestler's (1972) theory of coincidences attempts to generalize the "supernatural lawfulness" of the world of ESP to coincidences. If we inspect the nature of the coincidences described, we find that—much like the gambler who is surveying the sequence of outcomes—Koestler's attention is drawn to runs or clusters that appear nonrandom.

Likewise, a detailed study of the concepts employed by other researchers of coincidences reveals that a run or a co-occurrence of similar elements is often referred to. Such is the case with Kammerer's "recurrence" or "clustering" of events in time or in space, as well as with the term "synchronicity," coined by the psychologist Jung (with the physicist Pauli). Koestler (1972) clarifies his view on the extra-physical lawfulness of coincidences using the following description: "We thus arrive at the image of a world-mosaic or cosmic kaleidoscope, which in spite of constant shufflings and rearrangements also takes care of bringing like and like together" (p. 86). He imagined a prolonged process of random mixing and decided that the clusters he observed (i.e., the coincidences) could not be random. Ideal randomization should, in his opinion, completely separate similar elements. If they nevertheless appear in clusters, one must come up with a theory to explain such an odd occurrence—one cannot help but be reminded of the gambler's fallacy.

In conclusion, I have pointed out the possibility that cognitive biases—such as consideration of a too narrow event (especially with reference to oneself), a selection mechanism, and the gambler's fallacy—may be the source of erroneous supernatural theories; yet it was not proved. One can't completely rule out the possibility that the phenomena in question really are controlled by their own set of laws. One cannot but agree here with Koestler:

> Whether one believes that some highly improbable meaningful coinci-
> dences are manifestations of some such unknown principle operating
> beyond physical causality, or are produced by that immortal monkey at the
> typewriter, is ultimately a matter of inclination and temperament. [Hardy,
> Harvie and Koestler, 1973, p. 230]

References

Alvarez, L. W. 1965. "A Pseudo Experience in Parapsychology" (letter). *Science*
148: 1541.

Brown, G. S. 1957. *Probability and Scientific Inference*. London: Longmans
Green.

Diaconis, P. 1978. "Statistical Problems in ESP Research." *Science* 201: 131-136.

Falk, R. 1975. *The Perception of Randomness*. Doctoral dissertation (in Hebrew).
The Hebrew University of Jerusalem.

Fischhof, B. 1975. "Hindsight ≠ Foresight: The Effect of Outcome Knowledge on
Judgment Under Uncertainty." *Journal of Experimental Psychology: Human
Perception and Performance* 1:288-299.

———— 1980. "For Those Condemned to Study the Past: Reflections on Historical
Judgment." In R. A. Shweder and D. W. Fiske, eds., *New Directions for
Methodology of Behavior Science: Fallible Judgment in Behavioral Research*.
San Francisco: Jossey-Bass.

Gardner, M. 1957. *Fads and Fallacies in the Name of Science*. New York: Dover.

———— 1972. ("Mathematical Games" column) "Why the Long Arm of Coincidence
Is Usually Not as Long as It Seems." *Scientific American* 227 (4): 110-112B.

Hardy, A., R. Harvie, and A. Koestler. 1973. *The Challenge of Chance*. New York:
Random House.

Hintzman, D. L., S. J. Asher, and L. D. Stern. 1978. "Incidental Retrieval and
Memory for Coincidences." In M. M. Grunberg, P. E. Morris, and
R. N. Sykes, eds., *Practical Aspects of Memory*. London: Academic Press.

Kahneman, D., and A. Tversky. 1972. "Subjective Probability: A Judgment of
Representativeness." *Cognitive Psychology* 3: 430-454.

Koestler, A. 1971. *The Case of the Midwife Toad*. New York: Random House.

———— 1972. *The Roots of Coincidence*. London: Hutchinson.

Slovic, P., B. Fischhoff, and S. Lichtenstein (in press). "Informing the Public about
the Risks of Ionizing Radiation." In *Proceedings of the Public Meeting on a
Proposed Federal Radiation Agenda*. Bethesda: National Institutes of Health.

Snyder, C. R. 1980. "The Uniqueness Mystique." *Psychology Today* 13: 86-90.

Notes

Parts of this paper were included in one section of the author's Ph.D. thesis (carried out
under the supervision of Amos Tversky): *The Perception of Randomness* (in Hebrew), the

Hebrew University of Jerusalem, 1975. Parts were also included in a Hebrew paper: Falk, R. 1978 "On Miracles and Coincidences." *Mada*, 22: 130-134. I wish to thank Raphael Falk, for editorial comments and contributions to all the drafts of this paper. I am also grateful to my friends at Decision Research: Baruch Fischhoff, Don MacGregor, Sarah Lichtenstein, Paul Slovic, Peggy Roecker, Maya Bar-Hillel, and others for their helpful comments.

1. The stories of version PU did not claim that "at least one" of a set of events did happen; rather, they told about one specific event that had occurred and casually mentioned the other possibilities. Possibly those hints are not sufficiently suggestive of the union sets (see the example of version PU given above). In contrast to that, version FU asked unequivocally about the probability of at least one event out of many.

6

Fooling Some of the People All of the Time

Barry Singer and Victor A. Benassi

> *I shall fear no man, nor spirit except for the Lord our God, for he can destroy you and your soul and cast you into the pit where there is no rest from the eternal flame. You are in bondage to Satan and nothing can set you free apart from Jesus Christ. Can you see that you are in bondage?*
> —An experimental subject, in feedback to our performer.

Why do many people attribute occult powers to charlatans who perform old-hat magic tricks and label them "psychic"? People are often impressed by a "psychic" performer who appears to dematerialize ashes, to perceive through a blindfold, or to make a compass needle twirl by passing her/his hand over it. Yet any magicians worth their salt could not only perform these stunts, they could levitate bevies of assistants with a wave of the hand or dematerialize whole elephants onstage. Elephants seem much more difficult to dematerialize than ashes. That is, although professional psychics claim awesome powers, their actual "psychic" performances reduce to stunts that any amateur magician could perform. The fact that such amateur stunts compel many people to strong occult beliefs, then, demands explanation.

We constructed an experimental paradigm to demonstrate and explore this phenomenon (Benassi et al., in press). A student accomplice, Craig Reynolds, a talented actor and amateur magician, was asked to develop a standard magic routine that included "psychiclike" stunts. Craig was not informed of the purpose or the varying conditions of our experiment. To increase the ambiguity of Craig's performance, we dressed him in a purple choir robe, sandals, and a gaudy medallion. Craig's performance was presented to six introductory psychology classes. The class instructors understood that they were cooperating in an experiment, but were unaware of its nature and purpose.

In two classes, which we termed the "psychic" condition, the instructors introduced Craig's presentation as follows (Craig was waiting outside the classroom, out of hearing):

Today I've agreed to allow one of our graduate students to give a presentation to you. His name is Craig Reynolds. Craig is interested in the psychology of paranormal or psychic abilities. He told me he's been working on developing a presentation of his psychic abilities and asked me if he could present it to you and get your opinions and reactions. I thought that would be interesting, even though I'm not convinced personally of Craig's or anyone else's psychic abilities, so I agreed to let him do it. He's a little late . . . I'm going to look outside to see if he's around.

In two other classes, termed the "magic" condition, the introduction was:

Today I've agreed to allow one of our graduate students to give a presentation to you. His name is Craig Reynolds. Craig is interested in the psychology of magic performances and stage trickery. He told me he's been working on developing a presentation of his magic act and asked me if he could present it to you and get your opinions and reactions. I thought that would be interesting, so I agreed to let him do it. He's a little late . . . I'm going to look outside to see if he's around.

Upon entering the room, Craig performed the following routine. He passed out blank sheets of paper, explaining that he wanted written feedback on his performance. Craig did not label, interpret, or explain his performance. After asking the instructor to blindfold him thoroughly with numerous strips of masking tape and cloth, Craig attempted while blindfolded to read 10 three-digit numbers that had been written by volunteers from the class on ordinary notepaper. Craig had the capacity to read them all correctly; however, we instructed him to read four or five perfectly, come close on three and miss widely on the rest. Next, he selected a female volunteer from the class, whom he asked to sit immediately in front of him. After asking her to hold out her hands and clench her fists, he requested some cigarette ashes from an ashtray in the room. Craig smeared the ashes on the back of the woman's hands, then asked her to concentrate on relaxing and becoming immaterial. When he wiped the ashes off with a kerchief, he asked her to open her hands toward the class, and some of the ashes were seen to have been transferred to her palms. Finally, Craig passed around the class a sturdy, stainless-steel rod, challenging anyone to bend it by force. No one could. He then held it gently between his fingertips, stroked it softly with his index fingers, and asked the class to join him in chanting the word "bending." After a minute or two, the rod melted in his hand, reaching a 60-degree bend. Craig passed it back to the class for inspection and another try at bending it by force. Again, no one could bend it. All of the stunts described are easy amateur tricks that have been practiced for centuries and are even explained in children's books of magic. Craig asked for written feedback, collected the papers, and left the room.

When we inspected our data (the written feedback to the performer) we received two large surprises. First, there was little difference due to the experimental conditions (see Table 1). In both the "magic" and the "psychic" classes, about two-thirds of the students clearly believed Craig was psychic. Only a few students seemed to believe the instructor's description of Craig as a magician in the two classes where he was introduced as such. Second, psychic belief was not only prevalent; it was strong and loaded with emotion. A number of students covered their papers with exorcism terms and exhortations against the Devil. In the psychic condition, 18 percent of the students explicitly expressed fright and emotional disturbance. Most expressed awe and amazement. We were present at two of Craig's performances and witnessed some extreme behavior. By the time Craig was halfway through the "bending" chant, the class was in a terribly excited state. Students sat rigidly in their chairs, eyes glazed and mouths open, chanting together. When the rod bent, they gasped and murmured. After class was dismissed, they typically sat still in their chairs, staring vacantly or shaking their heads, or rushed excitedly up to Craig, asking him how they could develop such powers. We felt we were observing an extraordinarily powerful behavioral effect. If Craig had asked the students at the end of his performance to tear off their clothes, throw him money, and start a new cult, we believe some would have responded enthusiastically.

TABLE 1

Relation between instructional set and percentage of subjects who believed that the performer was a psychic or a magician

	Instructional Set		
	Psychic	Magic	Strong Magic
Response:	%	%	%
Psychic	77	65	58
Magician	14	17	33
Undecided	9	18	9
Total	100%	100%	100%
N =	75	51	52

Obviously, something was going on here that we didn't understand. Further, many students were experiencing serious emotional disturbance as a result of what we had done. Although in retrospect we might have known better, we did not imagine that having someone dress in a choir robe and do elementary-school magic tricks would produce such fright. We therefore took three further steps.

First, although we had originally planned to obtain demographic and personality measures from our experimental subjects three weeks after the performance, and only then debrief them, instead the experimenters and the performer went back as soon as possible to the classes where Craig had performed and fully explained to the students that they had been taking part in an experiment and had only been seeing tricks.

Second, we added several more conditions to our experiment. We thought that perhaps the classes had simply not heard or understood the instructor's initial description of Craig. To check this possibility, in three other introductory classes different instructors cooperated in presenting the magic and psychic descriptions at the start of class, and then immediately asked the classes to write down and interpret what they had just heard. Results showed that virtually all students heard and understood the descriptions, could define the word "magic" appropriately, and could differentiate it from "psychic."

Next, we asked ourselves whether there was anything at all that their instructors could say about Craig in introducing him that would convince students that he wasn't psychic. We added a "strong magic" condition in two new classes, where the magic introduction was repeated with the following added: "In his act, Craig will pretend to read minds and demonstrate psychic abilities; but Craig does not really have psychic abilities, and what you'll be seeing are really only tricks." Thus the concept of "magic," "trickery," "pretend," are enunciated clearly six different times.

The use of this introductory description did succeed in reducing psychic belief slightly, but not to a level below 50 percent. Most of the students still believed Craig was psychic. We had two raters, ignorant of the purposes and conditions of the experiment, analyze all experimental protocols in random order. The results, shown in Table 1, indicate that, although our introductory cognitive "sets" did affect psychic versus magic beliefs at a statistically significant level, the most salient result was our inability to reduce psychic beliefs more than slightly through even strong and clear instructor descriptions of what was actually taking place. This says either something about the status of university instructors with their students or something about the strange pathways people take to occult belief.

We made a follow-up experiment with new introductory psychology students to get an indication of what was happening. After assessing the students' general psychic beliefs, we gave them a questionnaire containing a detailed written description of Craig's performance, which asked for their judgment of whether the performance demonstrated psychic powers. Fifty-eight percent of the students were fairly sure that Craig had psychic powers, and their judgment correlated with their prior psychic belief, which was generally quite strong.

The next question asked was whether magicians could do exactly what Craig did. Virtually all the students agreed that magicians could. They were then asked if they would like to revise their estimate of Craig's psychic abilities in the light of this negative information that they themselves had furnished. Only a few did, reducing the percentage of students believing that Craig had psychic powers to 55 percent.

Next the students were asked to estimate how many people who performed stunts such as Craig's and claimed to be psychic were actually fakes using magician's tricks. The consensus was that at least three out of four "psychics" were in fact frauds. After supplying this negative information, they were again asked if they wished to revise their estimate of Craig's psychic abilities. Again, only a few did, reducing the percentage believing that Craig had psychic powers to 52 percent.

To summarize: our students initially were fairly sure that Craig was psychic; immediately afterward they themselves furnished information in logical contradiction to this judgment. They agreed that magicians could do exactly what Craig did and that most people who, like Craig, called themselves psychic were actually phonies. One could not then logically estimate that Craig was the real McCoy. In several other studies we have found results essentially similar to those described above.

What does all this add up to? The results from our paper-and-pencil test suggest that people can stubbornly maintain a belief about someone's psychic powers *when they know better*. It is a logical fallacy to admit that tricksters can perform exactly the same stunts as real psychics and to estimate that most so-called psychics are frauds and at the same time maintain with a fair degree of confidence that any given example (Craig) is psychic. Are we humans really that foolish?

Yes, In fact, a plethora of recent psychological research indicates that we are typically inept at reasoning through even the simplest conceptual task involving alternative hypotheses or probability (Nisbett and Ross 1980). And once we've chosen an incorrect hypothesis, we can stubbornly persist even when hints to problem solution or the correct hypothesis are placed in front of our noses (Chapman and Chapman 1967; Kammann 1980). Our results differ from previous psychological research on reasoning deficiencies only in showing an especially stubborn and dramatic resistance to the obvious truth.

There are further hints in our data that information processing deficits were occurring. In searching the data protocols from our initial experiment, we found only a few cases out of almost two hundred where students acknowledged what their instructor had said, and only two cases in the magic conditions where they explicitly recognized that their psychic beliefs contradicted their instructors' assertions. Rather than resisting or denying the contradictory information, which we know they heard and understood, students blindly ignored it. This is a typical information processing effect.

There is more to the story. We believe other factors abetted the serious reasoning problems our subjects experienced. First, we know from previous surveys (Benassi and Singer, in press), and from surveys we made of subjects in follow-up experiments, that most students entered our experiments having existing psychic beliefs. These initial beliefs probably channeled their thinking and prevented them from reasoning clearly about alternative hypotheses. For instance, in our follow-up experiments we asked our subjects to write a justification of their final probability estimates of the psychic qualities of Craig's performance. The most frequent justification cited was prior belief in psychic phenomena, and in these cases subjects typically made mention *only* of prior belief. That is, they seldom mentioned anything at all about the particular case they were supposed to be judging. For instance, the complete justification of one subject who estimated that Craig was psychic was: "My justification for the answer is I am a Christian and I feel strongly that ESP or anything dealing with that is of Satan. Yes, I believe it could happen, *but* I, being a Christian, will have *no* part of it."

Second, especially in our initial experiment, the input from the instructor was brief, verbal, and cognitive, while Craig's actual performance was relatively long, live and in person, and dramatic. We would then expect the latter to be more vivid or "available" (Nisbett and Ross 1980) when subjects are fishing for interpretations of the experience. Since many subjects had prior psychic beliefs, and since we deliberately had Craig dress and act ambiguously—somewhere between a stage magician with top hat and tails and Uri Geller—the hypothesis that he was psychic would have been especially likely to be formed and supported.

Some alternative explanations that have been offered for our results are that the students were putting us on, or merely cooperating out of politeness with the performer's perceived desire to be judged as psychic. These explanations will not hold water. Neither in their protocols nor in their debriefing did a single subject hint that they knew that Craig's performance was part of an experiment. Their noticeable emotion during the performance showed that they were genuinely involved and taken in by the experience. Neither Craig nor the class instructors knew what the experiment was about. Craig at no point labeled himself as psychic or made explicit psychic claims. Far from being "polite" or supporting Craig's alleged self-ascription out of a desire to please him, those who were most convinced often vehemently accused Craig of being in league with Satan. In the week following the experiment, a number of direct observations and informal anecdotes revealed that students who witnessed Craig's performance were expressing their enthusiasm and amazement to other students who had not been in the classes.

Another alternative explanation for our results is the "California Hypothesis": California is chock full of nuts and anything could be believed there—it would never happen at Yale. To the contrary, the number one state in terms of cult membership per capita is Nevada (Stark and Bainbridge 1980). Students at California State, Long Beach, cluster around national norms in intelligence and academic achievement, and average several years older (24) than most undergraduate populations. Further, our experiment has been replicated by us and by colleagues at UCLA and Ohio State University, with very similar results (Padgett, personal communication). And a similar result, where bizarre beliefs about their own psychic abilities were experimentally induced in students, has in fact originated from Yale (Ayeroff and Abelson, 1976).

We believe that our results, as bizarre as they may be, are of wide generality and that the psychological processes we have tentatively identified as being involved in supporting psychic beliefs are present and active in the general population.

References

Ayeroff, F., and R. P. Abelson 1976. "ESP and ESB: Belief in Personal Success at Mental Telepathy." *Journal of Personality and Social Psychology* 34 (2): 240-247.

Benassi, V. A., and B. Singer (in press). "Occult Belief." *American Scientist.*

Benassi, V. A., B. Singer, and C. Reynolds (in press). "Occult Belief: Seeing Is Believing." *Journal for the Scientific Study of Religion.*

Chapman, L. J., and J. P. Chapman 1967. "Genesis of Popular but Erroneous Psychodiagnostic Observations." *Journal of Abnormal Psychology* 72: 193-204.

Kammann, R. 1980. "Self-perpetuating Beliefs." In D. Marks and R. Kammann,

eds. *The Psychology of the Psychic.* Buffalo, N.Y.: Prometheus, 1980.
Nisbett, R., and L. Ross 1980. *Human Inference.* Englewood Cliffs, N.J.: Prentice-Hall.
Stark, R., and W. S. Bainbridge 1980. "Secularization, Revival, and Cult Formation." Paper delivered at the annual meeting of the American Association for the Advancement of Science, San Francisco, January 3-8.

7

Misperception, Folk Belief, and the Occult: A Cognitive Guide to Understanding

John W. Connor

Of the many theories advanced for the study of folk belief and the occult, none I am aware of attempts to apply the findings of modern cognitive theory and research. I believe an understanding of cognitive theory will help explain how the elements of folk belief and the occult are formed and transformed.

Research in cognition by Berlyne (1965) and Neisser (1967) discloses that cognition is itself an act of construction:

> The world of experience is produced by the man who experiences it . . . There certainly is a real world of trees and people and cars . . . however, we have no direct, immediate access to the world nor to its properties. Whatever we know of reality has been mediated not only by the organs of sense, but by complex systems which interpret and reinterpret information. [Neisser 1967:145]

Neisser also states that our experiencing of reality can never be entirely attributed to the stimulus alone. It is always a construction based only in part on currently arriving information. Our knowledge of things is not determined by the things themselves but by our perception of them. Reality testing does not depend on clarity. We do not doubt the existence of a distant mountain on a hazy day. Reality testing involves questions of coherence, predictability, and sensibleness (1967:150).

Since humans do not come preprogrammed to live in an already prestructured world, they must create their own world of meaning and impose an arbitrary order on the complexity of eternal reality. This is the sociocultural reality established initially in childhood and maintained in continual interaction with others. This ordering of reality is an act of construction based upon perceptual sets. Perceptual sets are means by which we order, perceive, and interpret reality. The complex reality of a forest will be perceived quite differently by a botanist, an entomologist, a logger, a poet, and a little boy who is lost. A college campus will be perceived much differently by students majoring in architecture and those in police science, to say nothing of the different perceptions held by the administration, the professors, the custodial staff, and the newly arrived freshmen.

Since perceptual sets are the means by which we order reality, it also follows that visual and auditory hallucinations (or misperceptions) are endemic in any normal human population. Almost everyone has experienced the rather common auditory hallucination of hearing a telephone ring while running water in the bathtub. Here our folk belief that the telephone always rings while one is in the bathtub is a major contributing factor; it creates in us a perceptual set. The running water creates a masking noise into which we read or hear the sound of a telephone ringing. Similarly, a few years ago, clerks in a California state office building began complaining of a phantom caller who would always hang up before the call could be answered. A little research soon disclosed that it was an auditory hallucination. During the lunch hour a clerk would be left alone in the office with strict instructions that any telephone call was to be answered at once. At the same time the clerk was also instructed to receive all incoming messages on the teletype machines located in a far corner of the office. As soon as the teletype would begin to chatter, the clerk would rush over to interpret it only to be interrupted by the telephone ringing. The teletypes created a masking sound into which the conscientious clerk misperceived a telephone ringing.

An even more dramatic illustration was a bill introduced into the California Legislature in April 1982 on behalf of Christian fundamentalists to place warning labels on certain rock rongs that when played backwards allegedly contained satanic messages. A few words from the scientific community were enough to convince the legislators that when confronted with an ambiguous auditory background—as that of a record played backwards—a concerned listener would very likely hear anything that he or she was predisposed to hear. In this regard, it is essential to understand that most often such individuals are not intent on deception—they genuinely believe that what they have perceived has indeed taken place. Thus it would do no good to subject them to lie-detector tests or other means to establish their veracity.

Misperceptions sometimes occur among the ethnic groups in our society because of their perceived lower social status. A few years ago a young colleague boarded a crowded city bus on a hot summer day. In a few moments he was bathed in sweat and had to use his index finger to keep his heavy glasses from sliding off his nose. When he got off the bus, he was followed by three burly Mexican-Americans who wanted to know why he had been "giving them the finger." Only his obvious surprise and quick wit kept him from being roughed up.

Perhaps one of the best examples of the influence of one's perceptual set comes from experimental studies. In 1962 Schacter and Singer administered epinepherine (an adrenaline extract) to groups of students who were then asked by the experimenter's "stooges" to fill out forms describing the effects of the drug. Although the physiological reaction to the drug—a heightened state of arousal—is the same for all groups, the group whose stooge pretended to be happy reported that the drug made them happy, while those influenced by an "angry" stooge reported that it made them angry (Schacter and Singer 1962).

An even more dramatic experiment was reported by C. R. B. Joyce (1976), who tells of two rooms in which there were ten people. In one, nine were given an amphetamine and the tenth, a barbiturate. In the other room the experiment was reversed. In both cases the "odd man out" behaved as did his companions. In one room the lone barbiturate taker behaved as though fatigue had been banished, while in the other room the lone amphetamine taker fell asleep (Joyce 1976:322).

The influence of a collective perceptual set is such that natural phenomena may be quite easily interpreted as being *ipso facto* evidence of the occult. Thus, with a collective perceptual set of the world being a vast battleground between the forces of Good and Evil, it is easily understandable that a fifteenth-century European peasant would interpret the sounds of nocturnal animals as being the movement of demons. Dimly seen in the evening twilight, a flight of birds could quite honestly be perceived as a coven of witches on their way to Sabbath (Connor 1975:372). From my own experience I recall a Japanese peasant who stoutly maintained that the luminous will-o'-the-wisp sometimes seen on summer nights near Japanese villages was tangible evidence for lantern-carrying badgers who figured largely in Japanese folklore. In the same fashion, Solon Kimball once reported his conversations with an Irish peasant who was convinced that a geological formation was the work of the "little people," and nothing could be done to convince him to the contrary (personal communication).

There also exists in humans a drive to reduce conceptual conflict. Berlyne (1965:255) defines conceptual conflict as a "conflict between incompatible symbolic patterns, that is, beliefs, attitudes, thoughts, ideas." Berlyne further states that the major types of conceptual conflicts, such as incongruity, contradiction, confusion, and irrelevance, can be reduced by suppression and conciliation. Suppression is simply the reduction of conflict by suppressing thoughts about the subject matter or by avoiding stimuli that tend to evoke such thoughts. Conciliation, on the other hand, is a way of reducing incompatibility by the use of new information that reveals that the beliefs are not incompatible (Berlyne 1965:256-60).

Perhaps one of the most influential and productive figures in the field of conceptual conflict, or cognitive dissonance, as he prefers to call it, is

Leon Festinger. Along with two collaborators he did a study of a group who predicted the end of the world by forces from outer space. Those who were the true believers would be saved by a flying saucer. When the appointed time came and went, there was indeed a considerable amount of conceptual conflict or cognitive dissonance. The problem was solved in part by "discovering" that the world had been saved because of the faith of the flying-saucer cult and that the earth was to be given a second chance (Festinger, et. al. 1956).

Discussion

In their attempt to order and make sense of the world, humans will interpret and reinterpret information to make it compatible with what is already known. In a classic experiment conducted over half a century ago, Bartlett (1932) had experimental subjects at Cambridge University read a translation of an American Indian folktale. Reproductions of the tale were made at various intervals ranging from 15 minutes after the event to several weeks, months, or even years later. In each subject there was a gradual transformation of the tale so that those elements alien to English culture were dropped while others more compatible were added.

An inadvertent experiment similar to Bartlett's is reported by Dundes (1965:269-276) in which Frank Cushing told the Italian tale of "The Cock and the Mouse" to his Zuni informants. About a year later he heard the same story in a Zuni version; as in Bartlett's experiment, the Italian elements had been dropped and were replaced by those more meaningful to the Zuni.

In another inadvertent experiment, near the end of the past century, Cushing told the story of Cinderella to his Zuni informants and later found that it had been thoroughly transformed into "The Poor Turkey Girl," with settings and incidents that were typically Zuni (Greenway 1964:227-234).

Similarly, when European folk tales like "Goldilocks and the Three Bears" have been added to Japanese children's readers, they have been changed to fit Japanese culture. In the traditional version of the tale, Goldilocks eats the porridge, sleeps in the bed, and then leaves suddenly when the bears appear. To the Japanese, such rude behavior is unthinkable. The Japanese version therefore has Goldilocks apologize profusely for her rude behavior (George DeVos, personal communication).

Another illustration of the importance of one's perceptual set may be found in the saying "A rolling stone gathers no moss." Most Americans would interpret this to mean that one must keep moving both socially and geographically; in the American mind a moss-covered stone is one that has stagnated. The Japanese, on the other hand, interpret the saying to mean that, in order to acquire experience, expertise, patience, and the patina (*Sabi*) that comes with age and experience, one must stay put.

The importance of one's perceptual set has long been known to influence the memory of an event. In another classic experiment, Allport had subjects memorize a picture of a subway scene in which several seated figures are depicted. Standing in the foreground are a black man in a business suit having an argument with a Caucasian male dressed in work clothes and holding a straight razor in his hand. In subsequent retelling of the story, there was a gradual shift in the razor from the hand of the Caucasian to that of the black (Buckhout 1974).

From my own experience, I recall the home of a black family that caught on fire in the summer of 1941. Because we often played in the burned-out shell, I can clearly recall that it was quite obvious that the fire began on a kitchen stove in the basement and roared up a stairwell. Yet, when I returned to the area in 1968, during the height of the civil rights movement, a story was widely circulated that the black family had been deliberately "burned out" by angry whites. Even though I pointed out the real cause of the fire and that sympathetic whites had helped with food and clothing, there was nothing I could do to scotch the story. Here was yet another sad example of the manner in which local history is rewritten.

Similar retelling of events often strips away the less dramatic elements of a story and substitutes those that are felt to be more suitable. Thus most of the tales told of the sinking of the *Titanic* include the story of the noble band playing "Nearer My God to Thee" as the ship went down. This story apparently originated a few days after the disaster. According to Tom Burnam (1980:216), the band played ragtime until the bridge went under, at which time the band master led the musicians in the Episcopal hymn "Autumn."

A less dramatic example, but still germane, was the recent Australian film *The Man from Snowy River* (1982). From my own experience, more than half of those who had heard of this film reported it to be "The Man from Snowy Mountain," because they associated snow with mountains and not with rivers.

An additional example can be seen in the corruption of the German word "Deutsch" (German) to "Dutch" in the "Pennsylvania Dutch," which has become so embedded in the language that many items sold in eastern Pennsylvania are stamped with tiny windmills to certify that they do indeed come from the Pennsylvania "Dutch" country and are therefore authentic.

The belief that St. Bernard dogs search for snowbound victims with little casks of brandy around their necks is so firmly embedded in the popular mind that it has been the subject of countless cartoons and even a Swiss chocolate candy-bar wrapper. Unfortunately, there is no record of a St. Bernard dog ever carrying a brandy cask (Burnam 1980:192).

Exceptionally dramatic events, especially those containing an element

of the unknown or danger, are extremely conducive to the rapid spread of
those urban legends that border on mass hysteria—or "collective behavior,"
as it is called by sociologists. In terms of the previously discussed experiment by Schacter and Singer (1962) in which subjects were administered
epinepherine and then experienced a "heightened state of arousal" that
was then interpreted for them by an experimenter's stooge, any dramatic,
unknown event that is potentially dangerous will evoke in a large segment
of the population a "heightened state of arousal." In turn, this "heightened
state of arousal" can be quickly defined or exploited by almost any dramatic event that symbolically contains an element of the unknown dangers.

Thus, in October 1938, the famous Orson Welles "War of the Worlds"
broadcast, which caused considerable panic, came after the Munich crisis
of the previous month and a growing fear of war. The initially unknown
cause of Legionnaires' Disease in the summer of 1976 created an epidemic
of plant shutdowns in various sections of the country as people reacted to
what many thought was a communist plot to spread germ warfare in the
United States. The Tylenol scare in the fall of 1982 created several incidents in which students at football games—already in a heightened state
of arousal – rapidly became ill when a rumor spread that a "Coke" machine
had been laced with arsenic-laden Tylenol. (*Sacramento Bee*, October 24,
1982). In like manner the "heightened state of arousal" that accompanies
international crises also creates perceptual sets for interplanetary intervention to save the earth, and the number of reported UFO sightings
begins to increase. One is also tempted to speculate that the enormous
success of Stanley Kubrick's film *2001* (1968) was a catalyst that created in
the popular mind a perceptual set that permitted the widespread acceptance
of the von Däniken books in the late 1960s and early 1970s.

The drama, excitement, and heightened state of arousal accompanying
the stock-market crash of 1929 created in the popular mind a belief that
Wall Street was littered with bodies because of the suicides of brokers and
businessmen whose fortunes had been wiped out in the crash. Yet, if one
remembers that the generation of the 1930s was in the midst of making a
psychological transition from the values associated with rural America to
those of the modern industrial, technological society, the story has all the
earmarks of a comeuppance tale; indeed, according to Burnam (1925:249),
there was no increase at all in the suicide rate.

In many ways we are still making the psychological transition to an
urban, industrial society. In the folk mind there still exists more than a
lingering mistrust of mass marketing and big business. There are many
cautionary tales that harken back to the pre-Food and Drug laws and the
dread suspicion that one really doesn't know what goes into one's food.
Despite the reassurance offered by such fast-food chains as McDonald's,
where the gleam of sterile stainless-steel kitchens and neatly groomed
attendants are deliberately designed to relieve anxiety, such persistent tales

as doctored hamburgers and Kentucky-fried rats recall an earlier generation's attitude toward the food industry that was expressed in the delightful doggerel: Mary had a little lamb/ And when it began to sicken/ She sent it off to the Packing Town/ And now it's labeled "chicken" (Bailey 1966:662).

The shock of the Arab oil embargo of 1973 and rising gasoline prices reawakened the old mistrust of big business, especially of the oil industry. As gasoline prices soared, so did the many versions of the tale of the "miracle carburetor," which promised unbelievable mileage. The product of some backyard genius, the carburetor was allegedly purchased by the oil companies to keep it off the market. Actually, the tale had been around since the 1930s.

In a similar manner, the widespread shock and dismay that followed the assassination of President Kennedy was marked by many attempts to find some sort of mystical connection between the martyred president and Abraham Lincoln. Thus it was shown that the names of both Kennedy and Lincoln have seven letters, while those of their assassins, Lee Harvey Oswald and John Wilkes Booth, have fifteen. Lincoln was shot in a theater and the assassin ran to a warehouse. Kennedy was shot from a warehouse and the assassin ran to a theater. Both Lincoln and Kennedy were succeeded by men named Johnson, and so on.

A more recent example of the influence of one's perceptual set during a heightened state of arousal is to be found in the story of a man who, while praying for his critically ill son, looked up and saw a likeness of Christ's face in the grain on the wooden door to the recovery room. Since then, hundreds of people have been jamming the third floor of the hospital to see the door (*Sacramento Union*, April 17, 1983).

On a less dramatic note, the great New York blackout of 1964 reawakened in the popular mind an adolescent perception of what one does when the lights are out. Thus, with smiles and naughty nods of insight, it was widely held that there was a significant increase in the birthrate some nine months later. Actually, the birthrate went down slightly (Burnam 1980:20).

A more recent transformation of an urban legend to fit a popular perceptual set is that of the young girl kidnapped at a shopping center by a man who had sedated her with either an injection or a cloth soaked in an anesthetic. First reported by Brunvand (1981:183-84), the story has appeared in many parts of the country. The most recent version reported by my students as having occurred in a nearby shopping center describes the kidnapper as a black man.

Another area in which existing folk taxonomies are reflected in one's perceptual set is in our use of the number three, as in three strikes in baseball, the number of the trinity, the branches of government, and so on. This usage is so widespread in our culture that professors automatically give three examples to support or illustrate their points. It is also seen in the folk beliefs that disasters or deaths occur in threes and that it is unlucky to light three cigarettes from the same match. There is also a carryover in such fairy tales as "Goldilocks and the Three Bears," "The Three Blind Mice," to say nothing of the propensity of many writers to produce trilogies (Dundes 1968:401-424).

Similarly, in the Near East the number forty is considered auspicious or magical. In the flood of Noah it rained for forty days and forty nights; Christ went into the wilderness for forty days; and in the realm of legend we have "Ali Baba and the Forty Thieves." In Japanese society, on the other hand, the number five is considered auspicious. Japanese work and the neighborhood groups were organized on the principle of *goningumi*, or a five-person group. Japanese doctors give prescriptions that will last for five days. In indigenous psychotherapies a "breakthrough" is supposed to take place on the fifth day, and tableware for domestic use comes in sets of five.

Summary and Conclusion

In this paper we have attempted to apply the findings of cognitive psychology as an aid in understanding the manner in which folk beliefs are begun, changed, and reinterpreted over time. Cognitive psychology shows that the ordering of reality is an act of construction based upon perceptual sets. Perceptual sets may be classified into those that are perhaps innate— as, for example, our response to a smile, scowl, or tears (Eibl-Eibesfeldt 1975:442-443); those that are cultural, as in the American propensity to think of three as a lucky number, while five is considered normal or auspicious in parts of Asia, and forty in the Near East; and, finally, those that are idiosyncratic, such as an artist's perception of color and form.

In every society, there exist "ideal" models of heroes, behavior, dress, and so on. These fluctuate through time and from event to event. All have in common elements that have been recognized from the time of Plato's cave to that of the six o'clock news. These elements could be called fitting or dramatic, or "neat." They sum up or encapsulate in dramatic, symbolic form the beliefs, values, ideals, fears, and prejudices of a group of people. It is for this reason that stories with dramatic potential are told and retold so that they "fit" the perceptual sets of a group.

Thus, as Dundes points out in his analysis of the life of Jesus, "I have tried to show that the life of Jesus must be understood as a version, a very special version, of the standard Indo-European hero pattern. . . . The life of Jesus cannot be understood in isolation from the cultural context . . . nor understood without reference to the social and psychological factors inherent in circum-Mediterranean and Near-Eastern form of the family" (1981:78-79).

For these reasons one cannot expect one's perception of reality to be the same as those of a different culture. Thus, during World War II, a U.S. Navy health officer attempted to explain to the natives of a Pacific island the health problems associated with the common houselfy. As a training aid he used a foot-long model of a fly. After seeing the rapt attention on his listener's faces, he thought he had made his point, until a chief stood up and said, "I can understand your preoccupation with flies in America. We have flies here, too, but fortunately they are just little fellows," and he held up his thumb and forefinger to show the size (Foster 1962:122).

In a campaign to encourage dental hygiene in Fiji, health officials reproduced a New Zealand poster depicting a whale jumping out of the water after a tube of toothpaste. The response was enthusiastic: Fiji fishermen sent in rush orders for the new fish bait (Foster 1962:138).

Cognitive theory cannot explain all the means by which misperceptions, folk belief, and the occult are created and transformed, but it can help us to better understand the process. By way of analogy, those rough fragments of dramatic experience that survive their initial immersion in the cultural stream are shaped by the various forces, eddies, and currents until the rough edges are removed and they are rounded and retained. On the other hand, it often happens that some loosely constructed tales simply break up and muddy the water.

References

Bailey, Thomas. 1966. *The American Pageant.* Boston: D. C. Heath.

Bartlett, F. C. 1932. *Remembering: A Study in Experimental and Social Psychology.* Cambridge, England: University of Cambridge Press.

Bascom, William. 1953. Folklore and anthropology. *Journal of American Folklore* 67:283-290.

——. 1954. Four functions of folklore. *Journal of American Folklore* 67:333-349.

Berlyne, Daniel. 1965. *Structure and Direction in Thinking.* New York: John Wiley and Sons.

Boyer, L. Bryce. 1980. Folklore, anthropology, and psychoanalysis. *Journal of Psychoanalytic Anthropology* 3 (no. 3):259-279.

Brunvand, Jan H. 1978. *The Study of American Folklore.* New York: W. W. Norton.

——. 1981. *The Vanishing Hitchhiker. American Urban Legends and Their Meanings.* New York: W. W. Norton.

Buckhout, Robert. 1974. Eyewitness testimony. *Scientific American* 231: 23-31.

Burnam, Tom. 1975. *The Dictionary of Misinformation.* New York: Thomas Y. Crowell.

——. 1980. *More Misinformation.* New York: Lippincott and Crowell.

Connor, John. 1975. The social and psychological reality of European witchcraft beliefs. *Psychiatry* 38: 366-380.

Dorson, Richard. 1963. Current folklore theories. *Current Anthropology* 4:93-112.

Dundes, Alan. 1965 (ed.). *The Study of Folklore*. Englewood Cliffs, N.J.: Prentice-Hall.

———. 1968. *The Number Three in American Culture*. Englewood Cliffs, N.J.: Prentice-Hall.

———. 1981. The hero pattern and the life of Jesus. In Werner Muensterberger, L. Bryce Boyer, and Simon Grolnick, eds. *The Psychoanalytic Study of Society*. New York: Psychohistory press.

Eibl-Eibesfeldt, Irenaus. 1975. *Ethnology: The Biology of Behavior*. New York: Holt, Rinehart and Winston.

Fisher, J. L. 1963. The sociopsychological analysis of folktales. *Current Anthropology* 4 (no. 3): 235-295.

Foster, George. 1962. *Traditional Cultures and the Impact of Technological Change*. New York: Harper & Row.

Greenway, John. 1964. *Literature Among the Primitives*. Hatboro, Pa.: Folklore Associates.

Joyce, C. R. B. 1976. Quoted in Steven Rose, *The Conscious Brain*. New York: Penguin Books. p. 322.

Neisser, Ulric. 1967. *Cognitive Psychology*. New York: Appleton, Century Crafts.

Sacramento Bee. November 24, 1982. Mass illness linked to Tylenol paranoia.

Sacramento Union. April 17, 1983. Man says the face is Christ's.

Schacter, S. and J. E. Singer. 1962. Cognitive, social, and physiological determinants of emotional state. *Psychological Review* 69:379-399.

Testinger, Leon, et al. 1956. *When Prophecy Fails*. University of Minnesota Press.

8

The Great Stone Face and Other Nonmysteries

Martin Gardner

Clouds often take the shapes of animals and human faces. The same is true of rock formations, such as the Great Stone Face in the White Mountains of New Hampshire, made famous by Hawthorne's tale. Draw a wiggly vertical line. It's easy to find spots where you can add a few more lines to make the profile of a face. On the left and right sides of the maple leaf on the Canadian flag you'll see the faces of two men (liberal and conservative?) arguing with each other. A few decades ago the Canadian dollar bill had to be re-engraved because the face of a demon accidentally turned up in the Queen's hair just behind her left ear.

This tendency of chaotic shapes to form patterns vaguely resembling familiar things is responsible for one of the most absurd books ever written about advertising: *Subliminal Seduction,* by journalist Wilson Bryan Key (Prentice-Hall, 1973). The Signet paperback had on its cover a photograph of an ice-filled cocktail with the caption, "Are you sexually aroused by this picture?" It was the author's contention that hundreds of advertising photographs are carefully retouched to "embed" concealed pictures designed to shock your unconscious and thereby help you remember the product. The hidden pictures include words ranging from *sex* to the most taboo of four-letter words, but there are also phallic symbols and all sorts of other eroticisms. In the ice-cube in an ad for Sprite, the author professed to see a nude woman cohabiting with a shaggy dog. It's hard to imagine anyone taking this nonsense seriously, especially since the author's many references to "recent studies" never disclosed where they took place or who the experimenter was. More amazing still, the Canadian Catholic philosopher Marshall McLuhan wrote the book's laudatory introduction. Key has gone on to write two even more bizarre books about the sneaky ways modern advertising is subliminally seducing us.

More recently, UFO enthusiasts have been playing the hidden-picture game with the moon and Mars. They pore over thousands of photographs of cratered surfaces until—aha!—they find something suggesting the pres-·

ence of alien creatures. An early anticipation of this pastime occurred in 1953 when H. Percy Wilkins, a retired British moon-mapper, discovered what looked like a man-made bridge on the moon. Frank Edwards wrote about it in *Stranger Than Science* (1959), and UFO cranks lost no time seizing on this as evidence of lunar life. Donald Keyhoe, in *The Flying Saucer Conspiracy* (1955), reported that spectroscopic analysis had identified the bridge's metal! When astronomer Donald Menzel said he couldn't see the bridge, Keyhoe called him an "army stooge" collaborating on a vast government conspiracy to conceal the truth about UFOs. (See James Oberg's article, "Myths and Mysteries of the Moon," in *Fate*, September 1980.)

As late as 1976, UFO buff George H. Leonard was claiming that bridges on the moon are among the "least controversial things about the moon." Alas, all bridges vanished when the Apollo photographs were obtained. The "bridges" were nothing more than illusions created by lights and shadows, yet the myth of moon bridges still persists in UFO fringe literature.

The same thing happened to mysterious spires on the moon. Photos in 1966 of the moon's surface showed objects casting such long shadows that UFOlogists decided they had to be rocket ships or radio beacons—at least *something* built by aliens. A Russian periodical called *Technology and Youth* featured a wild article about the spires in its May 1968 issue. The spires turned out to be ordinary boulders, their long shadows caused by the sunlight hitting them at extremely low angles.

George Leonard, in *Somebody Else Is on the Moon* (David McKay, 1976), carried this kind of speculation to such extremes that he managed to write one of the funniest books ever written by a UFO buff. Leonard is an amateur astronomer and retired public-health official in Rockville, Maryland. Photos of the moon's surface, he insists, show rims of craters sliced away by giant machines, jets of soil spraying out (caused by mining operations), and tracks of huge vehicles. "No, I do *not* know who they are," Leonard told the tabloid *Midnight* (February 8, 1977), "where they come from or precisely what their purpose is. But I do know the government is suppressing the discovery from the American people."

Leonard quotes an unnamed NASA scientist: "A lot of people at the top are scared." He thinks the aliens live underground and that seismic quakes on the moon are caused by their undersurface activities. "NASA is simply lying to the American people about UFOs," he told *Midnight*. He suspects the aliens are waiting patiently to take over the earth after we blow ourselves up.

Seeing familiar anomalies on Mars has been common ever since the invention of the telescope. Percival Lowell found the red planet's surface so honeycombed with canals that he wrote three books about how the Martians, desperately in need of water, built the canals to bring water from polar regions. Now, of course, we know the canals were only figments in Lowell's mind, distinguished astronomer though he was. Unfortunately, this has not deterred seemingly intelligent people from similar self-deception.

Here and there on Mars are formations with gridlike structures. "Did NASA Photograph Ruins of an Ancient City on Mars?" is the headline of a *National Enquirer* article (October 25, 1977). A photo of a region near Mars's south pole shows a series of squarelike formations called "Inca City" because they somewhat resemble a decayed Indian village.

In 1977, electrical engineer Vincent DiPietro came across a 1976 photograph taken by the Viking spacecraft that orbited Mars. At first he thought it was a hoax. The photograph showed a remarkably human-looking stone face about a mile wide. NASA had released the photo shortly after it was taken in 1976, and planetary scientists emphasized that it was a natural formation. DiPietro thinks it isn't. Computer scientist Gregory Molenaar used image-enhancement to explore details of the face, and in 1982 DiPietro and Molenaar published a 77-page book, *Unusual Martian Surface Features,* about their results. (It can be obtained by sending $9.00 to Mars Research, POB 284, Glenn Dale, MD 20769. "Face in Space," *Omni,* April 1982, was an excerpt from this book.) The authors concede that the face may have been produced by erosion but they suspect otherwise. They claim that computer enhancement shows an eyeball in the face's right eye cavity, with a pupil near the center, and what looks like a teardrop below the eye. "If this object was a natural formation," they write, "the amount of detail makes Nature herself a very intelligent being."

West of the big stone face, in the shadow of a pyramidlike formation, is a gridlike pattern suggesting a lost city with an avenue leading toward the face. (See "Metropolis on Mars," an unsigned article in *Omni,* March 1985). Skeptics have pointed out that the so-called pyramid is much cruder than scores of pyramids found as natural rock formations in Arizona.

Top drumbeater for the view that the stone face proves that an alien race once flourished on Mars is writer Richard Hoagland. He is completing a book about it that could make him lots of money, especially if he can tie the face into UFOs and get a chapter published in *Omni.* Fred Golden, writing the "Skeptical Eye" page in *Discover* (April 1985), ridiculed Hoagland's claims and ran a photo of another spot on Mars, where the topography resembles Kermit the Frog.

Let us not underestimate the public's scientific illiteracy. Dr. Emil Gaverluk, of East Flat Rock, N.C., is now lecturing around the country about the Martian face. A story in the Hendersonville, N.C., newspaper of February 16, 1985, reported that Dr. Gaverluk was speaking at the First Baptist Church on "the meaning of the gigantic face and pyramids and the laser of tremendous power that have been discovered on Mars." Why are these things on Mars? It's all explained in the Bible, Dr. Gaverluk told the newspaper columnist who wrote about him.

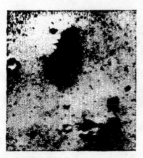

The Great Stone "Face." Photographed by the Viking 1 orbiter in July 1976. The feature is one mile across. Shadows in the rock formation give the illusion of a nose and mouth. The Jet Propulsion Laboratory points out that bit errors, caused by problems in transmission, cause the abundant speckles and comprise part of one of the "eyes" and "nostrils" on the eroded rock. JPL emphasizes that planetary geologists attribute the formation to purely natural processes.

Dr. Gaverluk was identified as an expert on communications science and the holder of a doctorate in educational technology, whatever that is. His lectures on science and faith are sponsored by the School Assembly Service, of Chicago. Dr. Gaverluk illustrates his talks with chalk drawings. He is a member of the American Association for the Advancement of Science and the Creation Research Society.

The great stone face can teach a serious lesson. If you search any kind of chaotic data, it is easy to find combinations that seem remarkable. Every page of a book of random numbers contains patterns with enormous odds against them if you were to specify the pattern before generating the random numbers. Every bridge hand you are dealt would be a stupendous miracle if you had written down its exact pattern before the deck was shuffled.

Let someone close his eyes and talk for 15 minutes about a scene he imagines. You'll have no trouble finding amazing correlations between his description and any randomly selected scenic spot. Let a psychic crime-solver rattle on for an hour about clues to a missing corpse. It's inevitable she'll have made some lucky hits if and when the body is found. If you don't have a tape of everything she told the police, how can you evaluate her accuracy? Jeane Dixon's few good hits seem impressive until you see a list of her thousands of whopping misses.

If hundreds of ESP tests are performed around the world during any given week, and only a few successful ones are published, the normal operations of chance are effectively concealed. J. B. Rhine was notorious in his belief that unsuccessful tests in his laboratory were not worth reporting and equally notorious during his youth in finding patterns in data to support correlations that the experiment had not been designed to find. Today's better parapsychologists are aware of such statistical pitfalls, but a failure to understand them casts a deceptively strong glow of success over the results trumpeted in the early naive years of modern parapsychology.

Let's take a closer look at that great stone face on Mars. Rotate the picture 90 degrees clockwise and what do you see? On the left is the nude torso of a woman, complete with dark pubic hair, small breasts, and an enlarged belly button slightly off center. I'm surprised Ken Frazier would allow such a picture in his family magazine.

9

Sir Oliver Lodge and the Spiritualists

Steven Hoffmaster

One of the major characteristics of science is its continual change, the ability to be self-corrective in the long run and to constantly increase in precision and application of theory. Considering pseudoscience in general and ESP in particular, when one wades through the verbiage one finds such progress largely missing.

An excellent way to show this is to discuss the spiritualist movement of the late 1800s and early 1900s. Many of the current fields of research were active fields of research then. Although the apparatus used has become significantly more sophisticated, the results have not become more impressive. A detailed critique of ESP research is beyond the scope of this article. I will attempt to describe one of the many influential participants in the spiritualist movement, Sir Oliver Lodge. By so doing, I hope the reader will obtain a better historical context in which to judge the issues and claims of modern ESP and also to perceive some of the difficulties of accepting such paranormal claims.

The Fox sisters are generally credited with starting the current revival of the spiritualist movement. In 1848 in the farmhouse of John D. Fox in Hydesville, New York, unusual noises were heard by Fox and his wife. These noises eventually became known as "rappings" and were witnessed by many neighbors. A code was worked out and many felt that, through its use, communication with the dead was possible. The rappings centered around two daughters of John Fox, Margaret and Kate, although it was only with the arrival of a stepsister, Leah Fox Underhill, that the girls began to turn their "talent" toward a more profitable end. Public demonstrations were given throughout the United States and England.[1] In the 1860s, spiritualism had begun to spread more quickly in England than in the United States; and as the demand for more unusual manifestations of spirit contact increased, they were certain to be met. The initial forms included rappings and table tiltings.

I then asked, "Where are Satan's headquarters? Are they in England?"

79

There was a slight movement [of the table].
"Are they in France?" A violent movement.
"Are they in Spain?" Similar agitation.
"Are they at Rome?" The table seemed literally frantic.[2]

They then progressed to spirit contact through séances and actual spirit manifestations. Most authors agree that the most convincing medium (a person who can contact the spirit world) was D. D. Home, although there were many others who were successful.[3]

The spiritualist movement was surprisingly popular around the turn of the century. There have been many speculations why. Certainly a major reason was the doubt being cast on fundamental Christian teachings as a result of archaeological and biblical studies. Those affected were primarily from the educated middle and upper class. For many, spiritualism offered a form of compromise, a way of keeping the spirit of Christianity and at the same time accepting the facts inconsistent with their former beliefs.[4]

Whenever an important belief-system is challenged, there will be strong feelings on both sides of the issue. The scientific community was very much a part of the debate about the validity of spiritualism. Men like Michael Faraday, Lord Rayleigh, and Sir William Crookes empirically investigated some of the phenomena. Others, such as Alfred Russel Wallace and Sir Oliver Lodge, were actively involved in séances and wrote and spoke extensively on behalf of spiritualism. Wallace, Crookes, and Lodge were believers in spiritualism (for at least part of their lives) and it is difficult to find a writer espousing the movement who does not mention one of these gentlemen in such a way as to at least imply the scientific consistency and feasibility of spiritualistic manifestations.

The scientific expertise and accomplishments of Wallace, Crookes, and Lodge are well documented. Wallace in the theory of evolution, Crookes in his work with thallium and X-rays, and Lodge with his studies on the ether and electromagnetism. In the remainder of this article I will concentrate on Sir Oliver Lodge and describe his interactions with the spiritualists and his attempts to reconcile his scientific and religious beliefs. In this way, I hope to show the dangers associated with assuming that expertise in one field can be carried over to another.

Oliver Joseph Lodge was born in 1851, the oldest of nine children, to middle-class parents. He obtained his doctorate in 1877 from the Royal College of Science. From 1881 to 1900 he was the first professor of physics at University College in Liverpool. It was at this time that his major contributions to science were made. Specifically, the case can be argued that he seemed to anticipate Heinrich Hertz's discovery of electromagnetic waves, using Leyden jars and conducting wires to investigate the propagation and detection of electromagnetic waves. His major contribution concerned the interaction of light and ether moving relative to one another.[5] These and other new effects would be explained in 1905, when Einstein published his first paper on what has become known as the special theory of relativity. With his assumption of more administrative duties and philosophical objections to the changes in physical theory, by 1900 Lodge's contributions to physics were in essence ended.[6]

Lodge's interest in psychical research began in 1883 and by the turn of the century he had twice served as the president of the Society for Psychical Research (SPR), the most prestigious of such organizations in England at the time.[7] Perhaps as his ability to contribute to physics decreased his enhanced interest in the paranormal helped to fill a void created by a changing interpretation of the natural world, an interpretation that upset many far greater physicists than Lodge.[8] In any event this interest in spirit phenomena continued until his death in 1940. Lodge's work in the paranormal can be divided into two parts: that occurring before September 14, 1915, and that occurring afterward. On that day his youngest son, Raymond, was killed in World War 1, a tragedy that deeply affected both Lodge and his wife.

In his 1909 book, *The Survival of Man—A Study in Unrecognized Human Faculty*, Lodge stressed that a scientific and objective study of psychical phenomena was necessary and implied that contemporary physics might be totally inappropriate to study it. Not unexpectedly, he tried to explain telepathy by means of the ether and speculated that some sort of resonance was necessary to send and receive telepathic signals. The general impression one gets is that Lodge was groping for an explanation of something he believed existed. He was much more hesitant than later in his life to accept the existence of the occult, but it is clear that he would have liked it to be genuine. Also obvious is that Lodge was only nominally aware of the possibilities of fraud and subconscious cues and biases. He explicitly stated that telepathy seems to work best when the receiver and sender are touching. In other instances he describes successful experiments in which the conditions for success had been set after the experiment was performed and the data analyzed.

What objectivity Lodge did possess regarding the paranormal all but disappeared after his son's death. In *Raymond, or Life After Death* (1916) and *Raymond Revised* (1922) Lodge used the first third of each book to reproduce the letters written by his son to the family. Lodge often spoke of his grief at Raymond's death. There are indications that Raymond was his favorite son. Certainly he was most like his father when young. Further, there are fairly strong implications of guilt feelings aroused by their son's death on the part of both Lodges. Mrs. Lodge is quoted as explicitly stating that she had not seen enough of Raymond while he was alive. At the time of his enlistment the Lodges were touring America, and one gets the impression that they felt they should have tried to discourage him from joining the British Army. One final note: In his letters home, Raymond, an officer of course, continually portrays the war as a glamorous adventure, something it was not for most. In my opinion, the circumstances of Raymond's death set the stage for Lodge's wholehearted acceptance of the validity of the spirit world and his contact with it. For his own peace of mind and that of his wife, he had to speak with Raymond and seek his forgiveness—something that he was quick to obtain. Having thus been positively reinforced, his entry into the world of séances and mediums was a natural step.

Soon after the death of Raymond, Sir Oliver was contacted by a medium, a Mrs. Kennedy, who arranged for a number of séances (at least the first six after Raymond's death). She professed ignorance of the Lodges' personal life and said only that Raymond had contacted a control (a spirit intermediary), who in turn had relayed Raymond's interest in communicating with his parents.[9] It should be noted that the Lodges were reasonably well known and much information about them was public knowledge. Further, the mediums operating in England at the time had an informal network that shared information. This, coupled with the possibility of some relatively easy research on the Lodges, makes Mrs. Kennedy's claim of ignorance of the family somewhat hard to accept.[10]

When reading Lodge's accounts of the séances, his metaphysics, and his description of the afterlife, it becomes quite apparent that he believed wholeheartedly in its validity and existence. His naiveté and his trusting nature were his undoing. A few examples will show this.

First, Lodge felt that the existence of spirits and their interaction with our world was demonstrated conclusively by William Crookes. Lodge stated several times that his major purpose was to build on Crookes's experimental foundation. His building was to be largely anecdotal. Crookes's experiments with the mediums Florence Cook and D. D. Home were at best questionable due to his lack of understanding of hydrostatics, magic, and human nature.[11]

Second, Lodge, like Crookes, was not at all aware of the methods involved in giving a cold reading.[12] This is a technique that has worked for hundreds of years and today is used successfully by astrologers and many others in the more interactive fields of the paranormal. In essence, it involves pumping the interested party for information without letting him/her know it and then feeding this information back in a slightly disguised form. As an example consider the following segment of a séance involving Lodge.[13] Feda is the medium's "control."

Extract from the Record of O.J.L.'s [Oliver J. Lodge's] Sitting
With Mrs. Leonard, 3, December 1915

FEDA—Now ask him some more.

OJL   —Well, he said something about having a photograph taken with some other men.

We haven't seen that photograph yet. Does he want to say anything more about it? He spoke about a photograph.

FEDA—Yes, but he thinks it wasn't here. He looks at Feda, and he says, it wasn't to you Feda.

OJL   —No, he's quite right. It wasn't. Can he say where he spoke of it?
FEDA—He says it wasn't through the table.

OJL   —No, it wasn't.
FEDA—It wasn't here at all. He didn't know the person that he said it through. The conditions were strange there a strange house. [Quite true, it was through Peters in Mrs. Kennedy's house during an anonymous sitting on 27 September.]

OJL   Do you recollect the photograph at all?
FEDA He thinks there were several others taken with him, not one or two, but several.

OJL   —Were they friends of yours?
FEDA—Some of them, he says. He didn't know them all, not very well. But he knew some; they were not all friends.

OLJ   —Does he remember how he looked in the photograph?
FEDA—No, he doesn't seem to think so. Some were raised up round; he was sitting down, and some were raised up at the back of him. Some were standing, and some were sitting, he thinks.

OJL   —Were they soldiers?
FEDA—He says yes— a mixed lot. Somebody called C was on it with him; and somebody called R – not his own name, but another R. K. K. K —he says something about K. He also mentions a man beginning with B (indistinct muttering something like Berry, Burney-- then clearly) but put down B.

A few items in this short segment should be of interest. The photograph mentioned was a pretty safe bet since it was typical of soldiers to have a company picture. Also note the vague answers given by Feda. Feda herself is an intriguing entity. Feda was claimed to be an Indian child. And only through Feda could contacts be made back and forth with Raymond. Raymond never talked directly with his father but always through a control, often Feda. Because Feda was a child, her manner of conveying knowledge would be that of a child and thus subject to misinterpretation. Add to that the language problems of a child speaking at best a second language, and the stage is set for a series of cryptic revelations. A few notable examples in this short passage (given by Lodge himself as proof of contact with Raymond) will demonstrate. In the picture, was Raymond standing up? No, he doesn't seem to think so. That's a pretty definitive answer. Who was standing or sitting? A mixed lot. Who was in this mixed lot? An R and a C. First name, last name, first letter, occupation? Why talk in riddles? At this point I am always reminded of T. H. Huxley's response to an invitation to join the Dialectical Society, the precursor to the Society for Psychical Research: [14]

Sir,—I regret that I am unable to accept the invitation of the Council of the Dialectical Society to cooperate with a Committee for the investigation of "Spiritualism," and for two reasons. In the first place, I have no time for such an inquiry, which would involve much trouble and (unless it were unlike all inquiries of that kind I have known) much annoyance. In the second place, I take no interest in the subject. The only case of "Spiritualism" I have had the opportunity of examining into for myself was as gross an imposture as ever came under my notice. But supposing the phenomena to be genuine—they do not interest me. If anybody would endow me with the faculty of listening to the chatter of old women and curates in the nearest cathedral town, I would decline the privilege, having better things to do.

And if the folk in the Spiritual world do not talk more wisely and sensibly than their friends report them to do, I put them in the same category.

The only good that I can see in a demonstration of the truth of "Spiritualism" is to furnish an additional argument against suicide. Better live a crossing-sweeper than die and be made to talk twaddle by a "medium" hired at a guinea a *seance*.

I am, Sir, &c.,
T. H. Huxley.

29th January, 1869.

There are other examples to be found in *Raymond,* but the major points are made with the one above. Walter Cook, in *Reflections on Raymond,* analyzed the inconsistencies and inaccuracies in much more detail.

As time goes on the questions asked by Lodge become more metaphysical. Eventually the next world is described to him in some detail, which he then relates to us in numerous books. Needless to say, it is much harder to verify information of such a world than it is to test descriptions of Raymond and his activities. No doubt the mediums were much relieved to be asked such questions. This is a segment of Lodge's *Phantom Walls,* published in 1929:[15]

The testimony so far obtained, or purporting to be obtained, from departed human beings is to the effect that memory continues after bodily death; for reminiscences are employed as one means of proving identity.

This, if accepted, shows that memory does not effectively reside in the brain, although habitual usage of certain nerve tracts no doubt makes recovery of memory more easy than when the material instrument has been lost.

Habits may be weakened by that loss, but memory need not be impaired. We find that incidents that have made an impression on the mind of the deceased personalities are remembered, and can be recalled under proper stimulus.

Intellect continues also. Literary quotations are often ingeniously applied, so as to convey information in a curious characteristic and evidential manner.

Aptitudes for learning and for artistic production seem also to continue. Innate, and probably also acquired, faculties and tastes belong to the individual, and are retained.

Indeed, the evidence is that the whole personality survives, with a character and powers similar to those displayed by the old bodily organism.

Above all, family affection continues strong; the desire to help friends and relatives is perhaps the most prominent feature, and indeed often constitutes the motive power that stimulated the effort to communicate.

Next, from his *Beyond Physics*, 1920.[16]

I postulate, then, as the one all-embracing reality on the physical side, the Ether of Space. And I conceive that in terms of that fundamental physical entity everything else in the material universe will have to be explained. To me the ether is a continuous substance, far more substantial than any matter. It fills all space; though we have no real knowledge of its constitution, for it is too fundamental to have its constitution expressible in terms of anything else. It seems to be analogous to a perfect incompressible fluid in a violent state of minute circulatory motion, imperfectly conceivable as ultramicroscopic vortices circulating or spinning with the velocity of light.

Unfortunately what these two quotations portray is a well-meaning but confused man. Scientifically he is trying to cling to a past that has been rejected by most physicists. Metaphysically he is creating an afterlife that for understandable personal reasons he hopes exists. This is a modified form of Christian belief, more worldly and less fundamentalist than previously held Victorian beliefs. The objectivity he had shown in his science, at least prior to 1900, is totally lacking in his spiritualistic endeavors. His need to believe is too strong.

There were a few other major developments that Lodge somehow had to discount in order to continue his belief in spirit communication. Periodically, the best known mediums would be shown to be using deceitful and fraudulent methods in spirit manifestations. Many were self-confessed. Of those the most significant were the Fox sisters, Margaret and Kate, who in 1888 publicly admitted that they had been cheating. They then proceeded to give a demonstration of their "rappings" at a public hall in New York. Incidentally, the rappings were the cracking of toe joints amplified by resonance with the legs and surfaces of tables. Their renunciation was widely circulated in the United States and Europe, with many spiritualists simply refusing to believe it. The spiritualists countered that the confession was a publicity stunt by the sisters and that they had done it to generate some income.[17] Many similar arguments are still presented when fraud is shown in a paranormal claimant.[18]

More convincing to a scientist should have been the work of Harry Houdini, who in addition to duplicating most of the "tricks" of the mediums was able to document the fraudulent nature of many others in the movement. He was able to do slate writing, table tilting and spirit photography, to mention a few. In his 1924 book, *A Magician Among the Spirits,* Houdini clearly presents quite normal means to produce these phenomena. He also makes a still valid observation that as the controls on an individual increase the paranormal incidents decrease. To me the most surprising and most distressing part of the book is a listing of murders of children and spouses, suicides, and insanity allegedly caused by belief in the information obtained in séances.[19]

A more detailed description of Lodge and the spiritualist movement I feel would provide examples similar to those already cited. I hope two points have been demonstrated. First, that the use of Sir Oliver Lodge (the same can be shown for William Crookes) as a reference to an early scientist involved in the spiritualist movement is valid. One cannot deny that. One should note, however, that the methods used and the conclusions reached by Lodge were questionable at best. His personal interest in the phenomena usually clouded his judgment and clearly forces one to reject him as an objective observer.

Second, because of his strong emotional involvement in the movement his scientific expertise was of little use to him. In fact, a scientist will generally assume the honesty of a colleague. This is a trait that simply is counterproductive in paranormal fields where frauds abound. As was pointed out by Houdini in the past, and by James Randi and Martin Gardner, among others, today, a group of individuals made up in part by scientists and magicians is best qualified to investigate occult claims.

Finally, in reading of the attempted scientific investigations of spirit phenomena, much of the general terminology and many of the techniques are still used today. Crookes felt that there was a psychic force. Lodge felt that telepathy would be easiest to show. People are still trying to prove the existence of such phenomena. Science progresses. Today most sciences are advancing at incredible rates, yet in many ways paranormal research is still where it was one hundred years ago. A cliché comes to mind concerning those who are unaware of the past being doomed to repeat it.

Notes

1. Harry Houdini, *A Magician Among the Spirits* (New York: Harper Bros., 1924), pp. 1-9.

2. Alan Gauld, *The Founders of Psychical Research* (New York: Schocken, 1968), p. 68.

3. Houdini. op. cit., Chap. 3.

4. Gauld, op. cit., pp. 75-76.

5. Charles Susskind, "Oliver Joseph Lodge," *Dictionary of Scientific Biography* (New York: Charles Scribner's Sons, 1973), pp. 443-444.

6. Oliver Lodge, *The Ether of Space* (London: Harper Bros., 1909).

7. Susskind, op. cit., p. 444.

8. Neither Planck nor Einstein were particularly pleased with the statistical nature of the new quantum theory.

9. Oliver Lodge, *Raymond, or Life After Death* (London: Methuen, 1916), p. 98.

10. Walter Cook, *Reflections on Raymond* (London: Grant Richards, 1917), pp. 25-31.

11. Houdini, pp. 200-205; E. E. Fournier d'Albe, *The Life of Sir William Crookes* (London: T. Fisher Unwin, 1923), Chap. 12.

12. Ray Hyman, "Cold Reading: How to Convince Strangers That You Know All About Them," *Zetetic* (now the SKEPTICAL INQUIRER), vol. 1, no. 2 (1977):18-37.

13. Lodge, *Raymond,* op. cit., pp. 106-107.

14. Houdini, op. cit., pp. 198-199.

15. Oliver Lodge, *Phantom Walls* (London: Hodder and Stoughton Ltd., 1929), pp. 232-233.

16. Oliver Lodge, *Beyond Physics* (London: Allen and Unwin, 1920), pp. 45-46.

17. Reuben Briggs Davenport, *The Death Blow of Spiritualism* (New York: G. W. Dillingham, 1888).

18. Martin Gardner, *Science: Good, Bad and Bogus* (Buffalo: Prometheus, 1981). See particularly his discussions of Uri Geller.

19. Houdini, op. cit., pp. 180-190.

Claims of Mind and Distance

10

Outracing the Evidence: The Muddled 'Mind Race'

Ray Hyman

Russell Targ is a physicist with patents in optics and laser physics. He also has devoted much of his adult life to research on psychic abilities. In collaboration with Harold Puthoff, in 1972 he originated the experiments on remote viewing at the Stanford Research Institute (now SRI International). He and Puthoff also gained considerable publicity for their experiments with Uri Geller. Targ and Puthoff summarized their earlier work on remote viewing and Uri Geller in their book *Mind Reach* (1977).

In many respects Targ's new book, *Mind Race: Understanding and Using Psychic Abilities,* co-authored with Keith Harary (1984), is an updated version of the earlier book.

Harary, a psychologist, is well known both as a parapsychologist and as an apparently successful percipient in parapsychological experiments. He and Targ founded Delphi Associates, an organization that sells psychic consulting services to individuals and businesses seeking advice on investments, exploration, or other important decisions.

Although Targ and Harary inform us that both the Soviet and the American defense establishments spend millions of dollars on psychical research, the "race" of the title does not refer to the competition for psychic superiority between the superpowers. Instead, as the authors put it:

> The Mind Race is a race to determine the future of your own consciousness before other forces decide that future for you. We must develop our ability to experience compassion and empathy with our fellow creatures, before we lose contact with our own humanity and exterminate one another over an ideological difference of opinion, or for some similarly foolish reason. The Mind Race is not a race between nations. Though the U.S. and Soviet governments are heavily involved in psi research, we are all in a more vital and personal race to determine whether we will be able to wake up to our deeper potential before we have exhausted the limited time available to us.

As a society we are in the process of making wide-ranging decisions about our evolutionary future. This decision is in our hands right now. The quality of future life on this planet will be determined for us by others if we do not choose to participate actively in determining our own destiny. We do not believe that any psychically sensitive human being would choose to live in a future that is dominated by robots, especially if we are to be the robots. We believe that our future must include psychic functioning if we are to achieve our full potential as human beings. We call this requirement the psi imperative. [p. 246]

The stakes are high. If Targ and Harary are correct, we have to enter the Mind Race and develop our psychic powers or end up as robots subject to the manipulation of others. But what if we lack psychic abilities? Or, if we have them, how can we develop them? Not to worry. The authors assure us that we all possess such powers. Furthermore, they supply directions for developing them. "It is past time for bringing psi into the open, where everyone can benefit from a realistic awareness of it. We believe it is time for all of us to claim our right to function psychically. You own your own mind. It is important not to give it away, or fail to use it to its full potential. So get going! You have to enter the Mind Race in order to win" (p. 246).

The authors' intentions are clear. They would like each of us to follow their directions and develop our inherent psychic faculties. They imply that some form of world utopia would automatically follow once each of us has heeded their advice.

But, as they see it, obstacles prevent many of us from making this commitment. Gurus, superpsychics, and occultists frighten and mislead many of us by depicting psychic functioning as special, abnormal, or available only to the initiated. The media portray psychic powers as weird, evil, or dangerous. Organized religion views such powers as satanic in origin. And critics, for what the authors assert are ulterior religious and philosophical motives, loudly proclaim that psi does not exist. "To those who refuse to develop their psychic abilities it makes little difference whether the force that manipulates them into repressing their human potential is organized religion, cults, materialistic critics, or the mass media. The end result of such repression is the same no matter where it originates" (pp. 245-246).

Targ and Harary's goals of creating a better world by helping us gain access to our psychic powers assume the truth of a number of propositions. I would list them as follows:

1. Psychic functioning, or psi, is real.

2. The reality of psi has been scientifically established beyond any reasonable doubt.

3. The individual reader can test the reality of psi by personal experi-

Promotion for *The Mind Race* and interview with Harary in *U.S. News* and *World Report*

ence in demonstrations suggested by the authors.

4. Psi is normal. It is a natural human function and does not depend on secret or occult rites, special states of mind, or abnormal circumstances.

5. Psi is universal. It is not a special gift. We all have the potential for psychic functioning.

6. Psi can be developed through simple exercises that help to discriminate valid psychic signals from "mental noise."

7. Psi can be put to practical use in any situation where decisions must be made with inadequate information, such as gambling, investing, finding parking spaces, etc.

8. It is important that we all develop and employ our psychic powers to the fullest.

The most basic claim, of course, is that psi is real. The arguments of the book make sense only if this claim is true. This "totally convincing book" (according to the dust jacket) employs a number of different grounds to convince the reader of the existence of psychic functioning. For the scientific justification they point to the list of 28 "published formal experiments" on remote viewing that they append to the book. Even more compelling, as far as the average reader is concerned, are the authors' accounts of the many impressive qualitative descriptions of targets by viewers and the successful applications of psi to predicting silver futures and the outcome of gambling choices. Readers are also urged to follow the directions and experience their own psychic success.

The Scientific Case for Psi

Let's look first at the scientific case they present. This is supported entirely by the published experiments on remote viewing. The phrase "remote viewing" was coined by Targ and Puthoff in 1972 as a neutral term to describe the phenomenon they believed they were capturing in their experiments at the Stanford Research Institute. These experiments employed at least three participants. A viewer, or percipient (the psychic), was isolated with an experimenter (the interviewer) in the laboratory. A second experimenter (the out-bound experimenter, or "beacon") then drove to a randomly selected geographical location (the target site) within a 30-minute drive from the laboratory. While the beacon was at the target site, the viewer described his or her impressions of the scene to the interviewer, and often made drawings as well. When the trial was over, the beacon would return to the laboratory and then all the participants would visit the target site to give the viewer feedback about how well the impressions had matched the actual target.

After a series of such trials (usually 7 or 9) had been conducted with a given viewer, the descriptions and drawings made by the viewer for each session were given to a judge, who then visited each site and ranked all the descriptions from best (a low score) to worst (a high score) according to how well each matched the target. If the agreement between the viewer's description and the actual target was simply a matter of guesswork, then, for example, with 9 possible target sites we would expect to find that the average rank of the descriptions would be 5. If the descriptions were actually related to the targets (by psi or some other means) then we would expect the rankings to be lower. In fact, this is what Targ and Harary claim the data from their own and other experiments on remote viewing have shown. In more than half the series the rankings have correlated significantly with the target sites.

Targ and Harary have no doubts that the scientific case for the reality of remote viewing has been established beyond all reasonable doubt. "In an examination of the twenty-eight formal published reports of attempted replications of remote viewing, Hansen, Schlitz, and Tart at the Institute for Parapsychology found that more than half of the papers reported successful outcomes." Part of this report is included as an appendix to the book. Hansen et al. compiled reports of remote-viewing experiments conducted during the years 1973 through 1982. They concluded: "We have found that more than half (fifteen out of twenty-eight) of the published formal experiments have been successful, where only one in twenty would be expected by chance."

To both the casual and scientifically trained reader the fact that 15 of 28 "published formal experiments have been successful" should seem rather impressive. But a more careful study of the list of experiments suggests that this data-base may not be as strong as implied.

The scientific literature in any given field consists of formal experiments published in scientific journals. Only those papers that survive a rigorous screening and revision procedure make it into print. In fact, many scientific journals reject more than half of the papers submitted to them. Rarely does a paper get published as submitted. Manuscripts are sent to two or more referees who are experts in the subject area of the manuscript. These referees advise the editor about whether the paper is of sufficient merit to be a candidate for publication. They also carefully scrutinize the manuscript for inconsistencies, unsupported claims, adequacy of the statistical analyses, unclear arguments, and so on. Typically, before a manuscript finally is accepted for publication, it has gone through several revisions as a result of this refereeing process. Such a screening process is not perfect and some defective papers do get published. But, for the most part, the process ensures that scientific reports have passed a number of tests.

Only 13, or less than half, of the "published formal experiments" meet the standards of having been published under refereed conditions. The remaining 15 were published under conditions that fall short of scientific acceptability. Some appeared as brief reports or abstracts of papers delivered at Parapsychological Association meetings or some other parapsychological conference. In addition to not having undergone the standard refereeing process, such abstracts present insufficient documentation for scientific evaluation. The same can be said for the other studies that appeared in print only as brief or informal reports in book chapters or letters to the editor.

The scientific case for remote viewing, then, rests upon 13 scientifically reported experiments, 9 of which are classified as "successful." Seven of these experiments were conducted by Targ and Puthoff. The remaining 2 came from two other laboratories. This harvest of 9 "successful" scientifically reported experiments emerging from just three different laboratories over the past 12 years hardly seems to justify the strong impression conveyed by the authors that remote-viewing studies have been successfully carried out in large numbers in laboratories all over the world. ("In laboratories across this country, and in many other nations as well, forty-six experimental series have investigated remote viewing. Twenty-three of these investigations have reported successful results and produced statistically significant data, where three would be expected" [p. 5].)

But even 9 "successes" out of 13 tries would not be bad if the successful studies met reasonable standards of adequacy. But all 9 suffer crippling weaknesses. At least 3, and possibly more, are what I would classify as "retrospective experiments"—experiments not explicitly planned in advance but apparently reconstructed from separate trials that were originally conducted simply as demonstrations. According to Kennedy (1979a), remote-viewing experiments have employed the wrong statistical test. When Kennedy applied a more appropriate statistical test he found, for example, that only 2 of 6 experiments reported by Puthoff and Targ were significant, whereas they had concluded that 5 were significant. This alone would reduce the total number of successful remote-viewing experiments to 6. Of these 6, all but one suffer from a "fatal flaw" that I first pointed out in 1977 (Hyman 1977b) and Kennedy (1979b) independently noted two years later. The one experiment that escapes this "fatal flaw" unfortunately suffers from another serious drawback. I will discuss these flaws later in this article.

Marks and Kammann (1978) raised serious questions about the validity of the findings on the remote-viewing experiments with Pat Price (Puthoff and Targ 1976). Marks obtained copies of the five unpublished transcripts from the series with Price. He found a number of clues in the transcripts to target sites without assuming the operation of psi. For example, in one transcript the interviewer mentions the nature reserve that had been the target for the previous day. Such a clue obviously helps the judge by informing him that the transcript in question should *not* be matched with the nature reserve. In addition, if the judge has information on the order of the target sites, it enables him to uniquely identify the transcript with its intended target. Using such clues within the transcripts Marks successfully matched each description against its intended target without actually visiting any of the sites.

Tart, Puthoff, and Targ (1980) responded to this critique with three rebuttals. Charles Tart, a parapsychologist who had not been involved in the original experiments with Price, reviewed the transcripts and removed "all phrases suggested as potential cues by Marks and Kammann" as well as "any additional phrases for which even the most remote *post hoc* cue argument could be made." The edited series was rejudged by a new and "qualified" judge who was able to successfully match seven of the nine transcripts. The parapsychologists argued that this successful rejudging refuted the "cueing-artifact hypothesis" as it applies to the Price series. Furthermore, they claimed that the hypothesis could not apply to their subsequent successful experiments because the transcripts were carefully edited to avoid such cues. Finally, they argued that the successful replication of their experiment in other laboratories confirms the reality of their psychic interpretation.

There is no need here to discuss the continuation of this controversy (Marks 1981; Puthoff and Targ 1981). Possibly this controversy as well as the critique of the statistical analysis being applied to nonindependent trials has helped to prevent the participants from realizing the full implications of the criticism raised by myself and Kennedy. Neither a more conservative test nor the editing out of obvious cues referring to previous targets can overcome the defect we have pinpointed. Once the viewer and the interviewer have been given feedback about a particular target, then *every* word and phrase in the subsequent descriptions of targets has been tainted. And it is not just the words and phrases that have been included but also those that have been excluded that create the problem.

The problem arises from the fact that the viewer is provided immediate feedback after each session. Say that the target for the first session was the Hoover Tower at Stanford. This will almost certainly influence what both the viewer and the interviewer say during the second and subsequent sessions in the same series. Almost certainly the viewer, during the second session, will not supply an exact description of the Hoover Tower. So, whatever the viewer says during the second session, a judge should find it to be a closer match to the second target site than to the first one. Now, assume that the second target site happened to be the Palo Alto train station. The viewer's descriptions during the third session will avoid describing either the Hoover Tower or the Palo Alto train station. We do not need to hypothesize something as mysterious as psi to predict that a judge should find this third description a better match to the third target site than either of the first two. As we add sessions, this effect of immediate feedback should continue to make the correlation between the viewer's descriptions and the target sites better and better.

Every experiment that has followed the original SRI protocols with immediate feedback is irrevocably flawed because there is no way of separating out a true psychic signal from the information in the transcripts provided by the fact that the viewer knows the previous target sites. So far as I can tell, only one of the nine "successful" experiments does not contain this fatal error.

This experiment (Schlitz and Gruber 1981) suffers from its own serious problems. Gruber, who was the beacon, also translated the viewer's target descriptions into Italian for the judging process. The translator knew which description went with a given target. With almost an infinite number of choices to be made in translating a description from English to Italian, and with the translator's task of trying to capture in the new language what the viewer "meant," it would seem inevitable that translations by the beacon would match the intended target sites. As just one example, assume that, as a part of her description, the viewer had mentioned "wood." One can translate the English word "wood" into Italian in a number of ways depending upon whether the translator believed the wood in question referred to the trunks of trees, the logs, the finished boards, the wood in furniture, or some other form of wood. If, in this case, the actual target site was a forest, then it seems reasonable that the translator would be strongly influenced to translate the English description to fit this known feature of the target. Given this blatant violation of controls, skeptics should not be surprised to learn that this experiment yielded the highest degree of significance of any remote-viewing experiment.

The foregoing considerations should make it clear that the scientific case for remote viewing rests on very shaky foundations. Further problems could be mentioned. For example, not one of the several skeptics who have seriously attempted to replicate the remote-viewing experiment has succeeded. I even know of two cases, neither yet published, in which a skeptic and a parapsychologist collaborated on a remote-viewing experiment with negative results.

Many problems involve inadequate documentation. In the early years of science, the ideal of a scientific paper was a report that was sufficiently complete so that any competent reader could both fully evaluate the results and repeat the experiment. The same ideal holds today, but with journal space costly and limited some practical compromises have to be made. Not all the data or complicated details of procedure can be included, but to the extent this is so the scientific community understands that the omitted details and data are publicly available and the authors are obligated, within the constraints of expense and practicality, to make them available to serious readers. A hallmark of scientific research is this public availability of the data for scrutiny by all interested parties.

The problem of public availability of the data is especially critical in the case of remote viewing. The raw data upon which the scientific case is built consists of the protocols or individual descriptions of targets provided by the viewers. It would take up a prohibitive amount of journal space to publish the complete set of transcripts from an experiment that consists of the typical nine or so trials. Without access to the original transcripts, the reader gets to read only those one or two exceptional transcripts selected by the authors. And, for the most part, only excerpts from the chosen transcripts are supplied.

The scientific public would never have been aware of the cues available in the Price transcripts if David Marks, overcoming strong resistance from Targ and Puthoff, had not obtained the original data. Because of the controversy that had arisen about those transcripts, Dr. Christopher Scott, an English mathematician and former parapsychologist, requested that Puthoff and Targ send him copies of the transcripts to signal to the scientific community that, in fact, these data were available for public inspection by responsible and qualified scientists. When his initial written inquiries failed to result in his getting the transcripts, Dr. Scott publicly repeated his request to Targ and Puthoff at the Parapsychological Association meeting in Cambridge, England, in August 1982. Puthoff admitted that Scott was entitled to inspect the transcripts and indicated that he would make them available for this purpose. Dr. Scott happened to be visiting California in the spring of 1983. Since, despite further requests, he still had not received copies of the transcripts, he made a special trip to SRI International to put his request to Puthoff directly. Unfortunately, Puthoff could not meet with Scott because of an illness in his family, but none of his associates would allow Scott to see the transcripts. Scott has persisted in his quest to see the data, but Puthoff and Targ, two years after they promised to make them available, still prevent public scrutiny.

Targ and Harary depict their critics as unreasonable dogmatists. They put all the blame for the failure of their work to gain scientific acceptance upon the religious fanaticism of blind materialists. Tragically, Targ does not realize how much of the blame must be attributed to his own unscientific behavior. By allowing only a small band of select initiates to inspect their raw data, Targ and Puthoff appear more like the leaders of an occult society who jealously guard their secrets rather than scientists who try to make their case in the public arena.

I do not have to develop my psychic powers to anticipate Targ and Harary's next reaction to the preceding critique. They preview their rebuttal, among other places, on pages 174 and 175 of *Mind Race*. Here they describe their reactions upon listening to the critics at the meeting of the Society for Psychical Research and the Parapsychological Association in August 1982:

> One question was repeatedly asked at this centenary conference: What has been accomplished in a hundred years of research? An answer that most of the scientists in the field would support is that as a result of thousands of laboratory experiments, comprising millions of trials, any fair-minded man or woman should be convinced beyond reasonable doubt that psi exists, and might possibly even be important. But many people at the conference did not share that view. Some were critics, and some were psi researchers. . . . It became clear from listening to these critics that any experiment, no matter how carefully carried out, may reveal a flaw in retrospect. There is always something that could have been done better. *This is true in every field of science*—and in recent years there have been many more examples of fraud in medical research than in psi research.
>
> Hearing what the critics have to say, we began to realize that psi may never be accepted into the mainstream of science on the basis of laboratory experimentation alone.

Like many other things the authors have to say, one can find circumstances and contexts in which the foregoing remarks apply. Some critics do fit this description. And not one will deny that after the fact we can always find in any experiment a defect or subtle variable that was overlooked. But the authors have again used an excuse that makes sense in some other context to avoid dealing with legitimate criticism.

The "critics" who gave papers at the 1982 conference were Chris Scott, Susan Blackmore, Piet Hein Hoebens, and myself. Scott is a former parapsychologist who has become a critic, but he is recognized by parapsychologists and others as scrupulously fair. He maintains good relations with psychic researchers and has written extensively for their journals. Susan Blackmore is a practicing parapsychologist. She has become skeptical of many claims in her field as a result of a decade of research in which she has failed to replicate many of the major findings. She remains in the field because she feels parapsychology badly needs friendly and constructive critics. Hoebens is a Dutch journalist who has gained an international reputation as a skeptic who leans over backwards to give the parapsychologists a fair hearing. All of us on the panel had agreed ahead of time that our task was to provide constructive and responsible criticism.

The flaws I have attributed to remote-viewing experiments in this article are definitely not flaws that are found retrospectively as new and better experiments emerge. They are the very same flaws I wrote about seven years ago (Hyman 1977). Kennedy (1979a, 1979b), a parapsychologist, complained about these same flaws along with others. Unfortunately, Targ and Puthoff, in their haste to dismiss any criticism as having ulterior motives, have kept repeating the same mistakes. Other researchers in remote-viewing slavishly followed their example. The tragic result is seven more years of wasted research.

The Nonscientific Case for Psi

The bottom line is that there is no scientifically convincing case for remote viewing. As the preceding quotations indicate, Targ and Harary, while insisting the scientific case for remote viewing is overwhelmingly strong, concede that they have little hope of convincing critics and the scientific establishment with such data. Consequently, the authors employ two other modes of argument to persuade the reader that psi is real. They supply many qualitative and compelling accounts of psychic successes, and they urge their readers to try experiencing psi for themselves.

Many of the qualitative accounts illustrate striking correspondences between portions of a transcript and the actual target during a remote-viewing session. In one example, the target was the Palo Alto Airport tower. The verbatim transcript and drawing made by the viewer, Hella Hammid, indeed seem to match the target well beyond any forced matching that one usually can achieve between a scenic description and a reasonably complex geographic site. But this particular session occurred after three preceding unsuccessful sessions. A skeptic might want to study all the transcripts in this series before jumping to conclusions about possible psychic correspondence.

This particular transcript obviously has been selected from hundreds available to the authors. Presumably it is presented in its entirety just because it appears to be a striking match. The authors present a number of other apparently striking matches between description and target, but in most of these cases only selected portions of the transcript are given. Again, the skeptic would want to study the entire transcript as well as all the other transcripts in the series.

Marks and Kammann (1980) employ the phrase "subjective validation" to label the subjectively compelling matches that viewers and judges discovered in their remote-viewing experiments. When they initiated their series of experiments in an attempt to replicate the remote-viewing experiments, Marks and Kammann first thought remote viewing was, in fact, occurring. Both they and their viewer, after getting the immediate feedback from the visits to the target sites, found amazing correspondences between the viewer's descriptions and the target. When the judging began, the judges also found amazing correspondences between the transcripts and the targets to which they matched them. Unfortunately, the judges' matchings of targets to transcripts did not correspond with the factual pairings in the experiment. Even when told of this, the viewers did not change their belief in the success of their remote viewing.

This tendency to find meaningful and compelling matches between verbal descriptions and arbitrary targets is quite pervasive. It helps account, for example, for the success of character readers and astrologers (Hyman 1977a). Furthermore, once an individual has found such a match, it is difficult to dissuade him or her from believing in the accidental nature of the correspondence (Nisbett and Ross 1980).

For such reasons as these, striking and "meaningful" correspondences between target and descriptions cannot be accepted as scientific evidence. This is why the elaborate blind-judging and complicated statistical methodologies have been devised. The scientific enterprise aims at separating out true correlations from subjectively compelling, but spurious, ones.

Unfortunately, the lay reader as well as the uncritical scientist will more likely be swayed by the colorful and vivid qualitative illustrations than they will by the abstract and quantitative scientific arguments. Nisbett and Ross (1980) cite abundant evidence to this effect. So we can anticipate that Targ and Harary will succeed in their tactic of bypassing the scientific case in favor of nonscientific arguments. They will succeed, that is, if their goal is to gain the adherence of their readers to their claims rather than to arrive at the truth.

Targ and Harary also describe successful applications of psi. An interesting example is the successful use of remote viewing by Elisabeth Targ to predict the winner of the sixth race at Bay Meadows. She picked a horse named Shamgo, and students from all over her college dormitory contributed to a betting pool. Shamgo won and paid six to one. As in other such accounts in this book, we are not told if this was Elisabeth's first attempt at predicting races or if she ever tried it again. Targ and Harary also retell the story of their venture into psychically predicting the silver-futures market. They claim to have correctly predicted both the magnitude and the direction of the change in all nine forecasts they made in the fall of 1982. Again they fail to tell us about any preceding or future forecasts (although on NOVA's program on ESP, the narrator casually mentioned that Targ and Harary's later attempt to repeat this feat failed).

The Proof of the Pudding Is Not in the Eating

Psychologists also will not be surprised if the readers who follow the authors' recipes for developing their own psychic powers become believers in the reality of psi. The authors write that readers can test the reality of psi for themselves. They supply general guidelines to follow to develop latent psychic abilities. The basic idea makes some sense in terms of general learning principles. If we accept their argument, then at any point in time our conscious experience consists of sensory impressions, memories, and inferences. In addition, some of this content may be impressions that have arrived psychically. If the viewer attempts to describe the psychic impression, the description is often contaminated and transformed by the viewer's expectations, memories, and current sensory impressions. The authors refer to this contamination as "mental noise." Developing one's psychic abilities involves learning to discriminate true psychic signals from "mental noise." This can be achieved, according to the authors' optimistic projections, by indulging in exercises in which immediate feedback supplies us clues as to which of our impressions were truly psychic and which were mental noise.

One exercise involves finding a parking place. Readers are urged to visualize a certain area of the city in which they want to find a parking place. When they get some sort of impression of a possibility, they drive to that spot. If the spot is occupied, they try again. They are to keep this up until they either find a parking spot or run out of gas. By repeatedly trying this exercise the learner, allegedly, can gradually improve the ability to discriminate between those impressions that work and those that do not. Exercises in playing black-jack, doing remote-viewing, anticipating traffic jams, and so on, are similar.

Targ and Harary confidently put forth such exercises as a way for the readers to find the truth for themselves. But we do not have to postulate psi to predict many of those who try such exercises will end up believing they are experiencing psi. For a sampling of just some of the enormous amount of psychological evidence for this expectation see Nisbett and Ross (1980). The authors do not bother to warn their readers of the traps that await them. Instead of forearming the readers, they *disarm* them. Consequently, instead of a path to the truth, they supply a recipe for self-deception.

Several things are wrong with such exercises. For one thing, one of them can succeed for reasons unrelated to psi. Indeed, the authors talk about developing intuition as if it is the same thing as psychic functioning. Some learners might actually improve their ability to find parking places. In some shopping areas the southern boundary, for example, might tend to have more unoccupied spaces than the other sections (because of prevailing traffic patterns). As learners practice trying to home in psychically on a parking space, they may gradually learn to follow impressions that lead them to the southern boundary. Such learning could take place without any conscious awareness on the learner's part. Very likely, the learner will attribute this increasing success to developing psychic powers. Other unconscious cues, such as hearing a motor start up as an auto vacates a parking place, could also become part of what the learner comes to rely upon as psychic abilities.

But even without any actual learning taking place, several psychological mechanisms can easily contribute to the illusion that psychic abilities are gradually leading to more and more successful outcomes. These are well-known distortions of memory, thinking, and other cognitive processes. And it is dismaying, especially when one of the authors claims to be an experimental psychologist, that Targ and Harary do nothing to protect the reader from such powerful pitfalls.

Furthermore, Targ and Harary provide no evidence that learning to discriminate psychic signals from "mental noise" according to their directions can actually occur. They refer vaguely to their experience with remote-viewing sessions. But they fail to hint at even one scientific experiment that suggests that such learning can take place.

Additional Problems

So far I have noted some serious weaknesses in the arguments for the propositions being asserted in *Mind Race*. My notes suggest a variety of other difficulties, but it would only make a long article much longer to try to list them all. In this section I want just to point to a few inconsistencies in the arguments of the book.

The key to developing psychic abilities, according to the authors, is learning to discriminate "mental noise" from true psi impressions. The authors repeatedly assert that their viewers become better and better at this with practice. Harary gives examples from his own experience as a viewer where he was actually able to indicate to the judge which of his statements were "mental noise."

Such a claim would seem simple to test, but the authors supply nothing but a few qualitative observations to back up their assertion. As a standard procedure, for example, each viewer, immediately after each trial, could review the transcript and indicate which statements are "mental noise" and which are true impressions. The experimenter could easily quantify such data and see whether the proportion of correctly identified statements increases with practice.

Furthermore, even if their claim is only partially true, it allows for an excellent control for the judging procedure. Each transcript could be divided into two transcripts—one part containing all the items identified by the viewer as "mental noise" and one part containing the items identified as true impressions. The judges should show high accuracy in matching the second transcripts and should be at chance in matching the first.

Nor do the authors face up to another inconsistency raised by their claim. It is the practice in remote-viewing experiments of employing independent judges to gauge the correspondence between target and description. The researchers who use this paradigm claim that they cannot get adequate results if they employ the viewers as judges because of "mental noise" that interferes with their seeing the correspondences. One inconsistency is that in the Ganzfeld experiments, which started at the same time as the remote-viewing paradigm, the reverse seems to be the case. In those experiments, the percipients are used to judge the correspondence of their own descriptions against the targets. This is done because apparently the results from independent judges do not work as well. Despite this odd reversal, the claims for success in the Ganzfeld experiments equal those for the remote-viewing experiments.

A second inconsistency is that, if in fact the viewer is learning to discriminate true psi signals from "mental noise," then the viewer should be a better judge than an outsider. The independent judge, after all, has to deal with the entire transcript and has no way of gauging which statements should be ignored in trying to match transcript with target.

The claims for gambling successes employing psi also hint at a variety of inconsistencies. Millions of gamblers over the years have presumably employed hunches and intuition in making their bets. They have gained enormous amounts of immediate feedback that should have taught them, according to the theory advanced in this book, which of their impressions should be trustworthy and which should be ignored. Even if such learning is only partial and even if it occurs in only some of the gamblers, it should raise the odds in favor of the players by some percentage. Yet the gambling industry rests on the assumption that odds that are only slightly (one or two percentage points) in their favor along with other restrictions upon the betting will ensure them against serious losses. If Targ and Harary are correct, the casinos in Las Vegas, Reno, Monte Carlo, Atlantic City, and elsewhere should have long ago gone bankrupt.

A third inconsistency arises from evolutionary theory. Targ and Harary assert not only that we all have latent psychic abilities that can easily be developed but that we *must* develop them if we are to realize our human potential and escape becoming robots. This first of all raises the question of the evolutionary forces that allowed such a capacity to develop and yet remain unused. Presumably, it must have had some survival value to have developed in the first place. But, then, we have to ask why it is now latent and needs special exercises to develop.

I can imagine possible answers to these inconsistencies, but it is strange that the authors have felt no obligation to deal with them.

Going Beyond the Data

In his review of Targ's earlier book *Mind Reach*, Robert Ornstein (1977) wrote:

> Throughout the book the authors state their hope that the study of parapsychology will become primarily a scientific one in which speculations are firmly grounded in the evidence. In their own writing, however, Targ and Puthoff almost always go beyond evidence and claim they have proven their case when they have done nothing of the sort. In writing this book, the authors have done more harm, perhaps, to their own position and to their field of study than they have helped.

These words apply with equal force to the current book. Targ and Harary's most conspicuous faults are hasty generalizations and overstated claims. On almost every page they make assertions based on inadequate or nonexistent evidence. I have already given samples in the preceding sections. But to back up my own assertion, I will list a sample of some of the more blatant cases·

1. As already documented, the authors overstate greatly the strength of the scientific support for remote viewing. They strongly imply that 15 of 28 "formal published experiments" from laboratories all over the world were successful. But, if we deal with papers actually published under accepted scientific standards, only 9 "successes" can be counted. Of these, 7 were conducted by Targ and Puthoff and only 2 come from other laboratories. All but one of these "successful" experiments suffer a fatal flaw that I pointed out in 1977, as did Kennedy in 1979.

2. On the basis of just two remote-viewing trials conducted with the viewers in a submarine, Targ and Harary conclude that ocean water provides no barrier to the psi signal and that remote viewing is unaffected by

seasickness. There is just no way such a conclusion can be drawn on the basis of just two data points. Even if the authors want to claim that remote viewing took place under these circumstances, they would need many more data points collected under the underwater conditions before they could say that no difference existed between this and land conditions.

3. Targ and Harary admit that they do not know if evil psychics can implant harmful thoughts in other people, but they do not hesitate to suggest steps that the readers can employ for "psychic self-defense." If we do not know if the disease exists, how in the world can we know if the cure will work or even if it may cause harmful side-effects?

4. As already indicated, the authors provide elaborate instructions for developing psi but cite not one piece of scientific evidence that suggests that such instruction works.

5. Targ and Harary freely accuse critics of fraud on the basis of undocumented or unsubstantiated allegations. They try to smear Martin Gardner by writing that he

> criticized the NASA-supported ESP-teaching-machine study carried out at SRI in 1974. He falsely alleged that the subjects in this experiment tore up their unsuccessful data tapes, and only handed in the successful ones. He said in his article, "I am not guessing when I say that the paper tape records from Phase I were handed in to Targ in bits and pieces." We now know the reason he could say that he "wasn't guessing." This is because he recently confided to a fellow reporter that he had just made it up, "because that's the way it must have happened." The reporter was so shocked at this disclosure, that even though he is not particularly sympathetic to our work, he felt compelled to call up the SRI researchers to pass on this remarkable piece of news. [p. 157]

On its face this vicious slander does not stand up. First, it is based entirely on an undocumented statement by an unnamed reporter. Second, it just does not make sense for a journalist whose profession is based on the integrity of its members to make such an obviously damaging admission. When I read it, I could tell it was completely false, not only because I know Martin Gardner but also because I am familiar with the circumstances of his having made the claim about the tapes. He felt he could make this statement with confidence because he was given the information by an informant whom he has every reason to trust. To further compound the damage of this slanderous accusation, it was published in *Fate* magazine as the "Quote-of-the-Month." The "fellow reporter" to whom Martin Gardner supposedly confided that he had deliberately lied about the ESP-teaching-machine experiment turns out to be Ron McRae, the author of the recently published book *Mind Wars*. McRae has written to *Fate* magazine that he never made such a statement to Targ or anyone else. Instead, he did happen to mention to Hal Puthoff, who was then Targ's colleague at SRI, that he had overheard another individual make such a claim but did not consider it reliable. As a result of McRae's letter to *Fate*, that magazine published an apology and retraction in its October issue.

Targ and Harary continue this reckless abandon by asserting that CSICOP "was recently caught conspiring to deceive the public about some research results that did not fit their expectations." They go on to say that Randi and the rest of the CSICOP members "were exposed when a member of their group defected and offered documented proof of the deception. . . . It is clear that the goal of the psi-cops is to *control your ability to access and interpret information* and to walk a beat in your mind" (pp. 157-158). The "documented proof" presumably refers to the charges made by Dennis Rawlins in his long attack upon three members of CSICOP published in *Fate* magazine. The incident refers to a controversy over the interpretation of data from a study by Michel Gauquelin. In cooperation with Gauquelin, three skeptics reanalyzed the original data and published an interpretation that was challenged not only by Gauquelin, but also, on certain obscure technical grounds, by Rawlins. All this was initiated before CSICOP was founded, and the project was never sponsored by the Committee. Randi, whom the authors obviously want to paint black, had nothing to do with the Gauquelin study. In addition, the debate involved complicated and subtle matters of how to interpret trends in the data, and no conspiracy to deceive or any other evils that Targ and Harary so carelessly invent were ever implicated.

6. The authors freely question the motives of those critics who disagree with them. "Some of these critics have ulterior motives for not wanting the public or the academic community to take normal psychic functioning seriously. In that, they are like anyone else who hopes to profit by misleading the public about psychic abilities. Critics, like cultists, can sometimes live off the controversy they generate; For example, one critic, now famous, was a minor entertainer until he began a nationwide crusade against psi research" (p. 156).

The "critic" they are talking about in the last sentence of the quotation is Randi, the magician. Randi was an established and well-known entertainer long before his attacks on Uri Geller. He is probably the best-known escape artist since Harry Houdini. It would be difficult to measure whether he profited or lost either professionally or financially from the publicity emerging from his critiques of psychical research. I know that he often gives up profitable engagements to attend conferences and give talks on his views of contemporary paranormal claims. From my personal acquaintance with Randi, I have little doubt that his motivations involve a love of his craft and a desire to prevent conjuring from being used to exploit scientists and the public. Motivations are of course complicated and elusive. I have been a critic of paranormal claims for at least 35 years. I am sure that my motivations have changed drastically over that period of time. And they are complex. Even today I would have a difficult time trying to give a full account of why I put so much time into it. But the complexities and multidimensionality of motivations do not deter Targ and Harary. Unencumbered by facts or proof, they freely and confidently assert which motivations guide the behavior of their critics.

If the critics were fair and honest, Targ and Harary believe, they would carefully scrutinize the parapsychological data and conclude that psi has been proved. But the critics, without having seriously examined the data, freely criticize the claims. This means, according to the authors, that they have ulterior motives for not wanting psi to be true and for keeping the public from believing. In the same breath, Targ and Harary acknowledge that some critics, both within and outside the field of psychical research, have examined the data and still debate the claims for psi. No matter. They say that these critics, too, have religious and philosophical motives and deliberately distort the facts so as to mislead the public.

These seem to be the kinds of rationalizations that enable the authors to cope with the fact that many critics deny that the case for psi has been proved. This rationalization seems to provide a protective shell. It keeps Targ and Harary from facing the reality that the case for psi is much shakier than they would like to believe.

In many ways their book does a disservice to the attempts of other parapsychologists to make their field into a respectable and serious branch of science. The authors boldly assert that the accumulated data are sound, consistent, and scientifically impeccable. Only prejudice and ignorance prevents the scientific establishment from recognizing this fact. They fail to realize that the parapsycholgists have much work to do in order to get their house in order before they are ready to withstand the scrutiny of serious scientists.

Some major parapsychologists, fortunately, do recognize this problem. John Beloff (1976), a past president of the Parapsychological Association, told his colleagues: "I think that one thing we have got to recognize is that our field is so much more erratic, anarchic and basically subversive than we like to admit when we are engaged in our public-relations exercises. . . ." And Martin Johnson (1976), who holds the Chair of Parapsychology at the University of Utrecht, wrote: "I must confess that I have some difficulties in understanding the logic of some parapsycholgists when they proclaim the standpoint that findings within our field have wide-ranging consequences for science in general, and especially for our world picture. It is often implied that the research findings within our field constitute a death blow to materialism. I am puzzled by this claim, since I thought that few people were really so unsophisticated as to mistake our concepts for reality. . . . I believe that we should not make extravagant and, as I see it, unwarranted claims about the wide-ranging consequences of our scattered, undigested, indeed rather 'soft' facts, if we can speak at all about facts within our field. I firmly believe that wide-ranging interpretations based on such scanty data tend to give us, and with some justification, a bad reputation among our colleagues within the more established fields of science."

Without a doubt, Targ and Harary's careless scholarship will contribute to the "bad reputation" that parapsychology still has among many established scientists. Perhaps it is equally unfortunate that this book may very well achieve the opposite of what the authors intend. They hope to demystify psychic functioning, put their readers in touch with themselves and the world, and to free them from false beliefs. Instead they have set the stage for new mystifications and self-deception.

References

Beloff, J. 1976. The study of the paranormal as an educative experience. In B. Shapiro and L. Colby (Eds.), *Education in Parapsychology*, 16-29. New York: Parapsychology Foundation.

Hyman, R. 1977a. "Cold reading": How to convince strangers that you know all about them. *Zetetic (SKEPTICAL INQUIRER)*, (Spring/Summer):18-37.

Hyman, R. 1979b. Psychics and scientists: A review of Targ, R., and Puthoff, H., *Mind Reach. The Humanist*, 37(May/June):16-20.

Johnson, M. 1976. Parapsychology and education. In B. Shapiro and L. Colby (Eds.), *Education in Parapsychology*, 130-151. New York: Parapsychology Foundation.

Kennedy, J. E. 1979a. Methodological problems in free-response ESP experiments. *Journal of the American Society for Psychical Research*, 73:1-15.

———. 1979b. More on methodological issues in free-response Psi experiments. *Journal of the American Society for Psychical Research*, 73:395-401.

Marks D. 1981. Sensory cues and data selection invalidate remote viewing experiments. *Nature*, 292:177.

Marks, D., and Kammann, R. 1978. Information transmission in remote viewing experiments. *Nature*, 274: 680-681.

———. 1980. *The Psychology of the Psychic*. Buffalo, N.Y.: Prometheus Books.

Nisbett, R., and Ross, L. 1980. *Human Inference: Strategies and Shortcomings of Social Judgment*. Englewood Cliffs, N.J.: Prentice-Hall.

Ornstein, R. 1977. A case for parapsychology (Review of *Mind Reach*). *New York Times Book Review*, March 13.

Puthoff, H., and Targ, R. 1981. Rebuttal of criticisms of remote viewing experiments. *Nature*, 292: 388.

Schlitz, M., and Gruber, E. 1980. Transcontinental remote viewing. *Journal of Parapsychology* 44: 305-317.

Targ, R., and Harary, K. 1984. *The Mind Race: Understanding and Using Psychic Abilities*. New York: Villard Books.

Targ, R., and Puthoff, H. E. 1977. *Mind Reach: Scientists Look at Psychic Ability*. New York: Delacorte.

Tart, C., Puthoff, H. E., and Targ, R. 1980. Information transmission in remote viewing experiments. *Nature*, 284: 191.

11

Remote Viewing Revisited

David F. Marks

The 1970s saw the emergence of an astonishing psychic phenomenon that the principal investigators called "remote viewing." This refers to an alleged ability to perceive information from remote sources not available to any known sense. Actually this product of the paranormal is not a new one—it is really good old faithful ESP in a new package and with a different brandname. The main promoters of remote viewing are two physicists at SRI International (formerly the Stanford Research Institute), Russell Targ and Harold Puthoff. The remote-viewing effect could apparently be obtained by anybody and it required no special training or unique abilities. The results were allegedly reliable and repeatable. In fact, remote viewing was every parapsychologist's dream come true.

This is how one observes remote viewing. Someone (let's call him the traveler, T) gets into a car and drives to a place some distance away (e.g., a park, a church, a city hall, a railway station, or a golf course). Someone else (let's call him P) waits with the subject (S) with a tape recorder and a drawing pad. At a prearranged time, when T will have arrived at the target location, S tries to describe the location using his or her imagination. Experimenter P may assist S by asking various questions to clarify the description. S may also draw a picture of the target location. T allegedly acts as a kind of "psychic beacon" beaming back information from the target using some unknown sensory modality. After a prearranged interval (say 15 or 30 minutes) of remote viewing by S, T returns and escorts S to the target site to provide feedback on how well S has done. The experiment is repeated a number of times (e.g., 9).

Generally speaking, S, T, and P are all highly delighted with the results and all kinds of matching elements are found between S's descriptions and aspects of the target sites. I can personally vouch for the reliability of the remote-viewing effect when evaluated by simple inspection of the target site immediately after S has produced his response. I have found the remote-viewing experience compelling and direct and, on occasion, eerie. However it is obviously important to validate the effect on a more objective level. To this end, S's taped descriptions are transcribed and, together with any drawings that may have been produced, the transcripts are given to a judge who tries to match them against the series of targets. The judge normally visits all the sites and ranks the transcripts at each site in the order of their degree of matching.

Targ and Puthoff (1974, 1976, 1977) have reported extremely successful results from this judging process, and so a compelling subjective phenomenon has apparently been quantified numerically in the form of probability values. The implications of remote viewing (if real) for science are enormous. Textbooks of psychology, physiology, anatomy, and physics would all need to be completely rewritten, as our knowledge of the bases of perception and psychophysics built up over the ages would have to undergo major revision. Before this revolution in science begins, however, it would seem prudent to examine the phenomenon of remote viewing more closely to determine whether there is any possibility of artifacts, flaws in the methodology, experimenter bias, or some other parsimonious explanation.

With this in mind I began an investigation in 1975, in association with Richard Kammann, the results of which were published in our co-authored book *The Psychology of the Psychic*. Like many other investigators we were unable to replicate the effect claimed by Puthoff and Targ, as none of our judges could match transcripts accurately. Following visits to SRI and multiple discussions with Targ, Puthoff, and their key judge, Arthur Hastings, it became clear that the SRI research program, promoted as well-controlled science, was actually a massive litany of fallacies and flaws. To avoid duplication, this article will take up the story more or less where our book left it two years ago. This review relates to the two series of experiments with Price and Hammid, and various attempted replications by other investigators.

Data Suppression

Science separates itself form pseudoscience along a number of dimensions. One of these dimensions is accessibility of the data. Scientific data are consensually validated by open inspection of the recorded observations or through replication of the relevant phenomena. Following publication of major observations it is an accepted practice in science for researchers to allow colleagues who are doing serious research in the same field to have access to their original data. When researchers consistently refuse to allow colleagues such access, something important is being signaled. Of course data may get lost or destroyed or be difficult or costly to retrieve in the form required. Or they may be classified information or have commercial value that a scientist may wish to exploit prior to their general release. However, when none of these considerations is applicable, a refusal to supply a copy of a set of data leads to the unpleasant inference that something is wrong, that the data do not support what is claimed for them, that the data are an embarrassment following an extravagant claim that cannot be substantiated.

Over the past few years I have made frequent requests to Puthoff and Targ for copies of their remote viewing (RV) transcripts obtained in their experiments reported in *Nature* (1974), *Proceedings of the Institute of Electrical and Electronic Engineers* (1976), and *Mind-Reach* (1977). Targ and Puthoff have consistently refused to supply this important information, as they have to all others I know who made this request. The only concession has been to supply a single transcript from the Price series (Experiment 7), which is published in *Mind-Reach*. Normally no explanation or reply follows such requests. However in May 1979, while preparing the manuscript of *Psychology of the Psychic* I did receive the following explanation from Puthoff:

> With regard to your request, at the present time we are subjecting the transcripts to a number of blind judging procedures in order to assess which of several approaches constitutes the best way to handle free-response text in remote viewing studies. As a result we are not ready to release the materials, since premature disclosure would prevent further blind analysis work. I am sure you can appreciate the fact that after spending years building up a data base, we don't want to do anything that would jeopardize our options with regard to blind analysis of this data pool.
>
> By the way, this work is proceeding quite well. We have several research analysts breaking transcripts down into concepts which are then individually rated against concepts generated for the targets. ...
>
> *When our blind analysis work has been completed, you may of course have access to the raw data.* I do not think it will be in 1979, however, as our analysis contracts extend beyond that; thus, I doubt it will be in time for your book." [Italics mine.]

Three years later I'm still waiting! Surely "several research analysts" are not still "breaking transcripts down into concepts." *I would like to publicly ask Targ and Puthoff to release all of their remote-viewing transcripts in their original unedited form.* This would enable members of the more skeptical scientific community to evaluate the data for themselves.

Targ and Puthoff supplied John Wilhelm with a single transcript (Price series, Experiment 4, Redwood City Marina), reproduced in Wilhelm's book *The Search for Superman* (pp. 213-18). Although this was purported to be a complete transcript, it actually had two pages missing. A copy of the whole transcript given to me by Arthur Hastings (the SRI judge) contained nine pages. Pages 6 and 7 are missing from Wilhelm's version, and one can only speculate on the implications of the fact that one of Targ's most blatant cues ("Nothing like having 3 successes behind you") was on the missing page 7. Who could avoid the conclusion after reading this cue that this experiment was number 4 in the series?

The refusal to release the data-base for their controversial claims is a disturbing feature of Puthoff and Targ's remote-viewing project. The task of presenting an accurate and fair evaluation of the research is made none the easier as a consequence.

Fortunately, the main judge in the SRI research team, Arthur Hastings, was more helpful. Hastings loaned me a complete set of nine transcripts from the Price series and showed me six of the nine Hammid transcripts. However, Hastings said he felt "paranoid" about releasing this information and soon requested their return as they had become "confidential." How much longer these data will remain confidential is anybody's guess. It may be forever.

Sensory Cues

Most of the information in the RV transcripts is descriptive material, honest attempts by the subjects to describe the remote locations. Examination of the RV transcripts supplied by Hastings, however, revealed an enormous array of extraneous information and cues. Basically these are bits and pieces of information about the experiments that enable the judge to place the transcripts in their correct sequence: dates, times, references to previously visited targets, statements made by the experimenter monitoring the subject (including leading questions), and other information of this type. For both the Price and Hammid series there are enough cues to place the transcripts in almost perfect sequence. I could hardly believe my eyes when I saw that four of the six Hammid transcripts actually carried dates. Hastings told me one transcript even carried the name of the associated target, although he returned this one to SRI!

Being able to place a series of transcripts in correct sequence is useful only if the target sequence is also available to the judge. I was dumbfounded to discover that Puthoff and Targ gave Hastings a target list for the Price series in correct sequence. Hastings gave me a copy of the actual list he received from SRI. However, on 10 July 1978, Puthoff wrote me a letter in which he stated: "With regard to your assumption that the judges knew the order of target visitation, that is incorrect. Neither the single judge (Hastings) nor the panel of five judges were given the order; they were kept blind." This information was quite false. Two years later, Puthoff retracted it: "You are correct that, in the case of the Price series, the published list of the targets was in the order of target usage, and that, furthermore, when the target list was provided to the judge, it was inadvertently (through an error) given in the same order" (letter from Puthoff, August 25, 1980).

Using only the extraneous information in each transcript, together with the same list of target names provided to Hastings, independent judges in New Zealand (about 12,000 km from the target sites) were able to blind-match the transcripts against the Price targets with exactly the same degree of accuracy as that achieved by Hastings. I have called this phenomenon "remote judging" (see Marks and Kammann 1980, pp. 29-31, for more details).

The same excellent results from remote judging have been obtained from the Hammid series. The Hammid series is a little more complex than the Price series but basically the remote-judging results are as good as those obtained by the original SRI judge, as shown in Table 1.

It is important to remember that the three remote judges had access only to the cues provided in the Hammid transcript together with the target lists and the map provided to the SRI judge. No site visits were possible and none of the descriptive material from the SRI transcripts was available.

To illustrate how the remote judging process works, a total of 24 cues found in six Hammid transcripts, together with the conclusions that can be drawn from them, are shown in Table 2.

In the light of Table 2 the following claim (Tart, Puthoff, and Targ 1980) is very difficult to believe: "In the extensive replication studies, which also yielded significant results, the Marks-Kammann criticisms do not apply in principle. Target lists and transcripts were separately randomized, and transcripts were carefully checked before judging to ensure absence of any phrasing for which even a weak post-hoc potential-cue argument could be made." How careful was a checking process that missed 24 cues in

TABLE I

Distribution of Rankings Assigned to Transcripts
Associated with Each Target Location in the Hammid Series

Target	Remote Judges			SRI
	A	B	C	Judge
Methodist Church	2	2	1	1
Merry-go-round	2	2	1	1
Parking Garage	1	1	1	2
Railroad Trestle Bridge	2	1	2	2
Pumpkin Patch	2	1	2	1
Pedestrian Overpass	1	1	1	2
Total sum of ranks	10	8	8	9
p value	.005	.001	.001	.002

TABLE 2

Sensory Cues Available in the Transcripts for the Hammid Series
and Conclusions That Can Be Drawn from Them about the Experimental Sequence

Experiment Number	Transcript[a] Number	Cue No.	Text Phrases Relevant as Cues	Conclusions
1 (Courtyard)	Not available[b]		(Almost certainly mentions date and/or time. Other cues?)	(First target?)
2 (Auditorium)	Not available[b]		(Date and/or time? Other cues?)	(Second target?)
3 (Merry-go-round)	3	(1)	The time is 11:07	Possibly first of the day.
		(2)	This is the 18th of September	Target early in series.
		(3)	Russell Targ has gone to the remote site	Target possibly outside SRI
		(4)	"My mind is unfortunately clouded by his parting words when he kissed me goodbye and said 'I'll see you in church.' I wish he hadn't said that."	Maybe the church target or a target to be visited near the church (merry-go-round?).
		(5)	"All the various subliminal images I discounted yesterday."	Not first target of series
		(6)	"I keep seeing him riding his motorcycle maybe because I know that he went there with a motorcycle."	Probably a target outside SRI (confirms cue 3).
		(7)	"His body isn't moving like it was yesterday."	Not first target of series (confirms cue 5).
4 (Church)	8	(8)	September 18th Russell Targ at the location	Target close to target associated with transcript 3?
		(9)	"Is this going to be one of those where you tell me or not?" (Hammid's question.)	Not first target of series
		(10)	"We haven't decided yet. We haven't given you any feedback today?" (Puthoff's answer and question.)	Second target of the day
		(11)	"No we only did one." (Hammid's reply.)	Target must follow target associated with transcript 3
5 (Pumpkin patch)	6[c]	(12)	October 11	Target must follow targets associated with transcripts 3 and 8
		(13)	a.m.	Target precedes target associated with transcript 9.
		(14)	Reference to Hal (Puthoff) as visitor to target? (Other cues?)	
6 (Railroad bridge)	9	(15)	"Our second experiment of the morning."	Target immediately follows target associated with transcript 6
		(16)	"It is a quarter to twelve."	
		(17)	"Hal has gone to a second target."	
7 (Overpass)	2	(18)	October 11, 1974	Target immediately follows targets associated with transcripts 6 and 9, but must have at least 2 afterwards—*must be the 7th target.*
		(19)	"Hal has gone to the first of three remote sites that he will visit in this experiment."	
		(20)	"At 3:45 Hal will be at his first target selected outside SRI."	
		(21)	"—I have something in mind too—maybe its because I noticed it on the way back when I looked at the pumpkin field."	Target must be after the pumpkin patch (Confirms cues 18-20.)
8 (Parking garage)	7	(22)	"Hella has made a drawing of Hal's first location."	Target immediately follows target associated with transcript 2.
		(23)	Four-fifteen to four-thirty	
		(24)	"Is it a new thing—or from the previous site?"	
9 (Bicycle shed)	Not available[b]		(Almost certainly mentions date and/or time. Other cues?)	Transcript must be the last of the series (see cues 19, 20, 22-24.)

Notes: a. Transcript identifiers were randomly assigned.

b. Dr Hastings would not consent to the release of these transcripts

c. This transcript was available for a cursory inspection only and therefore may have contained other cues unlisted in this table.

six transcripts?

The claim that the Hammid target list given to Hastings was randomized is also highly moot. It actually depends on which list one is talking about, because, although the SRI researchers appear unwilling to admit this, no less than *three* listings of targets in the Hammid series were given to the judge. "*I received three target lists*" (letter from Hastings, May 26, 1977). One of these lists was randomized; this is the one cited by Puthoff as *the* (implying *only*) target list given to the judge. The other two lists provided by SRI (described in detail by Hastings in his letter to me of May 26, 1977) were not random. One of the two nonrandom listings was "a series of pages, each with the name of a target and location. The order of those pages, as I now have them, is this: courtyard, auditorium, Methodist church, playground, overpass, parking garage, railroad bridge, pumpkin patch, bicycle shed" (Hastings's letter, May 26, 1977). The correlation of this target listing and the order of target usage is 0.833 ($p<.01$). This listing therefore would have provided an artifactual basis for correct matchings (see Marks 1981a).

The second nonrandom listing of Hammid targets given to the judge by SRI was a map of the Menlo Park/Palo Alto area indicating the location of the nine target sites. There is a strong correlation between map codes and pairs of targets visited outside SRI by Targ or Puthoff on different days and half-days over the series of experiments.

Here is a verbatim description from a remote judge of how he matched the six Hammid transcripts against the targets:

> The six transcripts fall into three pairs. The first pair are 8 and 3 [see Table 2]. Russell Targ is the traveller to both, on the same day, September 18. The give-away phrase in 3 is "I'll see you in church." Therefore I would say 3 is the Church and 8 is the Merry-go-round. The remaining two pairs are 6 with 9 and 2 with 7. Puthoff is the traveller. 6 and 9 were apparently visited one immediately after the other, on the morning of October 11. Looking at the map 6 and 9 either go with the Pumpkin Patch and the Railroad Bridge, which are very close together, or the Overpass and the Parking Garage, which are further apart. However this pair (6 and 9) must go with the Pumpkin Patch/Railroad Bridge combination because of the cue phrase in transcript 7 which mentions the "pumpkin field" as having already been visited. Therefore 6 and 9 go with the Pumpkin Patch/Railroad Bridge, and 2 and 7 go with the Overpass and the Garage. I am unsure of the order but I'll stick to the following: (6) Pumpkin Patch, (9) Railroad Bridge, (2) Overpass, (7) Parking Garage."

This remote judge (Judge B in Table 1) gained a sum of ranks of 8 giving a p value of 0.001. This result achieved at a distance of 12,000 km and with no ESP ability—just plain logical thinking—exceeded that obtained by Hastings. Clearly the 24 cues available in the transcripts, the presence of which Targ, Puthoff, and Tart have denied, are sufficient to account for the entire RV effect claimed for the Hammid series.

This review of the SRI RV research unfortunately remains incomplete owing to the unavailability of the majority of raw data. Real difficulties can occur in evaluating research on the basis of incomplete information, and misinterpretations and errors may occur. An example of this can be found in *The Psychology of the Psychic* in relation to a mistaken inference concerning the Hammid experiments, which were hypothesized to have been selected from a longer series. Although training experiments were run prior to the published experiments, the allegation of data selection has been retracted (see Marks 1981b). However, I remain undecided about whether all of the drawings produced by Hammid were submitted for judging, since drawings were missing from some of the transcripts given to me by the SRI judge. But this is a minor point in the light of the remote-judging controls. The SRI experiments with subjects Price and Hammid can be seen as providing no evidence whatsoever for remote viewing. Contrary to Targ and Puthoff's claims, the quality of the subjects' descriptions is extremely poor and, without the cues, cannot be matched against the targets.

Replication Attempts

Attempts to replicate RV fall into two distinct groups. First there is a group of carefully controlled studies that avoided the flaws present in the Puthoff-Targ experiments and that found no evidence of RV (Allen et al. 1976; Rauscher et al. 1976; Karnes et al. 1979; Karnes and Susman 1979, 1980; Marks and Kammann 1980). When all normal or artifactual methods for matching the subjects' descriptions against target sites are eliminated, it is apparent that the RV effect completely diappears.

A second group of experimenters claim to have confirmed RV. However, all of these experiments are flawed in a variety of ways, some of them reminiscent of the original SRI research and some new. The size of the RV effect claimed in these studies actually correlates quite well with the magnitude of the flaws present.

Hastings and Hurt (1976) reported a single RV experiment that used as a target "a circular play area, filled with sand, and containing a log structure with chains hanging from it and a slide on one side. ... There were swings next to the play area, and a jungle gym alongside the sidewalk." A group of 36 subjects attempted to describe this target, which was selected from a set of six possible targets all within a 10-minute drive of the "laboratory." The names of the targets were written on cards, which were sealed inside envelopes. One envelope was randomly selected by throwing a die, and David Hurt and a companion then departed with the sealed envelope. The envelope was opened in Hurt's car and the two travelers proceeded to the target site and spent 10 minutes there.

Apparently the other five envelopes containing the names of the nontargets were left behind with the other experimenter (Arthur Hastings) and the group of subjects. It is a matter of some speculation what Hastings could possibly have discerned from the five envelopes left in his control. However, before the remote team returned, no less than 20 of the 36 participants had selected the correct target site from the six available alternatives named and described by Hastings. This procedure seems incredibly lax and poorly controlled and the potential for cueing and bias in the subjects' descriptions is enormous. To quote the experimenters' report: "We think that the effectiveness of the experiment was partly due to the way we conducted it." Hastings and Hurt recommend that in future research "the target pool should be made up by someone not at the experiment." Who would not agree?

Whitson, Bogart, Palmer, and Tart (1976) reported a study not dissimilar from the Hastings-Hurt experiment just described. A target was randomly selected from a set of ten sites. A group of 27 art students attempted to draw this site while it was viewed by one of the experimenters. A judge was asked to match a first and second choice from a set of 10 slides of the sites projected simultaneously on a screen. A bias was observed in the judge's rankings favoring the target site, a bike tunnel. A second experiment was run using another target and another art class of 14. The results of the second experiment displayed a drawing bias but, on this occasion, favoring a different site, which was not the target. This study illustrates an important control for RV research—it is essential to control for biases operating in the subject's responses. The next study provides a potent illustration of what can happen when response biases are ignored.

Vallee, Hastings, and Askevold (1976) reported a series of 33 experiments using a computer-conferencing network with 12 participants in various locations across North America. At prearranged times the participants used portable computer-terminals located in their homes or offices to type descriptions of a mineral sample, selected from an available set of 10. Eight out of 33 descriptions were correctly matched by a panel of five judges, compared with a chance expectation of 3.3.

The implications of this result (if true) for human perception are absolutely mind-boggling. Apparently human beings can discriminate a baseball-sized object at a distance of 2,500 km, differentiate the object from nine others in an arbitrary and unknown set, through an incalculably huge number of physical barriers, and without interference from a huge landmass of rock and minerals in the North American continent. Before rewriting our textbooks on perception it would pay to examine the details of this experiment a little more closely. In fact we discover that although 12 participants took part in the experiment, results for only 6 subjects are included in the experimental report. Of the 8 matching descriptions, 3 were obtained for target D (opals) and 2 for target F (common salt crystals). The results become a little less surprising when one notes the strong bias in the subjects' descriptions (as rated by the judges) toward opals and salt, regardless of which mineral was the target. The results for these two minerals when they were targets gave 5 correct matches in 10 attempts (50 percent). However, for the remaining 23 experiments, when samples D or F were *not* the target, 8 (or 35 percent) of the descriptions also matched D or F more closely than any other. When the results are considered in terms of rankings instead of the peculiar rating system used by the Vallee team, we find that the average ranking of the subjects' descriptions against samples D or F when D and F were targets was 2.80, and when they were not targets, 2.96, a minuscule difference. The results for the other 8 targets gave 3 correct descriptions in 23 attempts, almost exactly what we would expect purely by chance. Therefore, having corrected for response bias, these results lack any real substance in their support of RV.

Dunne and Bisaha (1979) conducted precognitive RV experiments in which the subjects allegedly described the targets *before* they were selected. In the first series of seven experiments subjects worked in pairs but independently. Dunne and Bisaha reported results for seven subjects (S4–S10). For some unexplained reason the results for the three other subjects (S1–S3) are missing from the experimental report. Also ten targets were selected, but the results for only seven are presented. The protocol differed from the SRI research in that photographs of the targets were used by the judges rather than on-site visits. Apparently several photographs were taken at each target site and a selection must have then been made, as only one photograph of each site was given to the judges. Since the photographs used in judging must have been selected *after* the subject had described the target, unless the selection was conducted by an independent third party, the experiment is worthless. The selection could obviously have been biased by the experimenter's knowledge of the subject's descriptions.

Dunne and Bisaha randomly divided the total of 14 transcripts into two sets, Group A and Group B. Two judges were asked to blind rank Group A, two judges to blind rank Group B, and two judges matched Group A transcripts against Group B transcripts. John Bisaha sent me the transcripts for this experiment and they do contain a few cues, although not on the same scale as the SRI transcripts I have seen. For example, transcript A-5 is labeled "Barb 5/9," the subject's name and the date. Transcript B-2 is labeled "Steve B 5/9,". Needless to say these two transcripts were correctly matched. Evidence of post-hoc data-selection is also present in one of the transcripts. The experimenters claim to have attached any drawings the subject made to the transcripts for judging. Yet Transcript A-2 from subject 5 for the Madonna del Strada target included no drawings even though the transcript states in two different places that a drawing was done. Why was this drawing excluded? Was it a poor match?

An RV report claiming confirmation of the SRI research was published by Schlitz and Gruber (1980). A target in Rome was visited on each of 10 days at a prearranged time, while a second experimenter in Detroit sketched and wrote down her impressions as she concentrated on the distant target. Transcripts, translated into Italian, together with associated drawings, were given to five judges who visited the 10 sites and then ranked and rated the 10 transcripts against each target. The resulting scores were highly significant for four out of the five judges.

Schlitz and Gruber state that no cues were found in the transcripts so that no editing was necessary. However the drawings were attached to the transcripts following their translation into Italian. Were the drawings and transcripts dated or coded? If not, how could they have been put together correctly following the translation? What sort of coding was used? How were the targets listed when given to the judges? Were they in the correct order or in random order? Unfortunately we are not told this vital information and are left with an uncomfortable feeling of déjà vu.

Conclusion

Well-controlled experiments never find the RV effect, while poorly controlled experiments nearly always do. Data suppression, flawed methodology, and lack of replication lead to the conclusion that remote viewing is a cognitive illusion, an artifact of human error and wishful thinking.

References

Allen, S., P. Green, K. Rucker, R. Cohen, C. Goolsby, and R. L. Morris. 1976. "A Remote Viewing Study Using a Modified Version of the SRI Procedure." In J. D. Morris, W. G. Roll, and R. L. Morris (eds.), Research in Parapsychology 1975. Metuchen, N.J.: Scarecrow Press.

Dunne, B. J. and J. Bisaha. 1979. "Precognitive Remote Viewing in the Chicago Area: A Replication of the Stanford Experiment." Journal of Parapsychology 43:179-204.

Hastings, A., and D. Hurt. 1976. "A Confirmatory Remote Viewing Experiment in a Group Setting." Proceedings of the IEEE 64:1544-45.

Karnes, E. W., and E. P. Susman. 1979. "Remote Viewing: A Response Bias Interpretation." Psychological Reports 44:,471-,479.

Karnes, E. W., J. Ballou, E. P. Susman, and P. Swaroff. 1979. "Remote Viewing: Failure to Replicate With Control Comparisons." Psychological Reports 45:963-73.

Karnes, E. W., E. P. Susman, P. Klusman, and L. Turcotte. 1980. "Failures to Replicate Remote-Viewing Using Psychic Subjects." Zetetic Scholar 6:66-89.

Marks, D. F. 1981a. "Sensory Cues Invalidate Remote Viewing Experiments." Nature 292:177.

_____ 1981b. "The Assessment of Parapsychological Studies on Remote Viewing: A Reply to Morris." Journal of the American Society for Psychical Research 75.

Marks, D. F., and R. Kammann. 1978. "Information Transmission in Remote Viewing Experiments." Nature 274: 680-81.

_____ 1980. *The Psychology of the Psychic*. Buffalo, N.Y.: Prometheus Books.

Puthoff, H. E., and R. Targ. 1976. "A Perceptual Channel for Information Transfer Over Kilometer Distances: Historical Perspective and Recent Research." *Proceedings of the IEEE* 64: 329-54.

_____ 1981. "Rebuttal of Criticisms of Remote Viewing Experiments." *Nature* 292:388.

Rauscher, E. A., G. Weissman, J. Sarfatti, and S. P. Sirag. 1976. "Remote Perception of Natural Scenes Shielded Against Ordinary Perception." In J. D. Morris, W. G. Roll, and R. L. Morris (eds.) *Research in Parapsychology: 1975*. Metuchen, N.J.: Scarecrow Press.

Schlitz, M., and E. Gruber. 1980. "Transcontinental Remote Viewing." *Journal of Parapsychology* 44: 305-17.

Targ, R., and H. E. Puthoff. 1974. "Information Transfer Under Conditions of Sensory Shielding." *Nature* 252:602-07.

_____ 1977. *Mind-Reach*. New York: Delacorte Press/Eleanor Friede.

Tart, C. T., H. E. Puthoff, and R. Targ. 1980. "Information Transmission in Remote Viewing Experiments." *Nature* 284:191.

Vallee, J., A. Hastings, and G. Askervold. 1976. "Remote Viewing Experiments Through Computer Conferencing." *Proceedings of the IEEE* 64:1551-52.

Whitson, T., D. Bogard, J. Palmer, and C. T. Tart. 1976. "Preliminary Experiments in Group Remote-Viewing." *Proceedings of the IEEE* 64:1550-51.

Wilhelm, J. L. 1976. *The Search for Superman*. New York: Pocket Books.

12

Gerard Croiset: Investigation of the Mozart of "Psychic Sleuths"

Piet Hein Hoebens

The Dutchman Gerard Croiset, who died unexpectedly in July 1980, was undoubtedly one of the psychic superstars of the twentieth century. His mentor, Professor Wilhelm Tenhaeff, has called him the clairvoyant equivalent of Mozart or Beethoven. Tenhaeff's German colleague, Professor Hans Bender, recently admitted that Croiset had been instrumental in transforming his belief in ESP into "an unshakable conviction." The obituaries published in the European press reflected the sensitive's unique reputation. According to the Amsterdam weekly *Elsevier*, the deceased had heralded a "new awareness of cosmic solidarity." The German parascientific monthly *Esotera* ran a cover story lamenting the death of "the clairvoyant who never disappointed." A professor from the papal university delivered the funeral oration.

Croiset's career in the supernatural has been distinguished indeed. According to his biographers, he has solved some of the century's most baffling crimes, traced countless lost objects, and located hundreds of missing persons. His paranormal healing powers are said to have been on the Caycean level. He "excelled" at precognition and is credited with having accurately foretold future events on numerous occasions. Most of his remarkable feats, it is said, were performed under scientific supervision, which supposedly would make Croiset one of the most thoroughly tested sensitives since Mrs. Piper.

Gerard Croiset was respectable. Many educated Dutchmen who profess disbelief in ESP have managed to hold the simultaneous conviction that Croiset, for one, was genuine.

This miracle man is the subject of a full-length biography by the American journalist Jack Harrison Pollack, who claims to have spent five years checking and double-checking the psychic's record. Pollack's verdict: "Unbelievable, but true." Unbelievable, indeed. But true?

Psychic Detectives

The practical achievements of Gerard Croiset and other sensitives who claim to assist the police share most of the features of "spontaneous cases."* Such cases typically occur under uncontrolled conditions and are by their very nature unrepeatable. This means that the only evidence we have usually consists of whatever witnesses are able to remember or care to report. Before reaching a verdict, the critical investigator has to address two crucial questions:

1. Are the reports free of omissions, errors, and deliberate distortions?

2. Does whatever remains after the first question has been answered admit no more plausible an explanation than ESP?

The Sources

Studying Croiset has been the virtual monopoly of Wilhelm Heinrich Carl Tenhaeff, the Dutch parapsychologist who in 1953 was appointed to the first chair of psychical research ever to be established at a regular university (Utrecht). Tenhaeff's books and articles constitute the principal source of information on Croiset, whose case may be said to stand or fall with the reliability of his learned mentor. Unfortunately, little of Tenhaeff's work has been translated into English, which leaves Pollack's *Croiset the Clairvoyant* as the main reference in this language. Pollack is a journalist, not a scholar. Yet his biography may be regarded as an authoritative document, since it was written under the personal guidance of Tenhaeff himself. "He indefatigably double-checked the facts in my manuscript," Pollack states in his Acknowledgment.

In Tenhaeff's *Tijdschrift voor Parapsychologie* (the official journal of the Dutch Society for Psychical Research), the professor has proudly confirmed this. According to Tenhaeff, *Croiset the Clairvoyant* "was written on the basis of information which I supplied and also under my supervision."

Police Records

According to Pollack, Croiset won plaudits not just from parapsychologists but from policemen all over the world for his achievements in psychic detection. "I checked documents in case after case in police records," the biographer assures us. I am not quite certain what he means. Most of the documents he refers to must have been in Dutch, and I doubt that he ever familiarized himself with the language. The only Dutch expression I found in the book is the equivalent of "thank you," and even that solitary

---

*This article and the one to follow are exclusively concerned with Croiset's activities as a paranormal sleuth. In that role he became best known abroad. About the experiments with him I will have more to say later.

example contains an error. Presumably, Pollack relied on summaries or translations of the relevant documentation, prepared for him by Tenhaeff and other acquaintances in the Netherlands. He must have felt it was quite safe to do so. After all, his material would be double-checked by a distinguished scholar, a professor at a state university, a pioneer whom the American psychiatrist-parapsychologist Dr. Berthold Eric Schwartz had compared to Copernicus, Freud, and Einstein.

The Boy on the Raft

It is time to take a closer look at one of Croiset's most impressive successes. It is the "Boy on Raft" case, and will be found on pages 106 and 107 of the Bantam paperback edition of *Croiset the Clairvoyant*. This case has often been mentioned in the psi literature, and Tenhaeff himself has indicated more than once that it is one of the classics.

This is how Pollack reports it (italics added):

> Ten-year-old Dirk Zwenne left his home in the dunes city of Velsen near the North Sea canal on Saturday, August 29, 1953, at about two P.M. to play. When the boy had not returned home by early evening, his parents began to grow uneasy. They telephoned the local police without success. When no trace of the missing boy had been found in two days, Dirk's uncle telephoned Croiset, whose phone number and address in Enschede, 115 miles away, had been given to him by a police superintendent. Among the clairvoyant's *immediate images* was that Dirk had drowned: "I see a *small harbor, a small raft* and a little sailboat. The boy was playing on the raft. He slipped and fell into the water. As he fell, his head struck the sailboat and he received *an injury on the left side of his head.* I am very sorry. There was a strong current in the harbor. *The boy's body will be found in a few days in another small harbor which is connected with the first harbor.*" Unhappily, five days after he had disappeared, the body of Dirk Zwenne was found in this second harbor. And, *just as Croiset had seen,* the boy had a wound on the left side of his head. The raft and small sailboat were recovered in the first harbor— again *just as the sensitive had described.* "It is very likely that everything had happened as the paragnost had seen it," summarized Professor Tenhaeff.

This seems a striking case indeed. Oddly enough, until now nobody seems to have thought of comparing this version with a letter that was published in *Tijdschrift voor Parapsychologie* in 1955 (vol. 23, no. 1/2). It was written by Mr. A. J. Allan, the uncle who had consulted Croiset. From this report (embedded in an article by Tenhaeff) we get an idea of what really happened. On Monday, August 31, Mr. Allan phoned the clairvoyant, who was at that time living in the eastern Dutch town of Enschede. He acted on the advice of Haarlem police superintendent Gorter, who happened to be the second secretary of the Dutch Society for Psychical Research (SPR) and an acquaintance of both Tenhaeff and Croiset.

The sensitive, after having made clear that he knew what the call was about, told the uncle: "You must look near a *gasholder.*"

Allan: "A gasholder?"

Croiset: "Yes. It might be a tank or a boiler or something like that. I see a road and a small ditch. I also see a small bridge and a small water. Do I speak to the boy's father?"

Allan: "No, you are speaking to an uncle."

Croiset: "All right, I can speak freely. The child has drowned. He is dead. I also see a jetty and a rowing boat or something like that. That's where the body must be."

Allan: "Could it be the North Sea canal?"

Croiset: "No, that is too broad. I don't see so much water."

Allan: "Then where is it?"

Croiset: "I don't know Velsen, but you have to look near that gasholder or tank. It is to the right of it. *To know for sure I ought to come to Velsen.* Call me again if that's necessary."

End of conversation.

Holland being a country full of roads, ditches, small bridges, small waters, jetties, rowboats, and objects that could be described as gasholders, tanks, boilers, or "something like that," Croiset's impressions had hardly been specific. His description could apply to any number of locations.

According to Mr. Allan, the police, "after having considered several possibilities," decided that Croiset must have "seen" a small harbor near a water purification plant. This is a rather surprising interpretation, as that "harbor" (really a recess) is *part of the North Sea canal.* The psychic had been specific on only one point: the water was *not* the North Sea canal. To me, this strongly suggests that the police had reasons of their own to regard the small harbor as a likely place.

The police decided to drag the harbor the next day. On Tuesday, they heard that Dirk, shortly before disappearing, had told one of his friends about "having found a nice raft."

Croiset, who was phoned again later that day, *now started to receive impressions of a raft also.* Mr. Allan suggests that this was due to telepathy, but the skeptical reader may be able to think of a more naturalistic explanation.

Nothing was found in the small harbor, and the next day Allan asked Croiset to come to Velsen. The clairvoyant arrived that same evening, in the company of Tenhaeff.

The psychic was taken to the small harbor, and there he started to get "strong emotions." He stated that the boy had been playing with his raft, had lost his balance, and had bumped his head on a hard object. "According to him [Croiset] *this had been fatal,*" Allan notes. Croiset predicted the body would *not be found before Monday, September 7, or*

Tuesday the 8th, and would show an injury *"on the left side of the forehead."* The clairvoyant was then taken to a second small harbor that also forms part of the North Sea canal. *There, however, he felt "no emotions."*

The next morning, Thursday, September 3, the body of Dirk Zwenne was found in the canal near the entrance of the second harbor. The head showed bruises, *but not at the location Croiset had indicated.* Where and in what circumstances the boy had fallen into the water appears never to have been ascertained.

Now please compare this long and tedious story with Pollack's "summary" and be surprised at the magical metamorphosis an entirely unspectacular event has undergone in the process of summarizing. Croiset's impressions had been vague and for the most part wide of the mark, and yet this case is cited as a classic instance of successful psychic detection.

How could this fantastic distortion ever have survived the checking and double-checking by an experienced American journalist and a distinguished university professor? The answer to this question may contain the key to much of the Croiset mystery. For it was Tenhaeff himself who concocted the fake version. Having published Allan's account in his *Tijdschrift,* destined for the home market, the professor prepared a special version for export.

All Pollack had to do was to paraphrase the version Tenhaeff had already published in German (*Zeitschrift für Parapsychologie und Grenzgebiete der Psychologie,* vol. 2, no. 1, 1958) and in English (*Proceedings of the Parapsychological Institute of the State University of Utrecht,* no. 1, December 1960). This latter version reads as follows:

> When no trace of the child had been found by 31st August an uncle of the missing child rang up Mr. Alpha [Croiset's code name], whose name and address he had obtained from a police superintendent. According to the paragnost the child had drowned. Among the "pictures" which presented themselves to Mr. Alpha were a few *which concerned a small harbour.* In this small harbour he "saw" a *small raft* and a *little sailing boat.* According to the paragnost the child had been playing on the raft. He supposed him while at play to have slipped and to have fallen into the water. In doing so he appeared to have incurred *a wound on the left side of the head* where he struck the sailing boat as he fell. *In consequence of a current in the harbour, so the paragnost said, the body would be found in another small harbour which was connected with the first. On 3rd September, just as the paragnost had "seen,"* the body of Dirk Zwenne was *in fact* found in the second harbor with a wound on the left side of his head. [Italics added.]

On all essential points, this version is identical to Pollack's. In this form, the "Boy on Raft" case has become the "believer's favorite." It was featured in the cover story devoted to Croiset in the September 1979

Holland Herald (an English-language magazine mainly concerned with "selling" the Netherlands) and found its way into Ryzl's *Parapsychology: A Scientific Approach* and numerous other publications. It is clear that in this case, Professor Tenhaeff "cooked the books." His probable reasons for doing so will be discussed later.

Pollack as a Witness

Jack Harrison Pollack can hardly be blamed for the serious errors in his report of the "Boy on Raft" case. A journalist may be forgiven for accepting a university professor's word, but he must be held responsible for his reports of what he claims he personally witnessed.

In *This Week* of February 19, 1961 (a slightly elaborated version will be found on pp. 25-26 of the paperback edition of the biography), Pollack recalled being present when, on May 21, 1960, Croiset was phoned by a neighbor of an Eindhoven family whose four-year-old son had been missing for 24 hours. According to the article, the police "*had no clues.*"

"The outlook isn't good," Croiset is quoted as saying. "Search the area immediately. But I'm afraid in about three days the child's body will be found in the canal close to the bridge."

Pollack continues: "Three days later, I checked up. The police of Eindhoven had just found the child's body next to one of the piers of the bridge over the canal—exactly as Croiset had predicted."

Something seems to have gone wrong when Pollack checked up. In 1981, I made inquiries with the Eindhoven police. Mr. W. Jongsma of the Information Office kindly offered to check the original police report. These are the real facts: The victim, three-year-old Anthonius Thonen, while playing with a friend, fell into the Dommel River on May 20. The accident was witnessed by the other boy, who told Anthonius's mother about it when she came looking for him. Mrs. Thoonen saw something floating on the water. Presumably, this was the body. It had disappeared when the police arrived. On May 23 (two days after the telephone conversation), Anthonius's remains were found in the river, near the Gessel playground.

The police report does not mention Croiset. Neither does it mention a bridge. (There are so many bridges over the Dommel that there is always one nearby.)

The authorities from the very beginning knew that the boy had drowned in the river. Pollack's claim that the police "had no clues" is utterly misleading. No one needed a clairvoyant to say that "the outlook isn't good" or that the area should be searched immediately. Yet, by overlooking some crucial facts, Pollack is able to present this case as "an amazing demonstration."

Search for a Child

Pollack's book and numerous other English-language publications convey the impression that psychics are employed as a matter of course in Dutch police investigations. Some journalists seem to think that a special hot-line connects Tenhaeff's office with police headquarters in every major town. Perhaps the language barrier may have been responsible for this exaggeration. In fact, Dutch police authorities tend to be skeptical of clairvoyants. Their typical reply to questions about Croiset is something to the effect that ESP may exist but Croiset was never of any use to them. However, there are a few exceptions.

Notable among the exceptions is a report by Inspector G. D. H. van Woudenberg, published in *Algemeen Politieblad* (no. 13, 1964, pp. 297-300). Van Woudenberg, at that time serving with the Voorburg police, relates an apparent success Croiset achieved in searching for the body of six-year-old Wim Slee. The child was reported missing on April 11, 1963. A thorough search was organized the same day. A police dog led the way to a certain spot on the bank of a canal locally known as De Vliet. There were good reasons to assume the dog was right, as it was known that Wim often went there to play. No body was found, however. The next day the case was mentioned in the press and on radio and television. A number of psychics volunteered with perfectly worthless information. In the meantime, an uncle had rung Croiset's phone number, to be told that the psychic had gone abroad. The uncle did not get through to Croiset until the 16th. The clairvoyant then told him that the boy had drowned in De Vliet. The body would surface in a couple of days near a bridge, a sluice, "or something like that," to the left of the spot where the accident had happened. Croiset asked to be called back in case the child had not been found by Friday the 19th.

That Friday, with still no trace of Wim Slee, Croiset came to Voorburg. He had with him a sketch of the location where the boy had fallen into the water. He invited the police to get into his car and then drove to De Vliet. He stopped near the spot indicated by the dog and stated that he now experienced "strong emotions." Van Woudenberg noted "striking similarities" between the sketch and the actual location. Croiset said the child had drowned there but would surface on Tuesday morning near a bridge some 800 yards downstream. This was to the right of the indicated spot (as seen from Voorburg), but the clairvoyant explained that "to the right" really is the same as "to the left" if you look at it from the other side. Near the bridge, van Woudenberg continues, "we saw there were points [on the sketch] that corresponded with what the uncle had been told earlier that week."

As it happened, Wim Slee's body was found the following Tuesday near the bridge. Presumably, the remains had been tangled up in refuse on the bottom of the canal.

Unless we want to make unfounded conjectures about a possible lapse of memory on van Woudenberg's part, the verdict must be: a hit. Yet I wonder if ESP is the only plausible explanation. Croiset gave his first "impressions" on the 16th, five days after Wim Slee had been reported missing. The case had received considerable media coverage. The police suspected that the boy had drowned in De Vliet and the dog had even indicated a likely spot. However, Croiset's initial "images" were vague, and he did not specify what bridge, sluice, or "something like that" he meant.

His description (as reported by van Woudenberg) would fit a good many bridgelike structures. He stated that the body had floated "to the left" but did not say from what vantage point. Moreover, from his request to be called back in case the body was still missing on Friday the 19th, we may surmise that he expected that the boy would have been found by that date. Friday the 19th would have been eight days after Wim Slee disappeared, and van Woudenberg tells us that "most bodies come to the surface in a maximum of nine days."

Croiset scored a hit only when he tried again. The accuracy of the sketch he showed on the 19th is not surprising. He had simply drawn the area where, according to the police dog, the child had fallen into the water. The possibility that he had obtained his information by normal means should not be ruled out. Van Woudenberg (personal communication) thinks this hypothesis somewhat unlikely, as the sketch contained a few details of the location not visible from the public road. I venture to suggest that the inspector *may* have underestimated the resourcefulness of a highly experienced psychic.

What remains is that Croiset, in his second series of "impressions," received eight days after the accident, correctly predicted both the date and the spot where the body would be found. Striking enough, but I doubt whether the odds against such a hit arising from chance alone are really astronomical.

Van Woudenberg (personal communication) is still impressed by Croiset's success, although he does not think it falls into the "conclusive evidence" category. "The weakest part of the case," he told me in February 1981, "is that it seems to be pretty unique. It happened 17 years ago and continues to be cited as possibly the best case that ever happened in Holland. One cannot help wondering why there seem to be so few comparable successes."

Failures

As an isolated case, Croiset's achievement in Voorburg is fairly impressive. However, we must guard against a common fallacy in assessing such apparently compelling "proofs." The chance hypothesis can only be ruled out if we know the hit/miss ratio in the psychic's total score. On this point, no statistics are available, but there are a number of reliable indications.

In his English-language *Proceedings* (1960) and in a number of other publications, Tenhaeff has admitted that the number of successful consultations (successful from a practical point of view) is limited. The bulk of the material in his *Beschouwingen over het gebruik van paragnosten* (1957), his major Dutch work on psychic detection, concerns cases where the psychics supposedly demonstrated ESP without actually solving any crime or finding any missing person. There are very few prize cases, and these are cited time and again. Some of these "successes," as I have shown, are striking only when the facts have carefully been doctored. In his book *Ontmoetingen met Paragnosten* ("Encounters with Psychics," 1979) Tenhaeff quotes Croiset as stating that he was consulted by relatives of missing persons on an average of 10 to 12 times a week. That is something like 500 times a year, and Croiset has been in the business since the forties!

All this strongly suggests that thousands of Croiset's attempts have ended in failures—even if we generously use standards that allow, for example, the "Boy on Raft" case to be judged a success. Given so many misses, an occasional lucky hit is hardly surprising. The miracle van Woudenberg thought he witnessed may simply have been one of those successful guesses we can expect once in a while if the number of trials is sufficiently large. Tenhaeff is remarkably reticent about the many failures, except when he feels able to explain them in terms of misdirected ESP. The complete disasters that cannot be rescued even by parapsychological special pleading are conveniently ignored. Examples, however, are numerous.

In May 1956 the public prosecutor in Amsterdam revealed that three psychics had earlier that year attempted to shed light on the disappearance of a 31-year-old inhabitant of Rossum. Croiset had stated that the man was alive and staying in Germany. Shortly thereafter, the body was found in a canal in the municipality of Ootmarsum, Holland.

In 1969, Croiset went to Viareggio, in Italy, to look for 13-year-old Ermano Lavorini. He "saw" that the boy had fallen into the water while playing. In fact, Ermano had been killed by a friend during a quarrel. The body was found in the dunes.

In 1966, Croiset journeyed to Adelaide, Australia, to search for three missing children. A local "committee" paid the expenses. The clairvoyant was sure the children were buried under a new warehouse. He advised demolition. The "committee" collected 40,000 Australian dollars to have the building pulled down. The soil under the concrete floor was removed to a depth of four yards. No bodies were found. Croiset urged them to dig one more yard, "and the children will be found." He was wrong. This costly mistake did not affect his reputation. Three years later the Amsterdam paper *Het Vrije Volk*, quoting an AP Telex, claimed that the

Australian authorities had "refused permission to search on the spot."
  In June 1973, Croiset was consulted by the relatives of a murdered Chinese from The Hague. The clairvoyant indicated that a Mr. Senf knew more about the crime. The relatives then abducted Senf and tortured him for three hours to get a "confession." Senf, however, had nothing to confess. He happened to be innocent. The following week, Croiset visited Senf, who was in the hospital recovering from his ordeal. He brought flowers and assured Senf that he now was quite convinced of his innocence.
  In the police journal *Algemeen Politieblad* of January 9, 1960, Utrecht Superintendent Th. van Roosmalen published a catalogue of psychic blunders. In December 1957, he revealed, the 14-year-old son of the E. family disappeared from his parental home in the Utrecht River district. The house was near one of several canals in that part of town. After a couple of days, the parents contacted Croiset. The psychic came and led Mr. and Mrs. E. to the quay, where he stopped and pointed. "This is where your son got into the water and drowned," he said. "I am desolate that I have to be the first to offer you my sympathy for having suffered such a grievous bereavement." The police learned from neighbors that the parents next day had contacted an undertaker to arrange for the funeral. A few days later the boy was found, alive and well and hiding in a haystack.
  In the light of such occurrences—and I could quote many more—*Esotera*'s description of Croiset as "the psychic who never disappointed" seems to contain an element of poetic license.

Note: While this article was in press, we were informed that Professor Tenhaeff died on July 9, 1981. Professor Tenhaeff had received prior notice of the results of Mr. Hoebens's investigations but declined several invitations to offer specific counterarguments. — Ed.

References

Brink, F. *Enige aspecten van de paragnosie in het Nederlandse Strafproces.* Drukkerij Storm, Utrecht, 1958.
_____. "Parapsychology and Criminal Investigation." *International Criminal Police Review 134* (January 1960).
Hansel, C. E. M. *ESP: Scientific Evaluation.* Scribner, New York, 1966, pp. 197-203.
_____. *ESP and Parapsychology.* Prometheus, Buffalo, N.Y. 1980, pp. 262-68.
Hoebens, P. H. "Vom Lob der Genauigkeit in der Parapsychologie." *Zeitschrift für Parapsychologie und Grenzgebiete der Psychologie 22,* no. 4 (1980): 225-34.
Pelz, C. "Herr Croiset, Sie konnen nicht hellsehen." *Kosmos* (1959/1960).
Pollack, Jack Harrison. *Croiset the Clairvoyant: The Story of the Amazing*

Dutchman. Doubleday, Garden City, N.Y., 1964.

Roosmalen, Th. van. "Ervaringen met Paragnosten en die zich zo noemen." *Algemeen Politieblad 109* (1960): 3-9.

Tenhaeff, W. H. C. "Aid to the Police." *Tomorrow,* Autumn 1953.

_____ "De Paragnostische Begaafdheid van de Heer C.H." *Tijdschrift voor Parapsychologie 23,* no. 2/3 (1955): 57-87.

_____ *Beschouwingen over het gebruik van Paragnosten.* Bijleveld, Utrecht, 1957.

_____ "The Employment of Paragnosts for Police Purposes." *Proceedings of the Parapsychological Institute of the State University of Utrecht,* no. 1 (December 1960): 15-32.

_____ *Ontmoetingen met Paragnosten.* Bijleveld, Utrecht, 1979.

_____ "Der Paragnost." *Esotera 31,* no. 9 (September 1980): 816-27.

Woudenberg, G. D. H. van. "Ervaringen met Paragnosten." *Algemeen Politieblad 113* (1964): 297-300.

Zorab, G. Review of Jack Harrison Pollack's *Croiset the Clairvoyant. Journal of the Society for Psychical Research 43* (1965): 209-12.

13

Croiset and Professor Tenhaeff: Discrepancies in Claims of Clairvoyance

Piet Hein Hoebens

According to Professor Wilhelm Heinrich Carl Tenhaeff *(Proceedings 1960)*, the majority of the hits scored by "psychic" detectives "appear to be of value solely from the parapsychological angle." They are of no use to the police, but to the experienced psychical researcher they constitute interesting examples of ESP. The psychic is supposed actually to have "seen" particulars relating to a given police case but to have been unable to get his vision into focus. Only post factum can the clairvoyant's impressions be declared hits. This, however, requires the facts to be subjected to a positively procrustean form of "interpretation."

An anecdote cited by American journalist Jack Harrison Pollack is an almost burlesque example of the lengths to which determined believers will go to make the outcome fit the prediction. Pollack is the author of a full-length biography of the Dutch clairvoyant Gerard Croiset, which Tenhaeff helped him with and vouched for. Consulted in a 1950 Arnheim rape case, Croiset "saw" that the rapist had "an abnormally big genital organ." When the police arrested a suspect, they had a good look at his private parts but found them to be standard size. Never mind, says Pollack, "They learned that he was a twenty-year-old cook who occasionally used a big, red basting syringe in the kitchen, which prompted Croiset's image of an abnormally large genital organ."

Both in the police cases and in the experimental "chair" tests, Croiset's typical ESP hit was on a comparable level. Of course the common willingness to believe that a post factum "explanation" reveals what the psychic really meant in the first place is at the bottom of the astounding success of hundreds of soothsayers, I Ching experts, tea-leaf readers, and other augurs. Tenhaeff, however, thought that those who make this objection suffer from "Gestalt blindness."

Professor Tenhaeff, who in 1953 was appointed to the first chair of psychical research ever to be established at a regular university (Utrecht), had a virtual monopoly on the study of Croiset. Tenhaeff's books and articles constitute the principal source of information on Croiset, and it may be said that the case for Croiset's clairvoyant abilities stands or falls with the reliability of his learned mentor. (Tenhaeff died on July 9, 1981, while this two-article series was in press. Professor Tenhaeff had received prior notice of the results of my investigations but declined my invitation to offer specific counterarguments.)

Critics

Th. van Roosmalen is not mentioned in Pollack's *Croiset the Clairvoyant* and neither are several other authors who have occasionally cast doubt on the psychic's achievements and their documentation. In the index to Pollack's book, one looks in vain for such names as George Zorab, the parapsychologist who first discovered Croiset and who could have told Pollack some interesting facts about both the psychic and the professor; Spigt, the historian who showed that Tenhaeff's inaugural address in 1953 was based entirely on a spurious source; Filippus Brink, the criminologist who wrote a major work on occult detectives; Pelz, the Hamburg police officer who in 1959 published a scathing report, titled *Herr Croiset, You Are Not Psychic;* and Ph. B. Ottervanger, the Dutch skeptic who in the fifties fired many a well-aimed shot at Tenhaeff and his protege.

Pollack may never have heard of these critics; presumably Tenhaeff did not encourage him to contact them. They might have persuaded the American journalist to correct at least a few of the most outrageous errors in his manuscript and to include some material that, while not flattering to the subject, might have improved the book.

In the same *This Week* article that we find the Eindhoven case (described in Chapter 12), Pollack praises Tenhaeff as "a stickler for complete scientific proof." Similar laudatory phrases are found in the biography. Pollack might have given a different description had he been familiar with van Roosmalen's article in *Algemeen Politieblad*. There, the Utrecht superintendent reports his meeting with Tenhaeff, which was arranged by the judge-commissioner in Utrecht. On that memorable occasion, van Roosmalen flatly told the professor that he did not believe in paranormal sleuthing. "Superintendent," Tenhaeff replied, "If you like, I will tell you of a few cases where the police failed and where Croiset was successful." Tenhaeff then related, in great detail, two ironclad cases. The first concerned a murder in the municipality X. After months of fruitless investigation, the police consulted Croiset. The psychic gave such a clear description of the murderer that they were able to make an arrest. The second case concerned a theft in a factory in the town of Y, where Croiset had identified the thief.

Van Roosmalen decided to check these claims. The police officer in X, when asked about the murder case, was puzzled. He said it was somewhat unlikely that Croiset had been successful in identifying the murderer because they had no record of such a crime having been committed! Van Roosmalen's colleagues in Y admitted that a suspect had been arrested on Croiset's advice. However, the alleged thief proved to be entirely innocent and had to be released with profuse apologies. Van Roosmalen was urgently advised not to mention the name of Croiset if he should visit the police in Y.

Pollack might also have found an interview with Filippus Brink enlightening. Brink, a police officer, in 1958 completed a doctoral thesis, *Enige Aspecten van de Paragnoise in het Nederlandse Strafproces* ("Some Aspects of ESP in the Netherlands Criminal Proceedings"), in which he reported the results of a series of experiments with occult detectives and of inquiries to police authorities in both Holland and abroad. Brink had tested four well-known "psychics"—one of whom was Croiset—by handing them photographs and other objects and requesting them to give their "impressions." Some of the materials were related to police cases, others were not. The experiments were extensive and lasted for more than a year. The results were nil. Looking at the picture of a murderer, the psychics clearly saw that the man was innocent; handling a weapon that came straight from the factory, they got visions of murder and hold-ups.

In "Aid to the Police," one of his few articles published in English (*Tomorrow*, Autumn 1953), Tenhaeff had assured his readers that Croiset "does not 'fish' for information." Brink observed nothing *but* fishing.

The results of Brink's police department inquiries were hardly more comforting to the proponents of paranormal detection. With very few exceptions, all Dutch and foreign authorities stated that psychics had never been successful in furthering any police investigation. (Incidentally, this was the reply even from the Haarlem police district, where Mr. Gorter had been superintendent. See Chapter 12.) The exceptions concerned highly ambiguous successes.

Brink recently told me: "I dare to say that, barring an occasional lucky guess, no clairvoyant has ever been able to solve a police case by paranormal means in the Netherlands." My recent inquiries to a number of Dutch police departments suggest that little has changed since Brink's 1958 publication.

Caught in Fraud

For a number of reasons that will be discussed later, these criticisms were not seen as fatal to Tenhaeff's reputation as a careful and honest scholar. His proponents privately admitted that the professor occasionally suffered from bouts of absentmindedness and might even sometimes have been led astray by his own enthusiasm. However, they insisted that the substance of his work was unassailable. Some of them began to lose faith only in 1980.

when I caught the professor red-handed in patent fraud. This time, it was difficult to think of innocent explanations.

In the course of my investigations into psychic claims, I have always concentrated on what Tenhaeff himself regarded as prize cases (to avoid the charge of biased data-selection). Given Croiset's international reputation as a psychic crimebuster, I was surprised at the scarcity of cases that would qualify him as such. Almost all the reports of his works with the police were about cases that did not result in the arrest of the actual culprit.

The Wierden affair (where Croiset is supposed to have identified the assaulter of a young girl by simply handling the hammer with which the crime had been committed) has been cited time and again by Croiset proponents, but it lost much of its appeal after C. E. M. Hansel reported that he made inquiries to the local authorities and was told that Croiset's endeavors had been of no use. Then, in the September 1980 issue of the German monthly *Esotera*, Tenhaeff published a report of a case that seemed ironclad. To summarize this report: On November 15, 1979, a state police officer, Commander Eekhof, had visited Croiset and asked him to help identify a mysterious arsonist who had terrorized the Woudrichem area for months and had completely escaped detection. Eekhof did this "in the hope that he [Croiset] would be able to provide the authorities with definitive information concerning the culprit."

A few weeks before Croiset's sudden death in July 1980, Tenhaeff had visited the Woudrichem police "to learn from Eekhof in full detail how successful his visit to Croiset had been." Tenhaeff wrote: "Everything Commander Eekhof told us was videotaped. The tapes were protocolled and the protocol was checked and signed by Mr. Eekhof."

According to Tenhaeff, Croiset was consulted at a moment when all official attempts to identify the perpetrator had proved fruitless. The clairvoyant described the arsonist as a man who "sometimes wore a uniform," "lived in an apartment building," and had "something to do with toy airplanes." Asked by Eekhof whether it could be "model airplanes," Croiset replied "in the affirmative." Eekhof allegedly "was shocked" by the clairvoyant's statements, "for Croiset's description fitted a quartermaster in his own police group."

According to Tenhaeff, at first the commander was incredulous, but "sometime later he saw himself compelled to admit that Croiset had been right: the quartermaster was a pyromaniac."

In order to check this remarkable claim, I contacted Commander Eekhof and showed him the *Esotera* article, which he had neither seen nor heard of. After having carefully read and re-read the report, he stated positively that it contained "outright falsehoods." He invited me to listen to the tape-recording of what Croiset had really said.

The grossest inaccuracies are corrected here:

● The consultation took place on November 15, 1977, a full two years before the date given by Tenhaeff. The quartermaster was not arrested until March 1980.

● Croiset at no time mentioned a "uniform," which would have been the most striking hit.

● Croiset did not mention "toy airplanes," although he did speak of "airplanes"—"sitting in airplanes," "airfields," and "airplane construction." When asked by Commander Eekhof whether it could be model airplanes, Croiset first said yes, maybe, but then retracted and said, "No, these are big airplanes." It is quite true that the quartermaster liked to build model airplanes. But these were first mentioned by Mr. Eekhof and not by Croiset. So, if this was an ESP hit, then it was scored by the policeman and not be the clairvoyant!

● Croiset, who in an earlier attempt to "see" the culprit had put the police on a false track, had finally identified the arsonist as a person in a photograph shown to him by Eekhof. This person—whom the police already considered a possible suspect—was not the quartermaster, who later admitted the crimes.

● Eekhof certainly was not "shocked" by Croiset's statements, because he could not possibly have recognized his fellow policeman in the psychic's confused "images." The quartermaster, who did not live in an apartment building, began to be suspected only months later, for reasons entirely unconnected with Croiset's "vision."

● Tenhaeff's claim that a protocol was "checked and signed" by Commander Eekhof is categorically denied by Eekhof, who told me he had not even seen a protocol.

Before exposing this quite extraordinary case of fraudulent reporting in two Amsterdam newspapers, *De Telegraaf* and *Courant Nieuws van de Dag* (October 18, 1980), I naturally invited Tenhaeff to comment. The "stickler for complete scientific proof" flatly refused to answer any questions, shouting a number of insults before slamming the receiver down.

The reader will not be surprised to learn that Tenhaeff's Dutch version of the report, published at about the same time in his *Tijdschrift*, does not mention either "uniforms" or "signed protocols." The worst distortions were prepared for export only. If I had not sent him a copy, Eekhof might never have seen the fairy tale in *Esotera*.

Methods

As I have demonstrated, at least to my own satisfaction, Tenhaeff's reports—our principal source of information concerning Croiset—are utterly unreliable. It is therefore hazardous to suggest possible "naturalistic" explanations for any "facts" presented in those reports. One may easily waste one's ingenuity on entirely spurious data. It is with this proviso that I will briefly offer a few conjectures about how the clairvoyant might have achieved what appear to be paranormal successes by perfectly normal means.

Croiset, a skilled hypnotist, was an expert muscle reader and a master of suggestive questioning. He was therefore well equipped for wresting shreds of information from unwitting clients and feeding these back as "telepathic" impressions.

It is possible that he occasionally resorted to cruder methods, such as using spies. In the pro-Croiset literature we find surprisingly little mention of the psychic's "assistants" and "secretaries," such as Dick West. Zorab (personal communication) has evidence that Croiset sometimes employed confederates in his experiments. Apparently the clairvoyant's professional ethics were not such as to forbid a little trickery now and then.

Even more important, Croiset knew how to make friends and influence people. He maintained very cordial relations with a number of journalists and law officers. (Some policemen were patients of "Dr." Croiset.) I do not need to point out the risks of such familiarity. Croiset was an engaging man who disarmed visitors with a convincing display of sincerity and simplicity. He was not the sort of person in whose company one felt the need to be on guard.

Discussion

The standard skeptical explanation for the alleged successes of psychic detectives is that these sensitives offer their consultants the verbal equivalent of a Rorschach test. Their statements are typically vague, rambling, and verbose. The accuracy of the readings is evaluated post factum: "Good sitters" retroactively interpret the ambiguous and often contradictory statements in such a way that they fit the true facts and obligingly forget the many details that were too wide of the mark. Complete failures are ignored or suppressed. The possibility that some of the paranormal information could have been acquired by normal means is quietly discounted. Occasional lucky guesses are enhanced by selective reporting and editorial embellishment.

The results of the present investigation suggest that this standard hypothesis does not need to be revised in order to explain the Croiset phenomenon. What always set Croiset apart was probably not the degree of his supposed paranormality but rather the success of the propagandistic efforts on his behalf. Unlike other psychics, Croiset had the extraordinary luck to find an impresario who enjoyed a fairly solid reputation as a scientist and scholar. The "Miracle Man from Holland" would never have achieved his status without the indefatigable help of Wilhelm Tenhaeff. The fact that his mentor was an authentic university professor has always protected Croiset. Croiset was widely believed to be the most honest of men because Tenhaeff said he was. As I have suggested, the key to the Croiset mystery lay with Tenhaeff. The question now is: Why did Tenhaeff act as he did? Why risk an academic reputation by engaging in palpable fraud?

The answer, I think, can only be that Tenhaeff *had to deceive* and thought that he would be able to get away with it. Soon after he made Croiset's acquaintance in 1947, the professor must have realized that the psychic's successes in occult detection were highly ambiguous. Straight and full reports of the "police cases" would never convince those not committed to a strong prior belief in the paranormal.

Tenhaeff, whose ambition had always been to become the Sigmund Freud of psychical research, devised something he was pleased to label a "theory" that enabled him to explain all but the worst failures in terms of ESP. The red basting syringe is a typical example. Yet the professor needed at least a few ironclad proofs to underpin this odd theoretical structure. The extreme scarcity of authentic miracles forced him to fabricate them.

What now remains to be explained is how Wilhelm Tenhaeff got away with this game for so many years. I can only offer a few suggestions.

1. His status as an "official professor": Tenhaeff was hailed as the first professor of parapsychology in history, and this carried much prestige. "He must be right or else they would not have appointed him, would they," his admirers were wont to say when confronted with an unbeliever. For many, such considerations effectively settled the matter.

2. The confusion he created with his publications: To the delight of the true believers, who find obscurity sure proof of profundity, Tenhaeff's writings are chaotic, verbose, and abstruse. A thick fog seems to exude from his books. Scanning these thousands of convoluted sentences for contradictions requires many hours of exceptionally dull work. Most skeptics sensibly give up after a few pages. I myself only embarked on this dreary task as the result of a challenge. After I had published some critical remarks on psychical research in 1978, local proponents defied me to come to terms with the best evidence they thought parapsychology had to offer: the "rigorously scientific work" of Wilhelm Tenhaeff.

Critical investigation is further complicated by the fact that the professor cleverly took advantage of the language barrier. The completely fraudulent versions of the "Boy on the Raft" and the Woudrichem cases were concocted for export only. To compensate, Tenhaeff occasionally did the reverse: in the German *Zeitschrift fur Parapsychologie und Grenzgebiete der Psychologie* (No. 4, 1980), I have shown how he dishonestly "edited" a report by Dr. Jule Eisenbud of an ambiguous success in a transatlantic ESP experiment with Croiset for home consumption in Holland.

3. The ambiguous feelings within the psi community toward cheating colleagues: The psi community has never completely freed itself from the

pernicious *idée fixe* that overt criticism of a colleague may damage the cause and play into the hands of the enemies of parapsychology. Some psychical researchers began to suspect Tenhaeff long ago. Seldom, however, did they voice their doubts openly. And, when they did, some sociological mechanism seems to have prevented an adequate follow-up.

4. Tenhaeff's mastery of propagandistic techniques: Tenhaeff has always been a master of propagandistic manipulation. He deftly used his excellent relations with the media to persuade a sizable segment of public opinion in Holland that he was a prophet of a new, nonmaterialistic science, who therefore had to suffer the irrational hatred of those whose world-view was threatened by the glorious discoveries of parapsychology. He never failed to remind his audiences of the religious implications of his work or to allude darkly to possible bolshevik influences in skeptical circles. His favorite trick was to tell the public that his critics were suffering from a Freudian complex and needed psychiatric treatment rather than to reply to their impertinent questions.

Ironically, Tenhaeff's bizarre behavior convinced a number of skeptics that he was a gullible victim of a devious "psychic" rather than a deceiver in his own right. His apparent credulity served as camouflage for his dishonesty.

Conclusion

My purpose has not been to deny Croiset's paranormal powers ex cathedra. Rather, I wanted to draw the reader's attention to some false notes in the "psychic Mozart's" scores, dissonants that seem to have escaped the notice of other biographers.

Of course *if* such a thing as ESP exists, it may well provide the most economical explanation for some of the coincidences that have been reported in connection with Gerard Croiset. On the other hand, it would be rash indeed to conclude the existence of a paranormal faculty on the basis of the Croiset material. If Croiset's amazing talents of clairvoyance had been genuine, then why would Tenhaeff have felt the urge to manipulate reports and present them fraudulently as prize cases?

I certainly do not wish to be dogmatic about psychic detectives: positive evidence of a much stronger nature may yet turn up. Until it does, the "believer" is well advised not to base his belief on the "Psychic Who Never Disappointed."

Acknowledgment

I am grateful to Mr. and Mrs. Ottervanger of Bussum, Holland, for having allowed me to make use of the extensive archives of the late Mr. Ph. B. Ottervanger.

References

Brink, F. *Enige aspecten van de paragnosie in het Nederlandse Strafproces.* Drukkerij Storm, Utrecht, 1958.

———— "Parapsychology and Criminal Investigation." *International Criminal Police Review 134* (January 1960).

Hansel, C. E. M. *ESP: A Scientific Evaluation.* Scribner, New York, 1966, pp. 197-203.

———— *ESP and Parapsychology.* Prometheus, Buffalo, N.Y., 1980, pp. 262-68.

Hoebens, P.H. "Vom Lob der Genauigkeit in der Parapsychologie." *Zietschrift fur Parapsychologie und Grenzgebiete der Psychologie 22,* no. 4 (1980): 225-34.

Pelz, C. "Herr Croiset, Sie konnen nicht hellsehen." *Kosmos* (1959/1960).

Pollack, Jack Harrison, *Croiset the Clairovoyant: The Story of the Amazing Dutchman.* Doubleday, Garden City, N.Y., 1964.

Roosmalen, Th. van. "Ervaringen met Paragnosten en die zich zo noemen." *Algemeen Politieblad 109* (1960): 3-9.

Tenhaeff, W. H. C. "Aid to the Police." *Tomorrow,* Autumn 1953.

———— "De Paragnostische Begaafdheid van de Heer C.H." *Tijdschrift voor Parapsychologie 23,* no. 2/3 (1955): 57-87.

———— *Beschouwingen over het gebruik van Paragnosten.* Bijleveld, Utrecht, 1957.

———— "The Employment of Paragnosts for Police Purposes." *Proceedings of the Parapsychological Institute of the State University of Utrecht,* no. 1 (December 1960): 15-32.

———— *Ontmoetingen met Paragnosten.* Bijleveld, Utrecht, 1979.

———— "Der Paragnost." *Esotera 31.* no. 9 (September 1980): 816-27.

Woudenberg, G. D. H. van. "Ervaringen met Paragnosten." *Algemeen Politieblad 113* (1964): 297-300.

Zorab, G. Review of Jack Harrison Pollack's *Croiset the Clairvoyant. Journal of the Society for Psychical Research 43* (1965): 209-12.

Claims of Mind and Matter

14

The Columbus Poltergeist Case

James Randi

March 1984 came in like a lion at the home of John and Joan Resch in the North Side district of Columbus, Ohio. Reporters who were called in to witness the evidence found broken glass, dented and overturned furniture, smashed picture frames, and a household in general disarray. The focus of all this activity seemed to be 14-year-old Tina, an adopted child who had shared the Resch home with some 250 foster children who came and went over the years.

Tina, a hyperactive and emotionally disturbed girl who had been taken out of school and was being privately tutored through the Franklin County Children's Services (FCCS), was interviewed by every media outlet who could get near the two-story frame house where these poltergeist activities were claimed to be taking place. Every day the street outside was jammed with vans and cars stuffed with television crews, reporters, and photographers who joyously tumbled over one another in their enthusiasm for what had become a circus.

Mike Harden, a reporter for the *Columbus Dispatch,* was the first on the scene. He had written an article on the Resch family some five months before, praising their work with foster children. He was aware that Tina was trying to trace her true parents—against the wishes of Mr. and Mrs. Resch, who felt it was not a good idea. One of their other adopted children had found his parents, and it did not turn out very well. In view of his previous encounter with the Resches, Harden considered himself a friend of the family.

During that first big press conference at the Resch home, more than forty persons were jammed into a 20' × 20' room. Participants described it to me as "rude" and "typically media." Comments from reporters we interviewed were:

"We didn't listen to what each of us was saying. We just jumped in."

There was "no development of questions."

"We tried to find the truth. We're obviously not equipped to do it."

"It would have been much more comfortable with two teams—or

three."

"It was a complete free-for-all."

"We were on her like flies on flypaper."

"Our attention was constantly diverted. When a reporter knocked something over by accident and took the blame for it, there was general disappointment."

Although the older Resches denied any prior belief in supernatural matters, they soon agreed that such goings-on probably resulted from a "poltergeist." This translates as, "noisy spirit," though some of the slightly less naive parapsychologists tend to ascribe these events to psychokinesis (PK) rather than ghosts. Since the record of past cases indicates that when these destructive phenomena take place very frequently an unhappy adolescent is in the vicinity and they cease when the youngster is recognized and satisfied, explanations other than supernatural ones immediately suggest themselves.

On March 5, photographer Fred Shannon, a 30-year veteran on the *Dispatch* staff, accompanied reporter Mike Harden to the Resch house to try to catch the elusive poltergeist events on film. By his own admission, Shannon was "afraid" of what might happen and was fully primed by Harden to witness miracles. During the first three hours of his visit, he took a remarkable series of photographs, but the actual story of how he and the public were apparently bamboozled by an adolescent girl is far more remarkable.

I have long believed that the major difference between the skeptic and the parapsychologist is one of expectation. The former does not believe that validation of paranormal claims is imminent; the latter depends upon that event for justification. Also, the skeptic will invoke parsimony—the simplest explanation consistent with the facts—where the parapsychologist eschews it. Personally, I find it much more reasonable, when objects fly about the room in the vicinity of an unhappy 14-year-old, to suspect poor reporting and observation rather than a repeal of the basic laws of physics.

It is true that the Committee for the Scientific Investigation of Claims of the Paranormal (CSICOP) was not invited to Columbus by the young lady at the center of these pranks, nor by her adoptive parents. But a call went out through the *Dispatch* for anyone who could help explain the phenomena. At that point, CSICOP chairman Paul Kurtz contacted me and asked if I would join Case Western Reserve astronomers Steve Shore and Nick Sanduleak in Columbus to look into the case. I arrived on March 13 and was met by a mob of generally hostile reporters at the airport. The official CSICOP statement released to them at the interview expressed the hope that we would be admitted to the Resch home to look into the events first-hand.

When I arrived in Columbus via Chattanooga, Tennessee, where I had been lecturing, and was joined there by the other two CSICOP investigators, I had no guarantee from the Resch family that we would be allowed to actually investigate anything. Upon reaching the house, Steve Shore asked Tina's parents whether we would be welcome. Mrs. Resch— as was her right, of course—said that the two astronomers could enter but not the "magician." She said it would be "sensationalizing" the matter to allow me access to the site. I did not see how she could honestly say that in view of the commotion brought about by the great number of press conferences and interviews that had taken place in the house. On one of those occasions there were, by actual count, more than 40 reporters, cameramen, and others rampaging about. The Resch case had become a major—though transitory—media event, featured all over the world.

But when we arrived the Resch home was already occupied by two investigators from the Psychical Research Foundation of Chapel Hill, North Carolina. William Roll and Kelly Powers had been enthusiastically welcomed to the house, and they had been living with the family for several days. Roll is the author of *The Poltergeist,* which J. B. Rhine referred to as "a book on . . . what to do with a poltergeist until the parapsychologist comes."

When asked why I had been refused admittance to witness the events—and we specified that we wanted to go in *after* Roll and Powers, so as not to interfere with them—Mrs. Resch told reporters that Roll had insisted that I not be admitted. Roll denied this, saying that it was her ruling, not his. Later Mrs. Resch said, "We have a circus going already, and I don't need a magic show as well."

Roll said he would have let me in "if the conditions had been completely up to me, and if there had been no problem about the health situation." He was referring to Mr. Resch's recently elevated blood pressure. But reporter Dave Yost of the *Columbus Citizen-Journal* told us that Roll had told him that he simply "didn't want Randi in there."

Following the departure of Roll and Powers, the Resch family told us they were leaving on a long vacation and would not be available for an interview with us. However, we discovered that they were still in their house two weeks after that announcement.

Because of the inaccessibility of the Resch home, the evidence we gathered centered around the film that was shot by photographer Shannon. One photo printed from the roll of 36 negatives (frame number 25) was published around the world as part of an Associated Press release. I had first become aware of the case from seeing the AP story in a Chattanooga newspaper. As a result of that one photo and its caption, much of the reading public now apparently believes that it represents a genuine example of either psychokinesis or spirit possession. This photo shows a telephone suspended in mid-air in front of Tina Resch while she cowers in fright. We called it "Attack of the Flying Killer Telephones" since the accompanying text said that the child was being assaulted regularly by these objects. The

Frame 25. This is an artist's rendition of the widely published photo sold by the *Columbus Dispatch* to Associated Press.

photo clearly shows that two telephones had been placed on the table at Tina's left side and that the handset of one of them is in motion in front of her. The cord is stretched out horizontally and shows transverse blurring.

Although it apparently had not occurred to any other investigators—including the parapsychologists—we asked to see the other 35 photos on that roll of film. We discovered seven flying telephones in all, and when the photographer admitted he had not been looking at the subject when taking the photographs, there was little mystery left.

Shannon had found that holding the camera to his eye and waiting for an event to occur was useless. It always seemed to happen just after he had relaxed and looked away. He referred to "The Force" when he spoke of the phenomenon. "It was tricky, and I would have to be tricky if I were to capture it on film. I decided I would outfox the force," he said. While Tina sat in a soft chair with two telephones within easy reach, Shannon looked away. When he saw a movement from the corner of his eye, he pressed the shutter. One result was the photo of Tina used by AP.

Now these photos were taken at 1/125th of a second, using a strobe flash. Shannon used a wide-angle lens, of 24 mm focal length, which subtends an angle of 84 degrees. Further evidence of the wide-angle lens is seen in the distortion of objects and persons at the margins of each frame. (For comparison, a "normal" lens of 50 mm focal length covers only 45 degrees.) Thus the cameraman would have to be much closer to the subject than might be supposed from the wide area shown by each photograph. If

Shannon was, as he has said, "turned away" from the subject, and yet seated as close as he apparently was, he could have had no appreciable perception of whatever action really took place.

None of Shannon's photos were taken in rapid succession, since the strobe flash (a Vivitar) would not have recycled in time. Only one of them, number 25, was published worldwide. Recall that this classic photograph shows transverse movement of the telephone cord. The others do not. In one of the other flying-phone photos, frame number 30, it can be seen that this blurring took place along the length of the cord. Any blurring shown in the photos was a result of the film registering ambient room light (from the nearby window) since the strobe flash lasts about one-thousandth of a second, freezing almost any rapid action. The conclusion is obvious: Frame 25 shows that movement of the telephone was caused in a different fashion than was the movement in frames 24, 30, 31 and 32—those in which telephones are actually frozen in flight. (Two others record the scene after the phone had fallen to the floor.)

What is different about the flying telephone in frame 25? Well, examination of frame 24, immediately preceding the photo published by the *Dispatch* and the AP, shows a strange situation. The phone cord is seen here already stretched out in front of Tina Resch, spanning the arms of the chair in which she is sitting, stationary, with the attached handset out of sight at her right, hanging down behind the side of the chair. Tina, typically, has her mouth open in a scream. (And there is a little girl—Miss

Frame 31. Tina appears to have been *holding* the phone base, and has *thrown* the handset out of frame. Note the foot of an (un-named) observer at the left margin.

"X"—standing and watching on the right. She was an eyewitness to this event and shows up in five other frames as well. We'll refer to her later.)

Simply by grabbing the phone cord at a spot near her right hand and yanking it hard, Tina could have caused the phone to fly up into exactly the position shown in the published photo, number 25. (We were able, at NBC-TV, to replicate this effect easily.) But looking at the other flying telephones that are revealed on the roll of film, we see longitudinal blurring, which indicates that Tina probably simply threw the telephone from camera right to camera left. In photo 12, Miss "X" is looking at the camera as if wondering whether that throw was convincing.

In photo 29, both telephones appear to have been in motion. Surely, it could be objected, if Tina simply tossed these phones around, she could not have tossed both of them without being caught at it! She surely would have been seen holding the phones, preparatory to throwing them when the "witnesses" were relaxed and looking away. A possible answer to this objection can be found when we examine frame 28, which shows Tina had a great deal of latitude in handling the equipment. She is freely holding the telephone apparatus in her hand while in animated conversation. It is evident that she, like other "psychics," is running the show her way, regardless of any requirements of security or control.

Frames 13, 22, and 28 show Tina similarly occupied, with the ubiquitous and mysterious Miss "X" once more present. During this entire photo session, it seems that an atmosphere of rather loose gaiety prevailed,

Frame 24. By grasping the cord at the arrow, Tina could cause the phone to fly up into the position shown in frame 25.

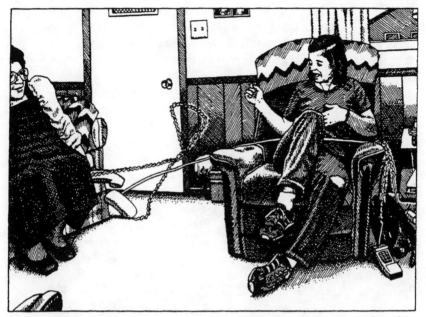

Frame 32. The major witness watches as the phone "flies" at her. Again, note the foot of the other observer of this miracle.

but we are told that that's the only way these things are expected to take place. Reporter Harden shows up in four frames, numbers 14, 21, 22, and 23. He is, in every case, either obediently not watching Tina, so as not to inhibit her performance, or paging through a phone book.

But there is another frame, number 32, in which a new witness appears at the left edge of the photo during a flying-phone event. We asked the *Dispatch* to identify this woman, but got no help at all. Finally, through a reporter in Cincinnati, we learned that she was Lee Arnold, Tina's caseworker. We contacted Ms. Arnold by phone, and she told us that she had been instructed by her employers, Franklin County Children's Services, that if she gave any information about her witnessing the events she would be in danger of losing her job. She told us nothing.

And what of that other frequent witness, little Miss "X"? She remains a mystery. Obviously, she could reveal a great deal about Tina's actions during the time those photographs were being taken. But try as we may, no one will inform us how we may contact her. That is most unfortunate, since her testimony might reveal very interesting data.

It is the last of the flying telephones on Fred Shannon's film that really asks a great deal of our patience. It shows Tina Resch seated in the chair, her pointing left hand extended to her right across her body. The telephone cord is horizontally stretched out and the telephone handset is so far away as to be out of the frame altogether. Tina is in a stance suggestive of a major-league baseball player completing a throw to first base. Now, with the simple principle of parsimony in mind, we must ask ourselves if we will choose to believe that this is a photograph of a girl being affected by poltergeist activities or a photograph of a girl simply pitching a telephone across the room.

As I have said, the Resch household was inundated by the media. And, as luck would have it, Tina was caught cheating by them—though not by the parapsychologists who were the officially sanctioned investigators. On the only occasion that she believed she was not being observed electronically while television equipment was present, she was caught red-handed—twice—as can be seen in a news tape obtained accidentally by WTVN-TV, Channel 6 (ABC), Cincinnati. It happened at the end of a long press visit on March 8. The TV crew was packing up their equipment, but had left a camera aimed at Tina. Seated at one end of the sofa, near an end-table, and believing the camera was no longer active, she watched carefully until she was unobserved, then reached up and pulled a table-lamp toward herself, simultaneously jumping away, letting out a series of bleating noises, and feigning, quite effectively, a reaction of stark terror. It matched other performances quite well. The lamp, on the first try, did not fall. Encouraged by the reaction, the girl then repeated the performance. This time, the lamp toppled to the floor. The TV crew hurried away to process their videotape for the next news broadcast, unaware that Tina's cheating had been recorded.

Discovering the imposture, WTVN-TV broadcast the tape, asking their viewers to make up their own minds about the event. Tina, confronted with the evidence after the broadcast, said coyly, and with much squirming about, that she was "only fooling" and did it because she wanted to get rid of the TV cameraman.

We may never see the rest of the TV tape as it was originally shown to the CSICOP team by WTVN-TV in their remote unit before it was edited. I pointed out to the TV crew at that time certain notable aspects of that tape and asked if I might have a complete copy of it. A copy was delivered to me, but I found that it consisted of only the portion that had been edited down for broadcast. When I asked about this I was told that I could not have the remainder, and when I reminded them of their promise to me they suddenly discovered that they had erased it—in error.

The missing portion of the videotape showed Tina Resch carefully and obviously setting up the trick. She edged around the sofa, glancing about her to be sure she was not being observed—not knowing the video camera on the floor was still connected, of course—and reaching up to test the height of the lamp shade. A moment later, thinking that she was safely unobserved, she is seen yanking at the shade and jumping away simultaneously, putting on her frightened act. Then she sets it up again and repeats the performance.

In frame number 20 of Fred Shannon's film, we see Tina doing the same trick. That same lamp falls in the same position, with her seated in the same place. Was this one, too, "only fooling"?

To take the edge off the cheating episode, after admitting that Tina had been caught, *Dispatch* reporter Mike Harden reassured his readers that WTVN-TV had witnessed a genuine miracle that same night: One of them had seen a table move mysteriously in the kitchen. But technician Robb Forest of WTVN told us that he'd caught Tina moving that table secretly with her foot, had accused her of it, and got only a horse-laugh from her for his trouble.

Similarly, Mary Anne Sharkey (a good friend of Mike Harden) of the *Cleveland Plain-Dealer* reported to us—but not her readers—that Tina's obvious hanky-panky with a candlestick had disillusioned her on the story. She had been there with three other reporters, two photographers, and a TV crew when Tina pointed out to her a candlestick located in a plastic wreath under a table. Sharkey wondered why it had been pointed out. One hour later, after they had all moved about the house, Tina announced that the candlestick had vanished, and she began looking for it. With Sharkey following her around, Tina "found" it under a chair. Sharkey was unimpressed, but this did not become part of her story. Instead, she reported another episode that seemed more convincing.

NBC-TV news reporter Bill Wolfson, at first fascinated by Tina's performances, changed his mind after prolonged exposure to events, contradictory reports, and reconsideration of what he actually saw—or didn't see. As for the press conferences, he said, "I thought the tone and quality of questions were somewhat less than poor. They were provocative and leading. The media were going crazy. One reporter asked Tina, 'Don't you feel guilty?' " Wolfson finally summed it up for us as "bunk."

While he snapped frame number 26, photographer Shannon, as he looked elsewhere awaiting a miracle, must have believed that something "psychic" was happening. It shows the footrest at the base of a "recliner" chair in the extended position, and Tina looking startled—as if it had suddenly popped out. I have one of these chairs at home. To make the footrest protrude, one need only grasp the arms firmly and push back. This is the only photograph among the 36 in which Tina is holding the chair arms. Her startled expression would indicate to me only what she has proven in the past (as in the videotapes)—that she is an excellent actress.

The *Dispatch* naturally had a fine time with this story. One Sunday edition contained a huge spread on the subject. Two new photos were shown. One (frame 17) shows Tina holding an already-broken picture frame. One can clearly see that the glass is broken, and Tina is holding it like a tray, with the shards of glass retained on it. The *following* photo (18) shows her tilting it forward to dump the glass. But in that Sunday newspaper account, we see that the caption for the *second* photo says, "Below, the picture shatters in her hand and falls to the floor."

This is not a responsible representation at all. Reporter Dave Yost, who attended the Resch press conferences and followed Tina all the way to North Carolina when Bill Roll took her there for further observation, was frustrated by the attitude of the media. He said, "The real story here, I suspect, is the reaction of a duped media." Added Yost, "In spite of repeated efforts, I have never seen these reported events."

To return to the Shannon photos: In frame 21 (with reporter Harden studiously looking at the phone book at the right) we see a rollaway couch "jumping out" from the wall at Tina, who is again startled at another wonder of poltergeistry. But examination of a small lower section of a previous frame, this time taken from 90 degrees away with Tina standing in front of that couch holding an object for photographer Shannon, reveals an interesting fact: Tina clearly has her right foot hooked under the edge of the couch! A sudden pull backwards and the couch would "jump out" at her easily. We don't know if that is how it was done when she was later "attacked" by the couch. But there is ample evidence here to believe that it might have happened that way, and none to show that it did not.

Admittedly, our team was not able to conduct a proper investigation of the Columbus poltergeist case. We were barred from the house and we never interviewed the girl involved. We could not trace one of two eyewitnesses to the photographed events, and the other witness was forbidden to tell us what she knew.

Witnesses we could identify were less than cooperative. Barbara Hughes, a neighbor and good friend of the Resch family and also a foster parent, spoke briefly with Steve Shore by phone, but refused to meet with us. She claimed to have seen one phenomenon while in the house. She said she addressed "The Force" out loud, demanding a demonstration. Something "fell," and she fled in terror. Drew Hadwal, working with WTVN-TV, "saw three chairs move apart" in the kitchen, we were told. I tried hard to reach him, but though the receptionist at Channel 6 said she knew he was in, when I gave my name over the telephone I was told that he was not going to be at work that day. Even electrician Bruce Claggett (to whom we will return in Part II) failed to return our calls and, although he was scheduled to be on the program with me at the annual meeting of the Parapsychological Association, he failed to appear.

Several reporters we did interview told us of damning details they had observed but never reported. One expressed his anger at the rewrite artists who had "fluffed up" his stories to the point where they were hyperbolic. The *Columbus Dispatch* gave endless excuses why we could not meet Fred Shannon to discuss the evidence with him during our stay in Columbus, and reporter Harden eventually was "out" to us when we tried to call on him.

On the other hand, Bill Roll actually stayed at the Resch house. He stated his professional conclusions twice during a press conference before taking Tina off to his lab in Chapel Hill. He said that, based upon stories told by witnesses, neurological and psychological tests of Tina, and his own experiences (during a half-hour period in the last hour of the last day of three he spent in residence) "when I felt I had Tina under close observation" he concluded that she had demonstrated "genuine recurrent spontaneous psychokinesis (RSPK)." Then he admitted that though he was "impressed . . . we are not dealing here with a controlled study, [but it was] sufficiently suggestive of RSPK." He added that his research was "in a very preliminary stage" and that he had come to "no definite conclusions."

Roll's evidence is based on a very short term of observation, with no other witnesses present and no direct experience of any event—only peripheral observations. His data consists of uncorroborated witnessing of a very few events, which he admits took place out of his direct line of sight at times when he was unable to anticipate them.

This is how he reported the events to the press conference: Immediately prior to the rush of phenomena, Tina had spent some 30 minutes upstairs, alone (only the two of them, so far as he knew, were in the house). Then she appeared at the top of the stairs screaming for him to rush up there and see miracles. A bar of soap, he reported, fell into the bathtub. Next, while they both were standing four feet from it, facing away, a picture fell from the wall. The nail had been pulled out of the wall. Roll and Tina rushed to it. Roll hammered the nail back with a pair of pliers. During this process, his small tape-recorder, which had been placed nearby on a dresser, flew to a position seven feet away. Roll and Tina went to it, Roll leaving the pliers behind. The pliers "moved from the dresser" to hit the wall near him.

Roll described his own observing abilities in such a way that we must place his performance in the paranormal category. Or, at the very least, he had to have rather remarkable sensory powers. Consider: (a) He was hammering a nail back into a wall using the edge of a pair of pliers (he called them "tongs")—an act that requires undivided attention, obviously; (b) he was "watching Tina carefully" (contrast this with his statement that he "felt he had Tina under close observation") and remember that the "possessed" girl was standing off to one side of him; and (c) he saw the tape machine fly away from a position directly behind him—a remarkable feat indeed, especially when the layout of the room is known. (See Figure 1.)

Questioned, he admitted he had not once seen any object in place as it began to move. I postulate that, since he could not see the tape recorder, Tina had ample opportunity to throw it along the dresser, from which position it fell to the floor. Then she picked up the pliers as the two of

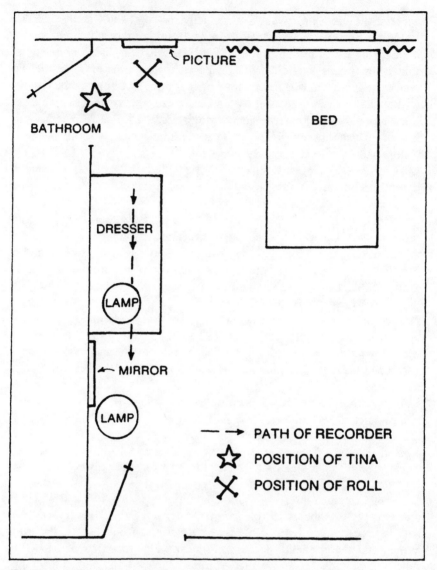

FIGURE 1. Layout of room in which Roll claimed tape recorder and pliers moved.

them went to recover the tape recorder and threw them against the far wall as Roll examined the recorder. It was an assumption on his part that the pliers "moved from the dresser." He said, "She wasn't doing anything with her hands that I could see." (Roll is myopic and wears thick glasses; he is a poor observer.)

An examination of the videotape made in that room shows that the dresser on which the tape recorder sat was directly behind Roll as he faced the picture on the wall! He could not have seen it move. It is an impossible scenario.

But why would Tina Resch want to smash up her home and allow others to believe it was a paranormal event? Factors are found here that suggest strong motivation on her part to create a sensation. She was admittedly under stress and had good reason to want to attract media exposure: she wanted to trace her true parents, against the wishes of the Resches. And her "best friend," Missy Johnson, had a fight with her and broke off their friendship two days before the phenomena began. She was a girl looking for attention. And she got it.

The evidence for the validity of poltergeist claims in this case is anecdotal and thin, at best. The evidence against them is, in my estimation, strong and convincing.

15

The Project Alpha Experiment

James Randi

Generous funding doesn't make scientists smart. . . . Nor are they able to detect trickery without help. . . .

When it was announced in 1979 that noted engineer James S. McDonnell, board chairman of McDonnell-Douglas Aircraft and devotee of the paranormal, had awarded a $500,000 grant to Washington University in St. Louis, Missouri, for the establishment of the McDonnell Laboratory for Psychical Research, it seemed the ideal opportunity to initiate an experiment I had contemplated for some time. It was designed to test two major hypotheses.

Parapsychologists have been lamenting for decades that they are unable to conduct proper research due to lack of adequate funding, but I felt strongly that the problem lay in their strong pro-psychic bias. The first hypothesis, therefore, was that no amount of financial support would remove that impediment to improvement in the quality of their work. Moreover, I have always been in accord with many others in the field—such as Stanley Krippner, current president of the Parapsychological Association—who insisted that qualified, experienced conjurors were essential for design, implementation, and evaluation of experiments in parapsychology, especially where deception—involuntary or deliberate—by subjects or experimenters, might be possible. So the second hypothesis was that parapsychologists would resist accepting expert conjuring assistance in designing proper control procedures and, as a result, would fail to detect various kinds of simple magic tricks.

U.K. parapsychologist Trevor Finch had even directly suggested that the skeptics try to introduce a conjuror into a lab disguised as a psychic. Certainly my plan seemed to be in accordance with the expressed needs of the parapsychological community.

The director of the McDonnell lab was physics professor Peter R. Phillips, who had a decade of interest in parapsychology behind him. He had declared in the press that the lab intended to investigate psychokinetic metal-bending (PKMB) by children. Accordingly, I asked two young conjurors who had been in touch with me by mail, and had expressed an interest in my work as a skeptic, to write the McDonnell lab claiming psychic powers. Our experiment was to be code-named "Project Alpha."

We learned that the lab had considered some 300 applicants who contacted them in response to notices in the media. Both my colleagues, Steve Shaw, an English immigrant hospital employee in Washington, Pennsylvania, and part-time magician/mentalist, and Michael Edwards, a student in Marion, Iowa, and well-known there as a magician, were the only McDonnell lab subjects chosen from that rather large group of applicants. They were 18 and 17 years old, respectively, when they began the project.

We had established well in advance of the beginning of Project Alpha that at a suitable date we would reveal the deception. Also, the subjects agreed that, if they were ever asked directly by an experimenter if they were using tricks, they would immediately answer, "Yes, and we were sent here by James Randi." They would then answer any and all questions concerning their involvement.

Even before the boys were tested at the lab, I sent Phillips a list of eleven "caveats" concerning tests done with human subjects. For example, I warned him not to allow the subjects to run the experiment by changing the protocol. Similarly, I suggested that capricious demands by subjects might well be the means of introducing conditions that would permit subterfuge. He was warned that reports of conditions should be very precise, assuming nothing. Above all, I urged that a conjuror be present. To that end, I offered to attend the McDonnell lab tests at my own expense, without any requirement that I be credited with any participation, or even attendance, in subsequent reports.

From the very beginning, the researchers ignored the rules I had suggested. As in other investigations, the "gifted subjects" took over running the experiments. They threw minor tantrums (inspired by similar events reported to have taken place at the Stanford Research Institute when Uri Geller was examined there in the 1970s) whenever conditions were not to their liking.

Though I had specifically warned Phillips against allowing more than one test object (a spoon or key, for example) to be placed before a subject during tests, the lab table was habitually littered with objects. The specimens were not permanently marked, but instead bore paper tags attached with string loops. Edwards and Shaw found it easy to switch tags after the objects had been accurately measured, thus producing the illusion that an object handled in the most casual fashion had undergone deformation.

During one type of telepathy test, a subject would be given a sealed envelope containing a picture drawn from a target pool. Left alone with the envelope, the subject would subsequently surrender the envelope to an experimenter, who would examine it for signs of tampering. The subject would then announce his selection from the target pool. This series of tests was quite successful—though not overly so, because the boys realized that 100 percent might be suggestive of trickery. They purposely minimized their success. The method was easy. Since the envelope was "sealed" only with a few staples, they removed them, peeked, then replaced the staples through the original holes! In one case, Michael lost two staples, and to cover this he opted to open the envelope himself upon confronting the experimenter. The breach of protocol was accepted. The subjects had been allowed to shape the experiment.

In other ESP tests, significant results were obtained only when one of the subjects was aware of the target drawing and was allowed to watch a TV monitor while the other tried to duplicate the drawing. The laboratory investigators decided, in their official report on the tests, that communication between the two by any means other than telepathy had been ruled out, since "Though it might seem suspicious that the most significant scores were obtained under just that condition which might have permitted collusion . . . we feel that any hypothesis of normal communication is very unlikely; even the best of our hits are not consistent with verbal cueing, but rather exhibit consistent resemblances of form without any semantic relation."

What the experimenters could have been told, if they had been willing to listen, was that the best of conjurors' "mind-reading" tricks are accomplished by a "hot-and-cold" system of communication having nothing to do with actual verbalization. Results obtained therefore appear much more striking in nature, and seem to be what might have been achieved as a result of "telepathic" transference. No amount of acting can simulate the actual difficulty experienced by the operators of such a system.

Though no communication took place during these tests, the lack of "consistent" hits referred to in the quote above would have provided just the required conditions for acceptance, had the experiment been successful. As it was, another common conjuror's ploy was utilized, giving an edge to the results that caused some excitement among the scientists. That "edge" would be eliminated, however, by proper double-blind evaluation of the data.

(A similar semblance of reality is accomplished by any truly competent "blindfold" performer who has learned the prime directive of that type of act. I recall Margaret Foos, who never did learn that subtlety. She was a "psychic" child who convinced TV host Merv Griffin and his audiences that she could see while her eyes were heavily bandaged. She appeared rather less than convincing as she walked briskly and confidently about the audience, simply because it was evident she really could see even while bandaged. Her act would have been improved immensely if she had really kept her eyes shut, thus genuinely stumbling about until she needed to peek.)

One rather naive experiment, conducted with Steve Shaw, involved a small slab of clear acrylic plastic in which a shallow groove had been cut. Into this groove was placed a thin (about 1/16") metal rod a few inches long that fit loosely, flush with the surface. It was believed, and so stated, that it was not possible to remove the rod from the groove by hand without either overturning the slab or using a tool of some sort. Steve was asked to stroke the metal with his finger and cause it to bend. He quickly discovered that the rod tilted up and out of the groove when he pressed down upon one end, the flesh of his finger having squeezed into the groove. He simply removed the rod unnoticed, bent it slightly, and re-inserted it into the groove, lying it on its side, since the groove was wide enough to accommodate the bend. Then he stroked and rotated the rod 90 degrees to make it appear to bend up and out of the groove. The feat was deemed impossible by trickery.

In another run of experiments, involving an electronic setup, the boys were asked to shorten the capacity of tubular electric fuses. A current would be passed through a fuse and gradually increased until the fuse blew out, and that value would be recorded. They obtained excellent results in this test, seeming to cause premature rupture of the fuses through mental influence. The reason for their success was simply that they were allowed to handle the fuses freely. They were able to re-insert the same blown fuse repeatedly! The circuit had been designed so that even if a fuse were "open"—already blown—the instruments did not reveal the open circuit until the current was somewhat advanced. Thus it seemed that the fuse had blown quite prematurely. Also, the boys found that by pressing down upon one end of the fuse in its holder, or merely by touching it briefly while passing a hand over it, the instruments recorded unusual results that were interpreted by the experimenters as marvelous psi effects.

Steve suggested that he might be able to affect a video camera and, while seated before one, staring into the lens, he gestured mysteriously over the instrument. The picture twice suddenly "bloomed" brightly, the image swelling and subsiding. This was recorded and subsequently shown in the official McDonnell lab film. It seemed to the researchers that this event was not possible by any but paranormal means; yet Shaw had simply reached forward and turned the "target" control on the side of the camera—twice—unseen by the lab personnel.

Steve, Michael, and I were in frequent communication by telephone and were thus able to plan what should be done to successfully get around the incomplete security measures that had been imposed on them.

The boys were also given small, transparent, sealed plastic boxes to take home with them, containing various objects they were asked to affect paranormally. The sealing was done by drilling small holes into the box and lid, passing fine wire through the holes so as to secure the lid, and sealing the joining of the wire loop with sealing-wax impressed with a symbol. The subjects found no difficulty in popping off the seal, opening the box, "affecting" the contents, and replacing the seal. In some cases they would re-melt the seal slightly to restore it. All three of us were astonished that the impression was formed by means of a standard 39-cent stationery-store seal bearing either an anchor or a bird symbol. For less than a dollar, we were able to defeat the security of a half-million-dollar project!

One box had a cork cemented into it and a fuse inserted into the cork, leaving one end in sight but the whole thing sealed inside. Rather than fuss with the wax seal, I defeated this one merely by sticking a heated pin through the mold-release mark on the bottom of the box and up through the cork until it contacted the concealed bottom of the fuse stuck in the cork. Another pin was inserted through one of the holes in the side and contacted the exposed fuse tip. A brief visit to my car battery, and in a flash the fuse was blown.

As in all cases where we had physically interfered with or altered the equipment, we purposely left positive traces of tampering. Although the hole in the mold-mark was resealed by pressing a hot pinhead against it, when the plastic box was viewed from the side the hole was easily detectable. But it was never noticed.

However, this interesting deception fell through for a strange reason. When the lab failed to express enthusiasm over the blown fuse, the boys asked why. They were told that, in the opinion of the scientists, a centrifuge could have been used to break the fuse wire! Now the small mass of the threadlike fuse wire made it unlikely to be broken by centrifugal force, and such a break would be very clearly distinguishable from a melted one such as the sample produced by our trickery — or one fused by mental power. So, although for once the scientists were right, it was for the wrong reason.

The McDonnell lab videotape showed the subjects causing a lightweight paper rotor perched atop a needle-point to turn — in either direction, at their will — while mounted inside what was called a "bell jar." The terminology was misleading indeed. A bell jar is a piece of laboratory equipment designed to be hermetically sealed to a base-plate, and usually capable of being evacuated of air. The actual item used was a glass dome, the cover of a cheap clock, placed upon a wooden base with a slot to receive the edge of the dome. A further seal was made by a layer of aluminum foil that settled inside the recess.

The boys demonstrated that a static-charged comb was not able to cause a deflection of the rotor because of a special anti-static coating sprayed upon the dome. But when they replaced the dome, it was an easy matter to drop a small ball of aluminum foil into the slot, causing the dome to tilt slightly and provide a gap. By simply blowing surreptitiously at the base of the dome, the boys could cause air to circulate within and thus turn the balanced rotor. Changing body position and blowing from a different angle changed the direction of the air movement, and thus the direction of rotor movement. Since the small ball of foil matched the layer of foil, it was invisible and could easily be removed after the experiment — especially since the boys were allowed to handle everything freely.

One device developed at the laboratory for testing the Alpha subjects consisted of an overturned aquarium bolted and padlocked to a stout table. Objects would be put inside and left overnight. Since the locks on the doors were of excellent quality, and Phillips wore the padlock and door keys around his neck, security was thought to be absolute. It was not. Edwards and Shaw simply left a window unlocked, and returned to the premises at night. There were several ways to open the sealed aquarium, and they were free to do anything they pleased with the contents, which were discovered in the morning by lab personnel to have been bent, twisted, broken, and moved about by mysterious paranormal forces.

A part of the aquarium test used a shallow box in which dry coffee-grounds were spread in a thin layer. Small cubes and other objects were placed therein, and were found to have spelled out strange cabalistic symbols when examined in the morning. This evoked much wonder among the investigators.

Later in the progress of Project Alpha, the amateur magician who originated the sealed-aquarium system, and who even now proclaims himself incapable of being flimflammed in his specialty of designing untamperable seated containers, tried to improve upon the aquarium by providing inviolable bottles. He failed even more spectacularly.

There is no question that the lab personnel believed that Mike and Steve actually were psychic. They did. It was this belief that made the deception exceedingly easy, and it was clear that, had the two entered the arena as conjurors, they could never have gotten away with all they did. The lab personnel further crippled themselves by referring to the kids as "gifted psychic subjects," even inventing the term *psychokinete* to apply to them. Simple tricks, performed under very informal conditions of control, were declared PK events, and careless descriptions of circumstances surrounding performances were written up. These factors certainly added to the sympathetic atmosphere in which the subjects were operating.

Another factor that led researchers down the garden path was their total, unquestioning acceptance of, and belief in, the work of their fellows in the field. Even the most doubtful results, seriously questioned and in some cases thoroughly denounced by colleagues, were embraced by the investigators when it matched their needs. It is apparent that many parapsychological investigators never do a house-cleaning to get rid of the obvious trash, and the clutter that results makes it impossible to obtain a clear picture of just what their problem is.

Any minor remark or claim made by the subjects that seemed to fit an outside researcher's notion of reality or fulfilled some expectation was further evidence to the laboratory investigators that they were dealing with the real thing. For example, Steve and Mike complained about electronic equipment putting out "bad vibes," not only to satisfy this established bit of mythology, but also to minimize proper video observation. Also, they were careful to mention that in early childhood both had experienced electric shocks, after which they had become aware of their psychic powers. Though not usable as strict evidence, acceptance of these tidbits further deepened the quicksand into which the researchers continued to sink.

All through the three-year period that Steve and Mike were at the McDonnell lab, I continued to write Professor Phillips offering to attend experimental sessions as a consultant. Phillips seemed quite confident that he could not be deceived, however, and did not accept my offer.

Then, in July of 1981, I "leaked" broad hints of Project Alpha at a magicians' convention in Pittsburgh. Eleven days later, I heard that some rumors had reached the McDonnell lab. This had been done in an attempt to alert the parapsychologists. Instead, the rumors were reported to Steve and Mike at the lab as great jokes. They were not asked if there was any truth to them.

Just previous to this event, Phillips had for the first time actually written to me for assistance. He asked if I would be prepared to supply him with a videotape of fake PKMB being performed, along with a revelation of how it had been done. He intended to show it at the forthcoming August meeting of the Parapsychological Association in Syracuse. I immediately agreed to do so, and within a few days I had excerpted a number of performances from my videotape library in which I was shown bending and breaking keys and spoons as well as doing some convincing "ESP" tricks. I supplied two sound tracks, one the original and the other a running commentary describing in detail the method used. I threw in, for educational purposes, an episode with Uri Geller in which he is seen to use exactly the same method in a spoon-bending performance, and is caught on tape doing so.

I felt that rumors of Alpha would reach Phillips at about the time he had my videotape and that he would be able to examine both his evidence and mine in light of the possibility that the collusion rumor was true.

In return for my participation, I asked Phillips if I might have a copy of the McDonnell lab videotape of the Alpha subjects that had been prepared for showing along with my tape at the upcoming PA meeting.

He agreed to do so; and, just days before the convention, I received his tape. I drew up a detailed analysis of the tricks shown there, pointing out that positively unmistakable evidence of deception was contained in their tape.

At the convention, Phillips showed my tape and his own. An active rumor began circulating that Phillips and I were working together to discredit the PA, and it was widely believed. It was no surprise that his announced findings were received by the parapsychologists with little enthusiasm — although some of them, Walter Uphoff and William Cox in particular, were ecstatic. Cox, never one to entertain any doubts, had written Phillips a month earlier objecting strenuously to his intention of showing the videotape I had prepared. He apparently felt that it would not be good to introduce any doubts whatsoever into the proceedings.

A formal report on the two subjects, prepared by the McDonnell lab and distributed at the convention, was hastily recalled, and modifiers ("apparently," "seemingly," and "ostensibly") were inserted at appropriate points in the text. It was reprinted and once again distributed. In somewhat a state of shock, Phillips was cornered by me after the workshop, and I insisted upon showing him and Mark Shafer, his principal researcher, where the tape showed evidence of fraud. Visibly shaken, the two thanked me for my efforts, and I parted from them reasonably sure that they had been impressed enough to change their ways.

Upon my return from the convention, I contacted Edwards and Shaw, and informed them that Phillips was now very suspicious, and that Project Alpha was probably about to end.

16

Lessons of a Landmark PK Hoax

Martin Gardner

The most significant recent event on the psi front was James Randi's Project Alpha. Since Randi himself gives the details in this volume, I will make only general comments.

Was it unethical? I think not, but before explaining why let's consider a few past instances in which deception was used to demonstrate the incompetence of researchers.

Early this century René Blondlot, a respected French physicist, announced his discovery of a new type of radiation, which he called N-rays, after the University of Nancy where he worked. Dozens of papers on N-rays were soon being published in France, but American physicists were dubious. One skeptic, a physicist at Johns Hopkins Univerity, Robert W. Wood, enjoyed playing practical jokes, especially jokes on spirit mediums. His humorous book *How to Tell the Birds from the Flowers* is still in print. Perhaps you have seen on TV a little pinheaded, bald creature with a huge flexible mouth that is produced by painting eyes and a nose on someone's chin, then viewing the chin upside down. It was Wood who invented this whimsical illusion.

In 1904 Wood made a trip to Nancy to observe N-ray research first hand. In one experiment he secretly removed from the apparatus an essentialism prism. This had no effect on what the experimenters said they were observing. In another test Wood surreptitiously substituted a piece of wood for a steel file that was supposed to be giving off N-rays. The imagined radiation continued to be reported by the Nancy scientists. Wood told his hosts nothing about either prank. Instead, he went home and wrote a devastating account of his visit for the British magazine *Nature*. It was a knockout blow to N-rays everywhere except at Nancy.

The reaction of the Nancy group to Wood's disclosures was well summed up by Irving Klotz in his fine article "The N-ray Affair," *Scientific American*, May 1980:

> According to Blondlot and his disciples, then, it was the sensitivity of the observer rather than the validity of the phenomena that was called into

question by criticisms such as Wood's, a point of view that will not be unfamiliar to those who have followed more recent controversies concerning extrasensory perception. By 1905, when only French scientists remained in the N-ray camp, the argument began to acquire a somewhat chauvinistic aspect. Some proponents of N-rays maintained that only the Latin races possessed the sensitivities (intellectual as well as sensory) necessary to detect manifestations of the rays. It was alleged that Anglo-Saxon powers of perception were dulled by continual exposure to fog and Teutonic ones blunted by constant ingestion of beer.

When N-rays became a huge embarrassment to French science, the journal *Revue Scientifique* proposed a definitive test that would settle the matter. "Permit me to decline totally," Blondlot responded, ". . . to cooperate in this simplistic experiment. The phenomena are much too delicate for that. Let each one form his own personal opinion about N-rays, either from his own experiments or from those of others in whom he has confidence." Like Percival Lowell, the American astronomer who drew elaborate maps of Martian canals, Blondlot could not prevent his strong desires from strongly biasing his observations. He lived another quarter-century. If he had any doubts about N-rays, so far as I know he never expressed them.

Move ahead to 1974. J. B. Rhine had appointed Walter J. Levy, 26, his successor as director of his laboratory. Levy was already famous in psi circles for his "carefully controlled" investigation of animal-psi. (One of them suggested that embryos in chicken eggs had psychokinetic [PK] powers.) Three older members of Rhine's staff were suspicious of Levy's string of successes. What did they do? They set a cruel trap. While Levy was testing the PK ability of rats to alter a randomizer, they watched through a peephole and saw Levy repeatedly beef up the scores by pulling a plug. Better yet, they installed another set of instruments, without Levy's knowledge, that kept an accurate score. The untampered record showed no evidence of PK. Levy confessed, and vanished from the psi scene.

To me the saddest aspect of this scandal was not Levy's deserved disgrace but the fact that it had never occurred to Rhine to check on Levy's honesty. Rhine himself was deeply shaken by the revelations. If the trap had not been set, Levy's papers would still be cited as strong evidence for animal-psi.

There are two reasons why traps to detect fraud are more essential in PK research than anywhere else. First, the claims are far more extraordinary and therefore require much stronger evidence. Second, the field has always been soaked with fraud. In the days when eminent physicists were convinced of the reality of floating tables and glowing ectoplasm, an enormous service to science was performed by Houdini and others who were willing and capable of setting traps for the mediums.

This brings us to the main moral of Randi's hilarious hoax. Paranor-

mal metal-bending is so fantastic a violation of natural laws that the first
task of any competent experimenter is to determine whether a psychic who
bends spoons is cheating or not. In England, when physicists John Taylor
and John Hasted were convinced that scads of children could twist cutlery
by PK, one would have expected two scientists to devise some elementary
traps, but they did not. The only good trap was set by two sociologists at
the University of Bath who did not even mean to set it. Puzzled by the fact
that no one ever *sees* metal bend—Taylor called it the "shyness effect"—
they put some spoon-bending youngsters in a room, then filmed them
through a one-way mirror. The purpose was not to embarrass the children,
but to record the shyness effect. To their amazement, they saw the children
cheating. Taylor soon became disenchanted, but such revelations had no
effect on Hasted's mind-set. Some spoon benders cheat, so what? Not in
his laboratory. You can read all about his naive experiments in his recently
published book, *The Metal Benders*.

Hasted and Phillips typify psychic research at its shabbiest. In spite of
many letters from Randi telling him that his two young subjects were
frauds, Phillips made no effort to check on their backgrounds. Not until
the very end, after Randi had severely criticized his videotapes, did he start
to tighten controls. Of course the wonders ceased. On many occasions
when controls were unbelievably lax, the two "psychics" suspected a trap.
It was never sprung. They overestimated the acumen of their monitors.

Think what the results might have been had the boys decided to
become professional psychics. They would have left Phillips's lab complaining that excessive controls were inhibiting their powers. Soon they
would be appearing on TV documentaries as wonder workers whose
powers had been validated by respected scientists. Uri Geller never tires of
talking about how the Stanford Research Institute (now SRI International)
validated his psychic abilities. Phillips's two young subjects are even better
than Geller. One of them invented a way to make one tine of a fork visibly
and unshyly bend that is superior to any of Geller's crude methods. When
Steven Shaw demonstrated this lovely illusion at Randi's Manhattan press
conference, the entire audience gasped. "Can you tell us how you did
that?" a startled reporter asked. Shaw walked to the mike and said, "I
cheat." It brought down the house.

It is to Phillips's credit that he had the courage to say (*Washington
Post*, March 1, 1983), "I should have taken Randi's advice." It is to the
credit of Stanley Krippner, a true believer in PK if ever there was one, that
he called Randi's project "a much-needed" experiment. It remained for
former CSICOP member and sociologist Truzzi to start the hue and cry
about entrapment. Truzzi had known about Randi's trap almost from the
beginning but had carefully kept his own trap shut. "Randi is hurting the
field with his gross exaggeration," Truzzi told the *New York Times* (February 15, 1983). "In no way will his project teach psychic researchers a lesson

and make them more likely to trust magicians' advice. Quite the contrary. This outside policeman thing sets up magicians as the enemy."

On this point Truzzi may be right. I, too, would be surprised if psychic researchers suddenly decided to study conjuring or to seek the active help of knowledgeable magicians. Conjurors are indeed the enemy. Their bad vibes alone are enough to kill any PK powers just by being there as observers; perhaps (as has actually been suggested by the sociologists at Bath) even their *reading* about the experiments afterwards influences the outcome by backward causality! But perhaps Randi's scam will have a salutory effect on funding. After all, the half-million bucks the McDonnell Foundation gave to Washington University could have gone to worthwhile research instead of down the drain to a group unqualified to investigate metal bending.

Am I saying that all psychic researchers should be trained in magic, or seek the aid of magicians, before they test miracle workers? That is exactly what I am saying. The most eminent scientist, untrained in magic, is putty in the hands of a clever charlatan. Without the help of professional deceivers—the conjurors—no testing of a superpsychic is worth ten cents of funding. That is the big lesson of Randi's hoax. That is why it is likely to become a landmark in the history of PK research.

17

Magicians in the Psi Lab: Many Misconceptions

Martin Gardner

Harry Collins, a University of Bath sociologist, is best known for his extreme relativistic philosophy of science (see my review of *Frames of Meaning,* which he coauthored, in the Fall 1983 issue of *Free Inquiry*), and for having caught a group of spoon-bending children at cheating. The *New Scientist* (June 30, 1983) printed his "Magicians in the Laboratory: A New Role to Play," in which he discusses what he calls the "vexed relationship" between magicians and psi researchers. His article contains many misconceptions about magic; but before detailing them, first a sketch of his views.

Randi's recent Project Alpha, Collins writes, has reminded us again of how easily psi researchers can be hoodwinked. Because the history of paranormal research has been riddled with fraud, Collins wisely recommends that, no matter how innocent a subject may appear, experiments must be designed on the assumption that the subject is "a notorious cheat." Unfortunately, he adds, completely fraud-proof tests are impossible because there is no way to anticipate new methods of cheating. Since magicians know standard ways, they can be enormously useful as advisors. But because they are not much better than nonmagicians in spotting new methods they are of little value as observers.

He feels that magicians should not be allowed to monitor experiments because they are usually unfriendly toward psi research and have a vested interest in seeing psychics discredited. Collins doesn't mention the belief of most parapsychologists that hostile observers inhibit psi phenomena, but even aside from this he thinks magicians would have a damaging effect on experiments if they were allowed to monitor them.

How, then, can conjurors help? One way, Collins says, is by breaking their code of secrecy and explaining to researchers how cheating can be done. If magicians are unwilling to do this, they should serve as "protocol breakers," by demonstrating the same paranormal phenomena under the same controls applied to the psychic. If they fail to break the protocol, this "would act as a certificate of competence in experimental design."

Misconception 1: Collins fails to distinguish stage performers from magicians who specialize in close-up magic. Throughout his article he repeatedly refers to "stage magicians" and "illusionists." The distinction is vital, because the methods used by psychic charlatans have almost nothing in common with stage magic. Although psychics like Uri Geller and Nina Kulagina may use a few concealed "gimmicks" (magnets, "invisible" thread, nail writers, palmed mirrors, and so on), for the most part they perform close-up magic that requires no apparatus.

Some stage magicians are knowledgeable about close-up magic, but not necessarily so. A stage performer is essentially an actor playing the role of a magician, relying for his miracles on costly equipment designed by others. Any good actor could easily take over Doug Henning's role in the Broadway musical *Merlin,* for example, and the stage illusions would work just as well. It is important for psi researchers to know this. Otherwise they might seek the help of a prominent stage performer who has less knowledge of close-up magic than do thousands of amateurs.

Misconception 2: Collins is persuaded that magicians are not much better than scientists in spotting new ways to cheat. He concedes that "skilled practitioners of deception" may be better than scientists in seeing loopholes develop in an experiment, but he adds, "I think it would be hard to demonstrate this."

On the contrary, it is easy to demonstrate. Collins could convince himself of this simply by accompanying someone like Randi to a magic convention at which dealers demonstrate new tricks for the frist time and see how he compares with Randi in figuring them out. It is true that magicians sometimes fool other magicians, but not often and not for long. The "magician's magician" who enjoys inventing tricks to fool his colleagues bears no resemblance to the psychic charlatan. The charlatan is usually a mediocre performer who has hit on some crude methods of deception all his own—methods that are transparent almost at once to any knowledgeable close-up magician who sees the charlatan perform.

When new tricks come on the market, dealers like to advertise them in magic periodicals with glowing descriptions that seem to rule out all standard methods. Magicians are often extremely good in guessing the modus operandi from the ad, without even seeing the trick performed. Of course, if they actually saw the trick demonstrated, it would be enormously easier. And if they saw it more than once, it would be a rare trick indeed that would resist unraveling.

A few years ago my friend Persi Diaconis, a statistician who is also a skilled card magician, telephoned to say that a certain Oriental conjuror was appearing that night on television and would be performing a sensational new trick with a silk. The silk is twisted like a rope, cut in half, the halves rolled into a ball, and when unrolled, the silk is restored. Persi had not yet seen the trick, but had heard it described by puzzled magicians.

After discussing several methods, we finally agreed on what we thought was the most probable technique. When we watched the show that night, our hypothesis was verified. The point is that we guessed the method before we even saw the trick.

Sometimes it is impossible to guess from a description. When I was a young man in Chicago, Joe Berg's magic shop advertised a miracle called the "none-such ribbon effect." A ribbon, the ad said, is cleanly cut in half and the ends widely separated. After the restoration, the ribbon is the same length as before. No ribbon is added or taken away, and no adhesives, magnets, or other secret aids are needed. I was unable to guess the method. A few days later, in Joe's shop, I asked him to demonstrate the trick. As soon as he did, I understood. I am free to give away the secret because this clinker of a trick has never been performed by a magician, and never will be. The "ribbon" proved to be crepe paper. It was genuinely cut, the halves folded into a parcel, one half palmed away, then the other half was pulled out of the fist in such a manner that it stretched to twice its original length!

New methods of deception are invariably based on ancient general principles that any experienced conjuror knows in his bones. No magician could have witnessed the none-such ribbon effect without seeing at once how it worked, even though no one had ever before thought of restoring a ribbon in this peculiar way. Scientists are helpless in the hands of a clever charlatan, whether he uses old or new methods, but knowledgeable magicians are far from helpless regardless of how unorthodox the new methods may be. Their ability to detect fraud by novel techniques is vastly superior to that of any investigator without a magic background, even if he has a high I.Q. and a Nobel Prize.

Misconception 3: The suggestion that magicians should advise but not observe is naive. Until a magician actually sees a clever psychic perform, he is in a poor position to know what controls should be adopted. It is no good to rely on a scientist's memory of what he saw, because such memories are notoriously faulty. Good magic is carefully designed to conceal a trick's most essential aspects, and even what a magician says is planned to make a spectator forget crucial details. The medium Henry Slade, for example, was once tested by a group of scientists. No one recalled afterward that a slate had "accidentally" slipped out of Slade's hands and dropped on the rug. Yet it was at just this instant that Slade switched slates. Magicians are alert to such misdirection. Nonmagicians are not. Incidentally, in Slade's day many scientists were totally convinced that his slate writing was genuine. Is it not curious that chalked messages appearing on slates have disappeared from the repertoire of modern psychics?

Conjurors obviously can be of great help in designing protocols, but if a charlatan is using new methods, or performing a feat never performed before (such as Ted Serios's trick with Polaroid cameras), it is almost essential that he be observed initially by a magician. True, in many cases a committee of magicians may, on the basis of a careful, accurate description of a psychic's performance, figure out how the psychic could be cheating and suggest adequate controls. In some cases, however, the memories of psi researchers are too vague and flawed to permit such reconstruction. Only by seeing the psychic do his or her thing can the magician make intelligent guesses and not waste the researcher's time by suggesting twenty different ways the psychic could have cheated. Of course it is essential that a psychic not know a magician is present. Psi powers have a way of evaporating even if the psychic only suspects a magician may be present. The reason D. D. Home was never caught cheating was that Home took extreme precautions to perform miracles only in the presence of persons he knew to be untrained in magic.

Suppose a club suspects a member of cheating at card games. How should members go about catching him? It is folly to ask an expert on card-swindling to design precautions, because there are thousands of ways to cheat. I can show you fifty ways to false shuffle a deck, and as many ways of getting secret peeks of top and bottom cards. Persi can demonstrate twenty different ways to deal the second card instead of the top one, some by using only one hand. There are dozens of subtle ways to mark certain cards in the course of a game. Nor is it feasible for a card "mechanic" to give club members an adequate course in cheating. It would require many months. Obviously nothing is gained by having the mechanic sit in on games if the hustler knows who he is. And how can club members be sure that the hustler doesn't know?

The fact is that there is only one good way to settle the matter. A trap must be set. Let the expert observe a game secretly, either through a peephole or a carefully constructed one-way mirror. This is such a simple way to trap a cheat that one of the great marvels of modern psi research is that the only researchers of recent decades who have used it seem to be Collins and the parapsychologists in Dr. Rhine's laboratory who set a peephole trap for their director, Walter Levy.

Consider the sad case of John Hasted, a Birkbeck College physicist who firmly believes that children can paranormally bend paperclips inside a glass sphere, provided the sphere has a hole in it and the children are allowed to take it into another room and do their psychic bending unobserved. A ridiculously easy way to settle this hypothesis would be to videotape the youngsters secretly, the way Collins did. If Hasted ever tried this, I haven't heard of it. It is passing strange that parapsychologists who become convinced that psychics can bend metal seem absolutely incapable of devising a simple trap. This augurs ill for the hope that they will ever seek the aid of magicians in any significant way.

Misconception 4: It is naive to suppose that most researchers are capable of setting up controls for a magician that are identical to those imposed on a psychic in the past experiment. If a videotape of an entire experiment is made, without breaks, it might be possible; but even here there are major difficulties. Take the case of Ted Serios. Suppose a tape had been made that showed Ted holding his "gizmo" (rolled piece of paper) in front of the camera lens and a picture of the Eiffel Tower appearing on the film. A magician asked to break protocols would ask: Was the gizmo examined immediately before the event was recorded? The researcher may honestly say yes; but unless a magician had been there, there is no way to rule out the possibility that Ted palmed an optical device into the gizmo *after* it had been examined. Even if the tape showed the gizmo being examined, if Ted were careful of camera angles nothing on the film would reveal palming. Similarly, an adequate tape would have to show the gizmo examined immediately after the camera snapped, and in such a way that it ruled out Ted palming a device out of the gizmo.

Jule Eisenbud, who wrote an entire book about Ted, has repeatedly challenged Randi to break his protocols. Why has Randi refused? Because Eisenbud, having learned from magicians how Ted could have cheated, now wants to impose Randi controls that were never imposed on Ted. Magicians think Randi has already broken Eisenbud's protocols; but Eisenbud does not think so, and neither do many top parapsychologists. Researchers typically demand of magicians that they repeat past miracles under conditions radically unlike those that prevailed when the "psychic" produced them. The fact is that there is no way to make sure controls are identical unless a magician has been there to see the psychic perform. Memories of researchers untrained in magic are far too unreliable. Of course one could ask that a magician and a psychic produce a paranormal event under identical controls, supervised by outsiders, but what psychic charlatan would ever agree to such a test?

Misconception 5: Collins makes much of his belief that magicians refuse to give away methods used by psychics. They do indeed refuse to give away secrets of tricks by which professional magicians earn a legitimate living, including entertainers like Kreskin who pose as psychics; but at the low level of prestidigitation on which psychics operate, magicians have never hestitated to give away secrets.

As Collins knows, Houdini constantly exposed the methods of fraudulent mediums. Randi has tirelessly explained the methods of Uri Geller and other mountebanks. The three magicians who investigated Serios for *Popular Photography* (October 1967) explained in detail how to produce all of Ted's effects with an optical gimmick. One of the three, Charles Reynolds (who designs illusions for Doug Henning and other stage performers) is certainly not going to tell how Doug vanished an elephant or how David Copperfield made the Statue of Liberty disappear, but he minds not at all telling any parapsychologist willing to listen how Geller bends keys. Surely Collins knows about my *Science* article (reprinted in *Science: Good, Bad, and Bogus*) that exposes the secrets of eyeless vision, except for Kuda Bux's method—and that was because Kuda made his living with it. Surely Collins knows of the two books by Uriah Fuller, on sale in magic stores, that give away all of Geller's basic techniques. Randi and I will happily tell anyone how Nina Kulagina uses invisible threads to move matches and float table-tennis balls, and how Felicia Parise could have moved a pill bottle for Charles Honorton. How Collins got the impression that magicians are reluctant to explain secrets of psychic fraud is beyond me. Even the secrets of legitimate magic are readily available to any psi researcher who cares to buy a few dozen modern books on the subject.

Misconception 6: Collins actually thinks that if magicians were routinely asked to observe psychic wonders it would wreck science. It is not just that fraud is possible in all experiments and there aren't enough magicians to go around, but psi research, like all research, is a vast social enterprise extending over long periods of time. It simply would not work, says Collins, if hostile magicians were perpetually underfoot.

What Collins ignores here are two all-important distinctions. One is between the operations of nature and human nature; the other is between ordinary and extraordinary phenomena. As I like to say, electrons and gerbils don't cheat. Even among psychics, very few claim such fantastic powers as the ability to bend metal by PK, translocate objects, and levitate tables. It is only when exceedingly rare miracles like these are seriously investigated that it is essential to call in an expert on the art of close-up cheating. And it is essential in many cases that the expert be there to watch, not just give advice at some later date to researchers who, more often than than not, in the past have paid not the slightest attention to such advice.

Some of the most widely heralded miracles are one-time events that the psychic never does again, such as the time Geller translocated a dog through the walls of Puharich's house, or Felicia moved a pill bottle, or Charles Tart's sleeping subject guessed the number on a card that Tart had put on a shelf above her line of vision. Since no expert on fraud was there as an observer, no one should take seriously the claims of Andrija Puharich, Charles Honorton, and Charles Tart that those events were genuine. Who can take seriously today J. B. Rhine's claim that Hubert Pearce correctly guessed 25 ESP cards in a row? Only Rhine observed the miracle, and there are 20 ways Pearce could have cheated. When a psychic produces events this extraordinary, it is impossible to imagine that he or she would ever submit to retesting under controls recommended by a magician, let alone being observed by a magician during the retesting.

In sum: if parapsychologists seeking the aid of magicians tried to follow Collins's naive guidelines, it is easy to predict the outcome. In a word—zilch.

18

How Not to Test a Psychic: The Great SRI Die Mystery

Martin Gardner

Writing in *Nature* (vol. 251, October 18, 1974) on their 1972-73 experiments with Uri Geller at the Stanford Research Institute, Harold Puthoff and Russell Targ (hereafter called P and T) described one sensational experiment as follows:

> A double-blind experiment was performed in which a single 3/4-inch die was placed in a 3 x 4 x 5 inch steel box. The box was then vigorously shaken by one of the experimenters and placed on the table, a technique found in control runs to produce a distribution of die faces differing nonsignificantly from chance. The orientation of the die within the box was unknown to the experimenters at that time. Geller would then write down which face was uppermost. The target pool was known, but the targets were individually prepared in a manner blind to all persons involved in the experiment. This experiment was performed ten times, with Geller passing twice and giving a response eight times. In the eight times in which he gave a response, he was correct each time. The distribution of responses consisted of three 2s, one 4, two 5s, and two 6s. The probability of this occurring by chance is approximately one in 10^6.

Surely this experiment deserves to rank with the famous test in which Hubert Pearce, a student at Duke University, correctly called 25 ESP cards in a row as J. B. Rhine repeatedly cut a deck and held up a card. In one respect, the die test with Geller is more significant because it rules out telepathy. Of course it does not rule out the possibility that Geller used precognition or that he decided on a number while the box was being shaken and then used PK to joggle the die to that number. In any case, the experiment seems to be a simple, foolproof, monumental violation of chance.

On the other hand, as in the case of Rhine's informal account of Pearce's equally miraculous run of 25 card-hits, P and T describe the die test with a brevity that seems inappropriate for so extraordinary a claim. We are not told who shook the box, where or when the test was made, who observed the trials, how long Geller took to make each guess, whether he was allowed to touch the box, whether there were earlier or later die-box tests with Uri, or whether the experiment was visually recorded.

When P and T released their official SRI film about their five-week testing of Geller, one of the die-box trials appeared on the film. It was accompanied by the following voice-over:

> Here is another double blind experiment in which a die is placed in a metal file box (both box and die being provided by SRI). The box is shaken up with neither the experimenter nor Geller knowing where the die is or which face is up. This is a live experiment that you see—in this case, Geller guessed that a four was showing but first he passed because he was not confident. You will note he was correct and he was quite pleased to have guessed correctly, but this particular test does not enter into our statistics.

The box is seen to be a metal one of the sort used for 3-by-5 file cards. The same box appears in two photographs that accompany an article on Geller in the July 1973 issue of *Psychic* magazine (now called *New Realities*). One picture shows Geller recording his guess, the closed box near his hand. The other shows Geller opening the box to check his guess.

John Wilhelm, in *The Search for Superman* (Pocket Books, 1976) reports that he was told by P and T of many other tests they made of Geller with a die in a box. Some of them took place in Geller's motel room, with Uri doing the shaking. "He's like a kid in that he had something that made a lot of noise and he just shook it," Targ told Wilhelm. Targ also said that during the experiment reported in *Nature* Geller was allowed to place his hands on the box in "dowsing fashion."

Targ also informed Wilhelm that they had a "good-quality videotape" of another die test in which Geller, five times in succession, correctly wrote down the die's number *before* the box was shaken. "We think it's precognition," said Targ. "We think maybe even on his original experiment it wasn't that he knew what was facing up, but that he had precognition as to what he would see when he opened the box."[1]

Wilhelm gives other details about the original test. Puthoff was the experimenter who usually shook the box. Many different dice were used, each etched with a serial number to guard against switching. To avoid ambiguity in guessing, Geller was asked to draw a picture of the spots rather than write a digit. "The experimenters also insist that a magician who examined the videotape of these performances found 'no way' in which Geller could have cheated."

Note that Wilhelm was told the experiment had been videotaped. In 1976, in a letter published in the *New York Review of Books* (reprinted on page 108 of my *Science: Good, Bad and Bogus*), I wondered if the test shown on the SRI film was part of the original test or a later one. If part of the actual test, I urged P and T to allow magicians to see the film of the entire experiment. James Randi, in *The Magic of Uri Geller* (the revised edition, *The Truth About Uri Geller,* has recently been released by Prometheus Books), suggested a method by which Geller could have secretly obtained peeks into the box. Seeing the film of all ten trials (not just a trial on which he passed; obviously if Uri had a way of cheating he would pass whenever circumstances made it impossible to make the needed "move") would provide valuable information concerning Randi's hypothesis. "Come, gentlemen," I concluded my letter, "let us see the entire tape! If we are wrong, we will humbly apologize."

After Randi's book appeared, P and T issued an 8-page "Fact Sheet" intended to correct what they considered serious errors in the book. They mention Randi's "elaborate hypothesis" about the die test and reply as follows: "Fact: Film and videotape show otherwise, and magicians examining this material have failed to detect a conjuring trick."

Now this statement clearly implies that the experiment had been videotaped or filmed in its entirety. Who were the magicians who examined the film? They could not have included Milbourne Christopher, a professional who visited SRI, because he has told me he saw no film of the die test. There are only two possibilities. One is Targ himself, who had a boyhood interest in magic. The other is Arthur Hastings, a close associate of P and T and a strong supporter of their work. P and T used him frequently as a judge in their remote-viewing experiments. Hastings claims some knowledge of conjuring techniques, but in my opinion his knowledge is extremely limited.

In the fall of 1981, almost ten years after the die test, Puthoff finally revealed an astonishing fact. No film or videotape was ever made of any of Uri's eight successful guesses!

This revelation came about only because Randi, in his latest book, *Film-Flam!,* concluded, on the basis of privately obtained information, that the episode on the SRI film, showing Uri passing, was a reenactment of the experiment. Both Puthoff and Zev Pressman, the research engineer who made the film, have since vigorously denied that it was a reenactment. In reply to an inquiry, Puthoff unequivocally told me in a letter (September 10, 1981): "Only one trial was filmed, and that is the one that appears on the film . . . the entire series of trials was *not* filmed."

Why? Because, Puthoff explained, Pressman's filming was done primarily to record PK efforts. As the Christmas holidays of 1972 approached, Puthoff said, they decided to "slip" in some die-box trials, "without making a big deal of it," to see if Geller could succeed in a pure clairvoyance test. These trials, Puthoff added, were "spaced-out over a few-day period" just before Geller left. When Puthoff saw they were getting hits, he decided that a film record of their protocols would be useful. Puthoff asked Pressman to make the record and he came over to do it. "We broke up for the holidays," Puthoff continued, "assuming that eventually we would get more trials on film, but we never came back to it, going on to other things. . . . I hope this clears it up for you."

Well, not quite. It seems passing strange that in a test of this importance P and T would see fit to film only the single trial on which Uri passed. Moreover, I was puzzled by the vagueness of the statement that the test had been spaced out over a "few-day period." I wrote again on September 14 to ask Puthoff if he could recall the exact number of days. Puthoff replied (October 5) that the experiment was spread over a "two or three day period, a few trials per day, sandwiched in among other experiments, until a total of ten trials were collected." He added that the length of time per trial, "from when I began the shake to when I opened the box, was relatively short—30, 40, 50 seconds. The one you see on the film is quite typical, and it is well under a minute."

I had asked for the "exact number" of days, but Puthoff's answer of "two or three days" was almost as vague as his "few-day period." D. Scott Rogo, writing about the die experiment in *Fate* (November 1981) said that P and T told him that Uri had found the die test difficult. It is hard to imagine that Uri would have considered the test difficult when he obtained eight hits in a row in trials that lasted less than a minute each. Nevertheless, Rogo continues, "he did only one or two trials a day over a period of a week. . . . He made a total of ten trials. . . . Only *one* of these trials was ever filmed. . . . This is the only SRI film ever made of any die-throwing tests."

Now there is a big difference between two or three days and a week. I was further mystified because Puthoff had also told Randi personally, at a parapsychology conclave in Toronto in 1981, that the die test had taken a week. Was it a week, or two or three days?

An incredible thought struck me. Could it be possible that P and T had not considered it worthwhile to keep a written record of the trials giving all the details about when and where each trial was made and who was present on each occasion? I sent Puthoff the following letter:

Dear Hal: 8 Oct 81

Your reply of 5 Oct was much appreciated. I did not even know, until I got your letter, that you were the experimenter in these tests.

May I assume from your statement of "two or three days" that the trials were not recorded and dated when they took place? It is the only way I can explain the ambiguity. (Rogo, by the way, in the latest issue of *Fate*, reports that he was told the trials took place during a period of a week, which only adds to the confusion.)

Perhaps I have regarded the test as more significant than it was considered at the time—especially since it ruled out the possibility of telepathy. If the die test was considered not important, and made more or less at random, with no keeping of records, then I can understand the confusion over the number of days. . . .

My assumption, Puthoff replied, was dead wrong. "Careful records were kept." He said he had now checked those records and determined that "the trials were carried out over a three-day period." Wilhelm, he added, confused two separate die tests. One was the test reported in *Nature*, of which only the passed trial was filmed. It used a red transparent die. Later a series of similar tests were carried out in a motel room in San Francisco when they were there for the *Psychic* article. These were videotaped. Puthoff closed by saying that he continues "to entertain Randi's hypothesis" but considers it ruled out by the SRI film. "Go back and view the film—that's what we have to deal with."

I found it curious that Puthoff would place any value on the filmed trial because, assuming Geller used a peek move for his hits, he obviously would not use it when being filmed. I wrote to Puthoff again (October 18) asking him if I could pay for the cost of having the original records photocopied. This is how I justified my request:

> In the interest of seeking the truth about this historic test (in which the results were so unambiguous and so overwhelmingly against chance), it would be enormously helpful to see these records. I want to be completely open. I know a great deal about dice-cheating techniques, and it is my belief that Geller did indeed peek by a method similar to the one Randi conjectured. The written records may cast no light on the matter, but at least they could be of help in pinning down the exact protocols.

The letter was never answered.

What conclusions can we reach from all this? The most important is surely the following. What seemed to any reader of *Nature* to be a carefully controlled die test has now become little more than a collection of anecdotes. At the very least P and T should make a full disclosure of all the details of the test, including photocopies of whatever records were made at the time. We also should be told the results of the videotaped tests made in San Francisco, and whether Wilhelm was accurate in reporting that a videotape was made of a successful precognition test with a die and box.

As it stands, the ten-trial test at SRI should not be called an experiment. There were too many ways Uri could have cheated (the peek move is only one)—ways that could be ruled out only if a knowledgeable magician had been present as an observer, or if a videotape had been made of all ten trials from start to finish, with no time breaks. In the absence of such controls for guarding against deception by a known charlatan, the die test was far too casual and slipshod to deserve being included in a technical paper for a journal as reputable as *Nature*. It belonged more properly in a popular article for *Fate*.

Notes

1. Both P and T are strong believers in precognition. Indeed, this was the topic of Targ's paper, "Precognition and Time's Arrow," delivered at the 24th annual meeting of the Parapsychological Association, at Syracuse University, August 1981. Targ gave his reasons for

thinking that precognition does not violate quantum mechanics and that it could be explained only by assuming time-reversed causality. He defended Helmut Schmidt's experiments that supposedly confirm backward causality, and cited William E. Cox's paper on precognition in the *Journal of the American Society for Psychical Research* (vol. 50, 1956, pp. 99-100), reporting a study of 28 trainwrecks that occurred between 1950 and 1955. Cox concluded that (in Targ's words) "significantly fewer people chose to ride trains on days when they were going to crash, than rode them on previous corresponding days of the week in earlier weeks or months."

2. Here is how Wilhelm reported what P and T told him about this die test (*The Search for Superman*, p. 95):

> "We only talk about the more conservative miracles," muses Targ. "We have another tape of Geller that's not reported because it's more outlandish. We have a very good-quality videotape in which Geller, on another visit, said, 'I don't want to repeat that, I have a new way of doing that dice experiment.' The new way is to write down on a piece of paper a number on the table. Then I [Targ] take the box and shake it vigorously. Then he takes my shaken box and he shakes it vigorously, dumps the dice out on the table, and it comes up the number he wrote down. We did that five times in a row."
>
> According to Puthoff, the dice was thrown "way up in the air, landing on the table, bouncing all over, and then coming up the [guessed] number." The die belonged to SRI.

In view of the fact that this entire test was videotaped, in contrast to the original test, which was not, it was a much better controlled test than the one reported in *Nature*. Does a tape of this test exist? If so, why has it not been made available to psi researchers?

Claims of Mind and Body

19

Iridology: Diagnosis or Delusion?

Russell S. Worrall

For centuries the eye has been said to be the mirror of the body and the soul. Terror and love are expressed through the eyes; and the general state of health can be reflected in the eyes, as in the vacant, glassy stare of the gravely ill. Today, a more precise analogy would be to describe the eye as a window rather than a mirror. The eye is an optically clear porthole that allows one to view body tissues, such as blood vessels and nerves, in their undisturbed state. The subject of this paper, iridology, proposes a more elaborate analogy—specifically, that the iris of the eye is a gauge registering the condition of the body's various organs; or, as Jessica Maxwell describes it in her book *The Eye/Body Connection*, the iris is "an organic Etch-a-Sketch" (Maxwell 1980, p. 12).

Iridology, pronounced "eyeridology," is the "science" of reading the markings or signs in the iris (the colored part of the eye) to determine the functional state of the various components of the body. It is not unique; other, equally sophisticated systems of belief exist utilizing the soles of the feet, the ear, the palm [see *Skeptical Inquirer*, Winter 1982-83], and the spine. A common theme unites these varied techniques. They are non-invasive (you do not have to be punctured or sliced into!) and they each involve a specific area of the body's surface, which when read by a "well-trained" practitioner reveals your innermost health problems. Elaborate charts (suitable for framing) that guide the practitioner and impress the patient are the centerpiece of each method.

Iridology may have its origins in antiquity, or more recently in Russia, as suggested by the *National Enquirer* (1978), but Dr. Ignatz von Peczely, of Hungary, is generally held responsible for developing and promoting the modern "science" of iris diagnosis. According to the story, the miraculous discovery of the iris-body connection came when, as a boy of ten, Ignatz accidently broke the leg of an owl and a black stripe spontaneously appeared on the owl's iris. He went on to develop charts based on his clinical observations (see Figure 1) and published a book on the subject (Von Peczely 1866).

In the United States, Bernard Jensen has been the most influential

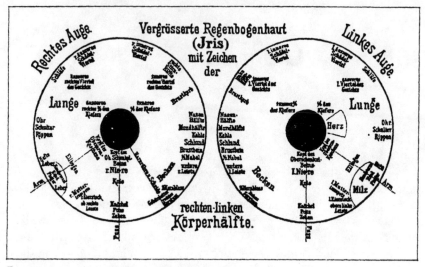

FIGURE 1. An iris chart developed by von Peczely.

proponent of iridology. His book, *The Science and Practice of Iridology* (Jensen [1952] 1974), has been the authority on the subject since its publication in 1952. His new book, *Iridology: The Science and Practice in the Healing Arts,* vol. 2 (Jensen 1982), is a 580-page epic detailing the history, "science," and application of iridology. Dr. Jensen's updated charts (see Figure 2) are the standard in the United States. European charts, though more detailed and exacting, follow the same general format as illustrated by Kriege's (1975) Bone Zone Chart (Figure 3) and Korvin-Swiecki's charts (Figure 4) shown in *The Eye-Body Connection* (Maxwell

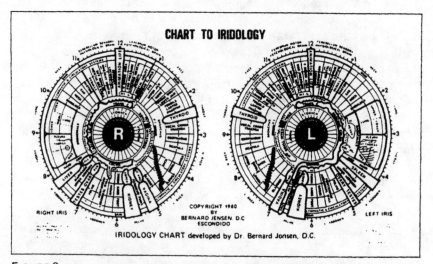

FIGURE 2. Iridology chart by Jensen (Jensen, *Iridology Simplified,* Iridologists International, 1980).

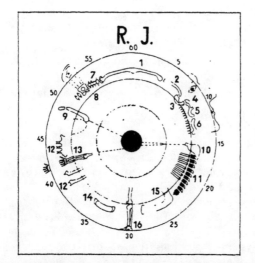

FIGURE 3. Schematic representation of the Bone Zone of the right eye by Kriege: (1) Cranial bone. (2) Frontal bone. (3) Orbit. (4) Nasal bone. (5) Upper jaw and teeth. (6) Lower jaw and teeth. (7) Cervical vertebrae. (8) Ear. (9) Shoulder and clavicle. (10) Scapula. (11) Spine and ribs. (12) Sternum and ribs. (13) Hand and arm bones. (14) True pelvis. (15) Pelvic crests. (16) Foot and leg bones. (*Fundamental Basis of Irisdiagnosis*, trans. Priest, Fowler Co., London, 1975, p. 102).

1980). This pseudoscience has been popularized by many recent books and articles, including a story in the *National Enquirer* (1978) with a typically dramatic headline: "Do-It-Yourself Eye Test That Can Save Your Life." All these reports are blandly neutral or, more commonly, outright

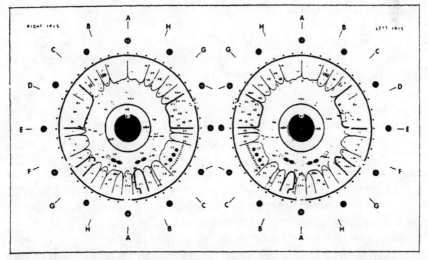

FIGURE 4. Basic European Iridology chart by Korvin-Swiecki. Some examples from the chart: (1) Cerebrum. (2) Cerebellum. (5) Ear. (6) Neck, throat. (8) Lungs. (9) Heart. (B) Aorta. (14) Liver. (16) Pancreas. (23) Kidneys. (27A) Uterus. (42) Larynx. (53) Autonomic nervous system. (54) Ascending colon. (58) Appendix. (59) Gallbladder. (60) Stomach. (61) Central nervous system. (62) Circulatory and lymphatic system. (Jessica Maxwell, *The Eye/Body Connection*, Warner Books, New York, 1980, pp. 60-63)

enthusiastic about this purportedly marvelous diagnostic procedure. Only two papers present a critical view based on controlled scientific studies, the results of which are not astonishing to the skeptical inquirer.

The philosophical, scientific, and clinical ramifications of iridology are succinctly stated by Harri Wolf (1979, pp. 7-8), founder of the National Iridology Research Foundation, in the introductory remarks of his *Applied Iridology*:

> Now wouldn't it seem *logical* [emphasis added] that through some creative design, or evolutionary process (whatever the reader's preference), the human body would be equipped with a metering device functioning as a gauge in regard to the health of the individual?
>
> Each of us is, *in fact* [emphasis added], equipped with just such a miniature recording screen —the iris. Via the direct neural connection of the surface layers of the iris with the cervical ganglion of the sympathetic nervous system, impressions from all over the body are conveyed to the iris. Thus is established the neuro-optic reflex.
>
> Iridology, as the study of the neuro-optic reflex is known, is the art/science of *revealing* [emphasis added] pathological, structural and functional disturbances in the body.

The Philosophy and Logic

The claim that iridology is a logical, natural system is central to iridology philosophy. As Jensen (1981a, p. 2) writes, "We must realize that iridology represents a law of nature that cannot be changed. I believe that it is just as immutable and unchangeable as any of the laws that govern the universe." When viewed from a critical perspective, the logic in iridology begins to fade. First, a gauge or metering system has to be read and understood to be useful. The iris of the eye is certainly inaccessible to all of earth's creatures, including man (unless he happens to have a mirror handy). Further, the iris signs are so complex as to be unintelligible to all but those who have been enlightened by von Peczely's theories. In short, there is no logical evidence to support a claim of functional utility for this complex biological system purported to exist in many diverse organisms, including man.

This apparent lack of utility to the organism exposes a more fundamental flaw in the logic of iridology when it is considered in the context of the evolutionary process. A physiological subsystem such as the suggested iris-body connection would be developed and refined under the gradual pressures exerted by natural selection. For such a system (more properly the gene pool that codes for the system) to have evolved in many diverse species, a distinct survival advantage had to be present for the organisms with this system. This is an assumption for which I can offer no logical arguments. Certainly the saber-toothed tiger derived little benefit from this amazing metering system, nor does today's modern owl

The Science

A second theme in iridology is that, as Wolf states, the neuro-optic reflex exists "in fact." The facts supporting the existence of the neuro-optic reflex are tenuous at best. It is postulated by Wolf (1979, p. 7) that the sympathetic division of the automatic nervous system mediates the iris response. D. Bamer (1982, p. 22) includes the parasympathetic division in his theory and offers an anatomical diagram to support his claim (see Figure 5). In a gross anatomical sense the autonomic system does interconnect and enervate almost every segment of the body, including the eye. However, anatomical interconnection does not imply functional connection any more than having a telephone in your home is proof of the proposition that you receive all of the calls intended for the president of the United States. Further, the autonomic nerves supplying the eye are of small caliber and would not seem to have adequate numbers of nerve fibers to handle the volume of information presumed to reach the iris. Anatomical, physiological, and clinical studies have eloquently demonstrated the functional neural pathways involved in many of the eye's control and response mechanisms, but published studies report no evidence in support of a functional iris-body connection (Moses 1975; Last 1973). Though these investigations were not specifically looking for a neuro-optic reflex, given the quality and quantity of information postulated to appear in the iris it is curious that even accidental detection

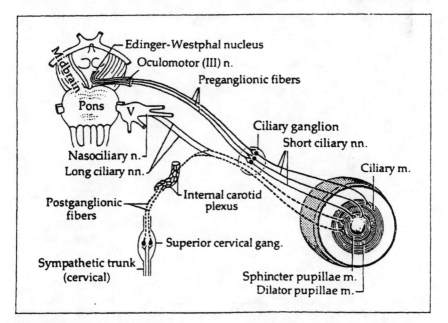

FIGURE 5. Anatomical diagram by Bamer (*Applied Iridology and Herbology*. BiWord Publishers, Orem, Utah, 1982, p. 22)

of this elaborate system has eluded researchers.

One aspect of the functional theory, as expressed by Jensen, is especially interesting in light of well-established neurological evidence. Anatomists and physiologists have long known that as a general rule the central nervous system is functionally split, with each side controlling and monitoring the opposite side of the body. In iridology it has also been "established" that each eye "sees" its own side of the body. Thus a conflict is created for iridologists in explaining the flow of information over the autonomic pathways that cross to the opposite side as they travel through the central nervous system. To explain this apparent difficulty, Jensen proposes that the optic nerve serves as the final link between the autonomic system and the iris (see Figures 6 and 7). Since the optic nerve crosses between the eye and the brain, information from an organ would make a second crossing on its way to the iris and register on the same side that it originated from. Jensen (1980, p. 3) also infers that the large size of the optic nerve would provide the needed transmission capacity to account for the flow of information to the iris. Thus, with Jensen's assumptions, iridology theory seems to agree with anatomical evidence!

The assumption that the optic nerve mediates the final leg of the neuro-optic response solves the iridologists' theoretical dilemma by creating a double cross, but at the same time the assumption raises serious questions by anyone familiar with the tremendous body of literature published on the visual pathways. The visual system (including the optic nerve) is probably the most intensively studied and best understood neural system in the body. The Nobel Prize recently awarded to Hubel and Wiesel

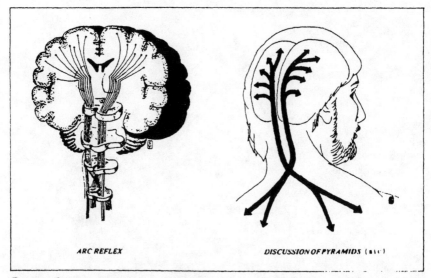

ARC REFLEX   DISCUSSION OF PYRAMIDS (sic)

FIGURE 6. The first crossing in the autonomic system according to Jensen (Iridologists International Manual for Research and Development. Iridologists International, Escondido, Calif., 1981, p. 25)

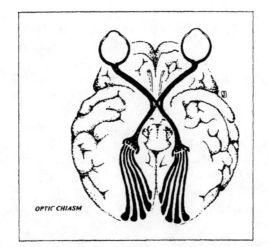

OPTIC CHIASM

FIGURE 7. The second crossing in the optic nerve according to Jensen (*Iridologists International Manual for Research and Development.* Iridologists International, Escondido, Calif., 1981, p. 27)

was the result of many years of work on this intriguing system. All of the accumulated research unequivocally demonstrates that the mammalian optic nerve is primarily an afferent pathway, that is, one in which the signals travel from the eye to the brain. There is no evidence suggesting that any fibers from the optic nerve make connections with the iris. This, combined with the fact that only half of the fibers in the optic nerve cross, makes the proposition that the optic nerve is the final link to the iris untenable (Moses 1975, pp. 367-405).

This double-cross hypothesis is characteristic of the "scientific" evidence presented in Jensen's new text. This volume contains countless misinterpretations of established anatomical and physiological knowledge and includes references to many pseudosciences, such as Kirlian photography and personology (Jensen 1982, pp. 88, 491).

This pseudoscience rises to the level of the ridiculous with the proposition that the iris can communicate information back to remote organs. D. Hall (1981, pp. 210-11) describes the "removal of function" following the surgical removal of a piece of iris tissue in glaucoma or cataract surgery. She says, "As sure as eggs, the iris zones affected will be down in function and maybe in structure too." As D. Stark (1981, p. 677) points out, "This can only be the case if the iris not only reflects but also controls bodily function, which is patently absurd."

Clinical Considerations

Though the gossamer theoretical structure that underlies iridology is apparent to the critical inquirer, proponents and practitioners continue to sell iridology as a "valid" clinical procedure. As Wolf (1979, pp. 7-8) states, iridology as a clinical tool reveals "pathological, structural and functional disturbances in the body." Jensen (1981a, p. 2) adds that "iridology is unique in its ability to make a subclinical evaluation, whereas the medical point of view only recognized lab test verification of dysfunction." Although Jensen mentions "diagnosis" in many of his writings, he qualifies this by stating (1981b, p. 16) that "iridology does not diagnose disease in the sense that Western medicine does, nor does it label combinations of symptoms with disease names." A paper by Fernandiz appearing in the same journal (published by Jensen) is titled "Hemicrania (Migraine) and Its Diagnosis by Means of the Iris"! Such double talk does not cover the reality of the fact that iridology is a clinical technique purported to determine functional states within the body for the purpose of recommending a course of corrective treatment.

Confusion is the first order of business in the clinical application of iridology; for, as Stark (1981, p. 677) notes, there are many iris charts (more than 19) and this "presents the first diagnostic dilemma—which chart to choose." Although most charts are in general agreement on major landmarks, such as the leg area being represented at the six-o'clock position, there are also many differences in both location and interpretation of iris signs (compare Figures 2-5).

Acknowledging the lack of objective clinical evidence, in a recent article Jensen (1981a, p. 1) says, "At the present time, we have no exact way of proving anything other than in a phenomenological manner that what we know and see is true." Thus, in support of the efficacy of iridology, proponents have published endless numbers of anecdotal case reports.

Medicine has long used clinical observation to support claims of observed phenomena even when detailed knowledge of the underlying functional processes are not fully understood. To ensure reliable reproducible results, clinical investigators have adopted strict rules in the form of controlled clinical studies. The controls are chosen to remove, to the extent possible, the inevitable bias of both the patient and the practitioner, to isolate the procedure or medication, and to provide subject groups of adequate size to make statistical comparisons. The controlled clinical study is not, as promoters of iridology suggest, a tool developed by Western medicine to attack unorthodox procedures (Jensen 1981b, p. 20), but in fact it is universally accepted and applied in all areas of scientific inquiry.

The design of a controlled clinical study of iridology would at first appear relatively simple, especially in light of Jensen's (1974, p. 2) comment that "iridology can diagnose a patient for the doctor if he has a perfect colored photograph showing three-dimensional depth of the patient's eye. The patient need not be present." Utilizing photographs in a study effectively isolates the iridologist from the patient, thus limiting the data available to only that obtained from the iris appearance. This eliminates the possibility that direct observation might provide information from general physical appearance and also prevents items of pertinent history from being inadvertently divulged, as is commonly observed in "cold readings" by psychics. The difficulty develops when a standard criterion for the diagnosis of the condition in question is established, as illustrated in the following discussion.

Establishing a standard diagnosis with which to compare the validity of iridology may well be impossible. First, as was quoted above, iridology claims to be able to make a subclinical diagnosis, that is, before symptoms or measurable signs develop. In addition, Jensen (1974, p. 12) adds: "Many times the conditions revealed in the iris today will not be apparent in the body for years to come, but time will inevitably show the analysis to be correct." Thus the proponents of iridology have an excellent but inherently unprovable explanation for the high rate of overdiagnosis (false positives) in controlled studies.

There is also little common diagnostic ground, because many of the conditions detected by practitioners of iridology are "diseases" whose existence has been disputed or discredited by scientific investigation. A common finding is a toxic bowel settlement (which is treated with procedures of questionable value, such as colonic irrigation!); however, the toxic settlement theory of disease was soundly discredited in the early part of this century (Ratclift 1962, p. 52). Thus from a critical perspective it would be difficult to agree on a standard diagnosis where the existence of the disease itself is in dispute.

Though the clinical application of iridology is widespread, the results of only two controlled clinical studies have been published. At the University of Melbourne (Australia), D. Cockburn compared iridology evaluations with known medical histories. The most interesting phase of his study had iridologists evaluate before-and-after iris photographs of subjects who developed an acute disease. He asked the iridologists to determine if a change in the iris had occurred and, if possible, to tell which organ was affected. The only set of photographs determined to have changes was a set taken as a control on the same subject two minutes apart! Cockburn (1981, p. 157) states, "It must be concluded that, at least for the subjects of the prospective trial and for the acute stage of the disease states represented, there were no detectable iris changes of the type depicted in the commonly used iris diagnosis charts."

At the University of California, San Diego, A. Simon, D. Worthen, and J. Mitas (1979) compared the accuracy of iridology based on the reading of color slides for the detection of kidney dysfunction. A blood chemistry test (creatinine level) was used as the standard for assessment of kidney function. Photographs of 143 subjects (48 with kidney disease) were read by three iridologists and three ophthalmologists. The overall record for hitting a correct determination was no better than chance when the number of incorrect and correct determinations were compared. In conclusion the authors state, "Clearly, none of the six observers in this study derived data of clinical importance or significance" (p. 1389).

Jensen, one of the iridologists participating in the San Diego study, has written several critical commentaries on the results. His first criticism was the poor quality of the photographs; however, at the time of the study he did not decline to read them. He also disputes the validity of the creatinine test as an indication of kidney function, though it is widely accepted and routinely used by orthodox practitioners. He asserts that the creatinine test has been around for 10 to 12 years, whereas iridology has been in use for more than 125 years (Jensen 1981c, p. vi). Jensen seems oblivious to the fact that the amount of time a test has been in existence has no relevance to its validity.

As further evidence of the clinical value of iridology, Jensen cites the work of Romashov and Velkelvor of the USSR (*National Enquirer* 1978), who reported a 95-percent accuracy in 1,273 subjects with diagnosed disease. He also states that Deck in Germany has reported a 92-percent efficiency in the detection of kidney disease through iris diagnosis (Jensen 1981c, p. 19). Jensen does not describe the details of these investigations, the nature of the controls, or the standards used for diagnosis. These are important, because one iridologist in the San Diego study also could boast of having correctly identified 88 percent of those with kidney disease. Unfortunately, he reported that 88 percent of the normal subjects included in the study as a control were also suffering from kidney disease (Simon, Worthen, and Mitas 1979, pp. 1387-88). Therefore, without specific details of the design, the use of these studies is of no value when offered in support of iridology.

To enhance the image of iridology by association with an accepted clinical technique, Jensen (1981c, p. i) writes: "The fundus examination, which has been accepted by ophthalmologists, reads the arterial circulation. Similarly, iridology reads the iris stroma . . ." The word "similarly" is loosely applied in this analogy. The fundus examination is a routine medical procedure using a special optical instrument (ophthalmoscope) that provides a view of the interior lining of the eye through the pupil. Body tissues, including blood vessels, can be studied undisturbed, an opportunity not afforded elsewhere on the body. The changes in arterial appearance in the eye represent a local manifestation of a more generalized vascular disease. No specific reflexive communication with remote body organs needs to be postulated to explain this or any of the many other observed phenomena in a traditional fundus examination.

Science or Rhetoric

In his rhetorical war with "Western" medicine Jensen (1980, p. 2) writes: "Iridology is based on scientific observation. It is the kind of science that cannot be related through scientific tests, for it does not provide clinical information." If this non sequitur does not deter all of those inclined to subject iridology to controlled clinical studies, Jensen (1981c, p. vi) adds the ultimate argument used by practitioners of all unorthodox procedures. He writes: "Iridology can only be judged by those who use it properly. Iridology has not been properly used by those who have criticized and say it fails the test." In other words, you have to be "sensitive" to the technique to ensure favorable results!

Even though proponents may have used iridology "properly" since Von Peczely published his theories in 1866, they have failed to publish even one well-documented study to support the validity of any of the information presented on their iris charts. Since efficacy has not been established, the ultimate question faced by practitioners of iridology is one of ethics in their relationship with patients.

Harmless Fad or Health Hazard?

It is clear from a logical, theoretical, and clinical perspective that iridology is a pseudoscience of no clinical value. Unfortunately, the use of iridology by unorthodox practitioners is all too common today, and the unsuspecting and often vulnerable patient in the clinical application of the "science" is the recipient of its presumed benefits. In my private practice and as a member of the faculty of the School of Optometry at the University of California, Berkeley, I have been increasingly alarmed by the growing popularity and acceptance of iridology as a diagnostic tool. This is an area of great concern to everyone in the health professions, because acceptance of this pseudoscience can lead an individual to delay needed treatment when a false-negative diagnosis is made (i.e., when a disease is present but not detected). This would appear inevitable given Jensen's recent advice on the differential diagnosis of appendicitis. He writes: "When trying to distinguish between appendicitis and cecal inflammation, we must carefully examine the area at five o'clock in the right iris. Many cases of cecal inflammation have been incorrectly diagnosed as appendicitis, but the iris reveals the location of the inflammation" (Jensen 1982, p. 235). (Remember Jensen claims not to diagnose!) A delay created while treating "cecal inflammation" could prove fatal if appendicitis is the correct diagnosis. On the other hand, when a false-positive finding is reported (i.e., when a disease is "detected" but not present) to a naive patient, extreme mental anguish can result. In addition, the patient may expend large sums of money on unneeded treatments or (if they are skeptical) on traditional diagnostic tests to confirm the reported nondisease.

The Formula for Success?

It would seem that the false-positive diagnosis of subclinical disease is the underlying key to the popularity and success of iridology. The bulk of diseases reported are vaguely stated conditions in organs, such as an "underactive" pancreas or "chronic weakness" in the lungs. Such vagueness permits clinicians to capitalize on any improvement in the way a patient "feels" as proof that the treatments are doing some good. Under those conditions the cure rate and patient satisfaction in a clinical practice can be very high.

Though the validity of the diagnosis and treatment may rest on false premises, many patients appear to experience a positive change in their health. This is understandable since these programs often include a good diet and moderate exercise, a formula that would do us all some good! However, as the following two cases will illustrate, the false-positive diagnosis can also have a negative impact on the patient.

A well-educated accountant, whom I have seen routinely for eye care, was experiencing lower back pain. He consulted a local chiropractor, and during the course of treatment an iridology workup was recommended. The results indicated, among many other health problems, the presence of cancer. Overwhelmed, the patient spent the day in torment. Unable to consult his family physician, who was out of the office, or his wife, who was at work, he finally sought my advice late in the afternoon. After a lengthy discussion I was able to allay his fears and he began to understand, in a more critical way, the complexities of a medical diagnosis. He wondered how an intelligent person like himself could be caught up in such a deep emotional web over such a diagnosis. This story fortunately had a pleasant ending. However, the outcome could have been much more serious since this patient is also suffering from a heart condition, which was not noted on the iridology evaluation!

Another patient in my office related her recent experience with an herbalist and iridology. Based on her iris photographs, she was given a list of herbs and advice supposedly needed to correct a long list of low-grade chronic conditions. The prescription for this long list was a total of over $200 worth of herbs. Considering herself to be healthy, she was skeptical and decided to save her money. This case also had a happy ending, but it leads me to wonder how many naive patients are investing in questionable treatments based on the results of questionable diagnostics?

A more humorous episode occurred recently when an investigative reporter had an iridology workup. She was told that "a whitish color emanating from the iris shows a lot of acidity and mucus throughout the body and could be from eating a lot of meat, bread and milk products. When told the reporter is a vegetarian, [the iridologist] said the acidity could be a reverse effect from eating too much fruit and vegetables." A classic example of clinical nonsense (Meyer 1982, p. 81).

Delusion or Diagnosis?

It seems that the pseudoscience of iridology has deluded both patient and practitioner alike. The surge in popularity that iridology and its fellow pseudodiagnostic sciences are enjoying is not surprising. Iridology is "amazing," relatively simple to learn, and painless and, most important, it has that mystical attraction on which unorthodox theories and practices have thrived over the centuries.

References

Bamer, D. 1982. *Applied Iridology and Herbology.* Orem, Utah: BiWorld Publishers.

Cockburn, D. 1982. "A Study of Validity of Iris Diagnosis." *Australian Journal of Optometry* (July).

Hall, D. 1981. *Iridology: How the Eyes Reveal Your Health and Personality.* New Canaan: Keats Publishing.

Jensen, B. 1974. *The Science and Practice of Iridology* (1952). Escondido, Calif.: Jensen.

_____. 1980. *Iridology Simplified.* Escondido, Calif.: Iridologists International.

_____. 1981a. "An Eye for the Future." *Iridologists International Manual for Research and Development* 2-11/12.

_____. 1981b. "Reply to Western Medicine's Study of Iridology." *Iridologists International Manual for Research and Development* 2-11/12.

_____. 1981c. "Answer to an Article Appearing in the Journal of the American Medical Association Evaluating Iridology." Insert to *Iridologists International Manual for Research and Development* 2-11/12.

_____. 1982. *Iridology: The Science and Practice in the Healing Arts,* vol. 2. Escondido, Calif.: Jensen.

Kriege, T. 1975. *Fundamentals of Iris Diagnosis.* Trans. A. Priest. London: Fowler Co.

Last, R. J. 1973. *Wolf's Anatomy of the Eye and Orbit,* 6th ed. Philadelphia: W. B. Saunders.

Maxwell, J. 1980. *The Eye/Body Connection.* New York: Warner Books.

Meyer, Norma. 1982. "Do Your Eyes Really Speak?" *Daily Breeze* (Torrence. Calif.), September 5.

Moses, R. A. 1975. *Adler's Physiology of the Eye.* New York: C. V. Mosby.

National Enquirer. 1978. "Do-It-Yourself Eye Test That Can Save Your Life" (May 23).

Ratclift. 1962. "America's Laxative Addiction." *Today's Health* (November).

Simon, A., Worthen, D., and Mitas, J. 1979. "An Evaluation of Iridology." *Journal of the American Medical Association* (September 28).

Stark, D. 1981. "Look into My Eyes." *Medical Journal of Australia* (December 12).

Von Peczely, I. 1866. *Discovery in the Realm of Nature and Art of Healing.* Budapest: Druckeree der Kgl.

Wolf, H. 1979. *Applied Iridology.* San Diego: National Iridology Foundation.

20

Palmistry: Science or Hand-Jive?

Michael Alan Park

He sealeth up the hand of every man, that all men may know his work.
—Job 37:7

Length of days is in her right hand, riches and honour in her left.
—Proverbs 3:16

You pay all your bills promptly.
—From the printout of an electric palm-reader

Along with reading tea-leaves and gazing into crystal balls, palmistry—or "Chiromancy"—is surely one of the forms of divination that comes to mind for most of us when "fortune telling" is mentioned. This occult art has become part of our culture, although for most of us it is a fairly benign form of the occult, and more than likely it is something we rarely think about. Palmistry is, however, an old and well-established method of divination and provides a good example of a claimed paranormal phenomenon for scientific examination.

Some of my own research (Park 1979) has been in the field of dermatoglyphics, which is the study of the patterns of ridges and furrows on the skin of the hands and feet—"fingerprints," although these patterns also appear on the palms, toes, and soles. The data utilized by palmists and by those of us in this scientific field of interest are similar, and in fact one area of dermatoglyphic research involves using finger- and palm-print data to "predict" certain aspects of people's lives. Thus we can examine two similar activities centered around the same general topic—one occult, the other scientific—to see how they compare in theory, method, and results. It is an opportunity not provided by other methods of divination such as those using tea leaves and crystal balls.

In scientifically examining the various "predictive" arts there are two potential approaches or levels from which one can work. The most straightforward is to examine the predictions and revelations to see if they are, in fact, accurate. This method has some obvious drawbacks. For one,

where predictions rather than revelations about personality are involved,
the study would necessarily be a long-range one—the lifetime of the
individual in fact. Secondly, as in any scientific study, certain controls are
needed. Here, for example. one would have to control for "self-fulfilling"
prophecies: the possibility that the subject takes a prediction or revelation
so seriously that, even if subconsciously, he causes it to come true.
Especially with such things as personality characteristics and social
matters, such self-fulfillment would be easy to bring about and hard to
control for in a study. There is also the matter of basing occult insights on
data that are not part of that "science." Scars, calluses, and stains on the
hands, as well as such factors as mode of dress and speech pattern, can be
interpreted à la Sherlock Holmes to "divine" some basic information
about an individual as a basis for "revealing" further information. (A
simple example: short nails and fingertip calluses on a person's left hand
are a pretty good indication of a right-handed stringed-instrument player.
From there, revelations concerning "musical talent," "creativity," and so
on, would be fairly safe.) In palmistry, this can be taken care of to some
extent by using ink-prints of palms rather than living subjects; but even
where use of prints is advocated by the palmists themselves (Gettings 1979,
p. 32) the necessity of also observing the living hand is made clear.

With predictive arts using tea leaves, crystal balls, tarot cards, and so
on, we are usually limited to testing prediction; the relationships claimed
by these arts find no analogues in any accepted scientific body of
knowledge. As noted, however, the palmist and the specialist in
dermatoglyphics are using some of the same data, and much dermatoglyphic research is concerned with the relationships between certain physical
traits of the hands and feet and other aspects of individual biology. We can,
then, test palmistry on the second potential level, that of examining
whether or not there actually are any direct, predictive, specific
relationships between the characteristics used by the palmist and any other
events or factors in an individual's life that may relate to the kinds of
information contained in the palmist's conclusions.

In attempting to define and describe the occult art of palmistry, some
difficulty is encountered. There appear to be a number of different versions
of the art that claim to be the "real" palmistry. The versions differ in the
number of characteristics observed, in the kinds of characteristics
observed, in details concerning the classification, nomenclature, and
nature of those characteristics, and in the meaning attributed to different
expressions of the characteristics. Since it is not my intent to examine
palmistry point by point but rather to discuss it more in terms of its general
theory, I will attempt to generalize from points that all or most of the
versions have in common or to find some sort of "average" statement that
seems to fairly reflect them all. I will, however, include in my examination
all the types of characteristics used in the most complex form of palmistry I
came across, that discussed by Gettings (1979), which uses hand shape and

fingerprints as well as the more typical lines and creases.

There seems to be no consensus as to just how old palmistry is. Recognition of individual uniqueness in fingerprint patterns goes back at least to China in the third century B.C., where a thumb-print impression on a pat of clay may have been used for personal identification (Cummins and Midlo 1961). As to palmistry itself, *The Encyclopedia of Occult Sciences* (1939) mentions ancient India and Egypt. In its modern form it can certainly be dated back to publications of the sixteenth century (de Givry 1958). Basically, palmistry is the use of the lines and creases of the palm to discern hidden information about a person's character and in some cases to foretell future events in the person's life. The lines used are not those familiar patterns of the fingertips (which upon close examination can be seen to extend to the palm), but those creases on the palms and inner surfaces of the fingers that become deeper and more defined when we flex our hands. These are in fact formally referred to as "flexion creases" and are described anatomically as "locations of firmer attachment of the skin to underlying structures" (Cummins and Midlo, p. 37).

These creases are the most important data for the palmist. There are three major lines (called the lines of life, head, and heart) that are almost always present and a number of subsidiary lines that may be absent in some individuals. The characteristics of these lines that are considered by the palmist include: the points at which the line begins and ends; the degree and direction of curvature; length; presence of forks; depth; and presence of a "chained" appearance. The lines are explained in various ways and at various levels of causality—from an astrological "influence of the stars" at birth (*Encyclopedia of Occult Sciences*, p. 150), to the lines as "conductors of energies from one part of the hand to another" (de Givry, p. 198). In any case, they are thought to be an outward manifestation of the inner state of the person possessing them, reflecting aspects of that person's personality characteristics, intellectual abilities, and even physique.

The question of the predictive potential of palmistry—the "fortune-telling" aspect—is not fully agreed upon. More traditional versions, like those discussed in de Givry, for example, accept this capability of the art. The best known example is perhaps the use of the "line of life" to determine an individual's approximate date of death. There are also recognized ways of foretelling such things as number of children, economic success, and other matters of life, and particularly love. More recent versions of palmistry, however, deny that specific events can be foretold. Gettings (p. 26) in fact calls such predictions the palmistry of "charlatans, gypsies, and popular articles in women's magazines." The best palmistry can offer by way of prediction, he says, are indications of "direction and tendencies." At the same time, however, Gettings does admit that on occasion external pieces of information can be arrived at—in one case the name of a subject's boyfriend and the fact that the boyfriend was married. Such information he says is derived "from unknown sources by the emotion" (p. 22), a process he calls "intuitive palmistry." There seems, then, to be some room left for the possibility of a clairvoyant capacity on the part of the palmist being set off by the palmar features. Finally, biologist Lyall Watson, who in *Supernature* (1973) proposes the possibility of a nonparanormal explanation for the connection between internal states and the features of the hands, says that "fortune telling by lines in the hands bears the same relationship to the serious study of [palmistry] as newspaper horoscopes do to true astrology" (p. 192). More on Watson later.

Especially in more current forms of palmistry, other features of the hand in addition to palmar creases are utilized. The form of the hand, specifically the shape of the palm in relation to the length of the fingers, is also considered to be a clue to personality traits, and Gettings links various "types" of hand-shapes with certain professions. The form of the individual fingers is also noted; the length of the little finger (the "finger of Mercury") and its individual phalanges, as well as the presence of any curvature, for example, are believed to give information about a person's honesty and dealings with the opposite sex and with money. Nail shape and color and the appearance of the fleshy parts of the palm (the "mounts") are of importance too. Last, and of special interest to me, is the use of fingerprint patterns. Again using Gettings as an example, it would seem that there is a recognized link within palmistry between various dermatoglyphic pattern types and certain personality traits.

It must be noted that, as practiced by many modern proponents, "real" palmistry is not the simple matter so often depicted in the movies. Several hours are said to be needed to adequately take into account all the features noted above and to evaluate their meanings—both individually and, more important, in relation to one another. Which particular finger a certain print type appears on is important, and that must be viewed in light of the length of that finger, the shape of the palm, and the indications given by all the other features. A "reading" is not a list of individual meanings but an interpretation based on a consideration of the balances and conflicts indicated by all the individual features as parts of the whole hand.

Dermatoglyphics may be generally defined as the study of the patterns of parallel ridges and furrows on the epidermis of the hands and feet. These are essentially the fingerprints so familiar to us in connection with law-enforcement work. However, whereas the police are interested in combinations of fingerprint traits that characterize individuals, the specialist in dermatoglyphics is interested in those traits that can be put into a finite number of categories, some expressions of which are exhibited by all persons. The most important and obvious of these traits is the pattern type. There are three main types: arches, loops, and whorls; there are also subtypes of each of these as well as some other, minor types. Every human finger and toe carries, with few exceptions, one of these types. In addition, there are areas of the palm and sole that also display these patterns. The other important trait used in dermatoglyphics is pattern size.

It is determined by counting the number of ridges between the center of the pattern and the triradius—the point from which the ridges that outline the pattern proper separate. In studying these two sets of characteristics, other traits of the hands and feet, including palmar flexion creases, have been noted, and in some cases have been incorporated into dermatoglyphic research.

Any human trait that shows variation is examined to try to determine to what extent its variation is a result of genetic differences between people and to what extent it may be environmentally explained. It became apparent during such investigations that pattern types and sizes were under some degree of genetic influence, although the specific details of the genetic mechanism have yet to be discerned. At any rate, the demonstration of at least a partial genetic basis led to two uses of dermatoglyphic data in other studies. One, of course, involved the examination of human "racial" differences. Dermatoglyphics were added to the growing list of traits whose expressions and frequencies help to distinguish and define the races. When it was seen that clear-cut boundaries between human races simply did not exist—and therefore that the whole concept of "race" needed to be viewed in a different light—the emphasis of studies of variable traits shifted to the search for explanations of the trait variation itself and to the use of the traits as genetic markers in research on the processes of human population genetics and evolution.

The second application of dermatoglyphic data, and the one most germane to the present topic, is in the field of medical genetics. Correlations have been established between dermatoglyphic features and certain human disorders, most of which have a known or suspected genetic basis. Here then is a scientifically testable link between some of the palmist's data and some other features of human biology, some involving behavioral characteristics.

The genetic mechanisms behind most human features are extremely complex; hence, we often get a clue as to genetic basis only when something goes wrong—when we can trace a specific disorder through family lines or can link one disorder with another or with some other particular physical feature. In this case, a good number of human ailments already established as having, or thought to have, genetic bases were seen to be statistically linked to particular features of the fingers and palms. It is not that specific, rare sorts of prints or lines are absolute "signs" of a particular genetic disease; it is just that samples of victims of certain disorders have unusually high or low frequencies of certain patterns or features when compared to the general population.

Several distinct kinds of disorders have been shown to have correlations with dermatoglyphic "abnormalities." (See, for example, Alter 1966.) A large portion of these ailments involve structural aberrations in the chromosomes—missing pieces, or chromosomes that are attached to part or all of another chromosome (translocations). Others involve too many or too few chromosomes (aneuploidy); this can occur either with the autosomal (nonsex) chromosomes or with the sex chromosomes. A number of disorders known or thought to be the result of mutations of single genes are also listed as having dermatoglyphic correlations. Finally, there are some conditions of uncertain genetic transmission and several disorders of external origin, such as rubella, thalidomide-damage, and cerebral palsy.

The dermatoglyphic "abnormalities" associated with these disorders include unusual frequencies of certain pattern types, unusually high or low ridge-counts for pattern size, unusual frequencies of a number of other dermatoglyphic features, the appearance of only one flexion crease on the inside surface of the fingers, and a particularly interesting feature known as the simian crease or line—where the two transverse flexion creases of the palm (the heart and head lines of the palmist) connect to form a single distinct crease.

As mentioned, the exact genetic mechanism responsible for dermatoglyphic features is as yet unknown. The correlations mentioned above, however, have shed some light on the problem. Other physical symptoms of many of the disorders studied in this context involve developmental abnormalities—aberrations that, in part at least, originate during the fetal period. Since it is known that the dermatoglyphic features develop between the sixth and twenty-first fetal weeks, it appears possible that these distortions in dermatoglyphic features have more of a mechanical explanation than a direct genetic origin. Other studies (Mulvihill and Smith 1969, for example) have suggested that pattern size may be under fairly direct genetic control, while pattern type is more the secondary result of the size and shape of the developing fetal fingertips. At this point one must conclude that the relation between dermatoglyphic features and genes is a complex one that also involves the effects of environmental factors, that is, factors relating to the development of other physical features and processes, factors perhaps of the internal environment of the womb, and maybe factors involving the complex interactions of differing genetic combinations. At any rate, whatever the mechanisms at work, there is a recognized relationship between certain features of the hands and other aspects of human biology that results in specific enough manifestations to allow dermatoglyphics, with limitations, to be used diagnostically for "strengthening diagnostic impressions" and as "screening devices" to select patients for further studies and tests (Alter 1966).

We may now examine some specific correlations to see if they resemble at all those proposed by the palmist. Again, using Gettings as an example of more recent versions of palmistry, we find the prints of the fingertips used as indicators of certain personality characteristics. Only the three basic patterns are used. Arches are said to indicate "crudeness," "practicality," and "rebelliousness." Loops point to "restraint," "lack of originality," and "coolness" of manner. Whorls are indicative of a person who is "creative," "restless," and "egocentric."

In dermatoglyphic research, an excess of arches (as compared with frequencies derived from large population samples) is associated with a number of disorders. These include trisomy 18—a condition where an individual possesses three rather than two of the eighteenth chromosome—a disorder that usually leads to infant death. Kleinfelter's syndrome, where males have two instead of one X sex chromosome (XXY), also is correlated with an excess of arches. Individuals with this syndrome exhibit such symptoms as underdeveloped gonads, sparse body hair, some breast development, unusually long legs, and some mental retardation. Other correlated disorders include certain forms of congenital heart disease, idiopathic mental retardation, epilepsy with retardation, and, according to some investigators, schizophrenia.

An excess of loops is associated with other forms of congenital heart disease and with another trisomy, trisomy 21, also known as Down's syndrome, or (unfortunately) mongoloid idiocy. This familiar disorder displays a number of characteristic physical features as well as mental retardation.

Whorls in excess of normal frequencies have been associated with another anomaly of the sex chromosomes known as Turner's syndrome. With this condition, outwardly physical females possess only one instead of two X chromosomes (XO). Most XO conceptions end in spontaneous abortions; but, when live birth does occur, characteristic symptoms of the disorder include very short stature, broad chest with underdeveloped breasts, webbing appearance of the neck, small uterus, and either no ovaries or those represented only by small "streaks" of tissue. There is no mental retardation. Other disorders correlated with excess whorls include Huntington's chorea, a neurological condition that results eventually in loss of mental faculties; some additional forms of congenital heart disease; and, according to other investigators, schizophrenia.

As I think can be readily seen, only with some semantic effort can any of these recognized correlations be construed as resembling those of the palmist. It is interesting, however, that so many of the disorders noted include among their symptoms some sort of behavioral manifestation.

The set of traits by far the most important in palmistry are the flexion creases of the palms. As noted above, various characteristics of these lines are said to relate information regarding an individual's personality, intellect, physical form, and even future. As might be suspected, interpretive systems differ enormously, but the kinds of information divined always fall into one of those basic categories.

Essentially the only interest the dermatoglyphic specialist has in flexion creases is in their medical relationships. In this regard, there are only two crease characteristics that have been shown to be well correlated with genetically based disorders. One is the presence of a single flexion crease, instead of two, on the inner surface of one or more fingers. This anomaly, so far as I can determine, is not treated by palmists at all. Anatomically it seems to be associated with a lack of mobility of the underlying finger joint. Single digital creases have shown correlations with three disorders that affect the development of an individual in numerous and rather disastrous ways. Two of these are known to have a genetic basis. They are trisomies 18, which is fatal, and 21 (Down's syndrome). The other disorder is known as the oral-facial-digital syndrome. It is suspected by some of being caused by a trisomy of the first chromosome, though this has not been proved. It results in various deformities of the regions of the body that give it its name.

The second important flexion crease feature is the so-called simian line or crease (so named because of its presence in some nonhuman primates). Found in under 2 percent of humans in general, this single line across the palm shows higher frequencies in victims of a number of genetically based disorders. Among these are: trisomies 18 and 21, Turner's syndrome, De Lange syndrome (various anomalies of the hands and feet), Ellis-van Creveld syndrome (dwarfism and polydactyly), psoriasis, Rubenstein-Taybi syndrome (broad thumb and great toe), idiopathic mental retardation, and two disorders of external origin— thalidomide damage and prenatal rubella.

The simian line is also used by palmists (under that name). It is said to indicate "a strong inner tension" (Gettings, p. 117). Depending upon the general personality traits shown by the other palmar and digital features, this tension can display itself in a number of specific ways: creative and artistic, destructive and criminal, or religious.

As with the fingerprint patterns, it would appear that these correlations accepted for the flexion creases and those proposed by the palmist bear little resemblance to one another. Again, it is of interest, however, that so many of the disorders mentioned have symptoms that involve behavioral manifestations.

Finally, a characteristic used in palmistry but not in dermatoglyphics (except with regard to physical deformity related to genetic disorders) is that of hand shape. Again, specific features and interpretations differ among various systems of palmistry and among palmists. The essential idea, however, is the same—that the shape of the palm and fingers is a clue to personality characteristics. This is reminiscent of the physique and temperament correlations espoused by W.H. Sheldon (1942). Sheldon developed a system for quantitatively describing an individual's overall body build that consisted of three components each scored on a scale of 1 to 7. Sheldon further suggested that these genetically based component expressions were linked with expressions of personality or temperament, also described by three components scored 1 to 7. This subject is complex and is really a topic for another paper; suffice it to say that Sheldon's ideas, especially with regard to temperament, are seldom used today, in part because of problems of interobserver uniformity and in part because, as I have suggested elsewhere (Park 1969), the evaluation of types of temperament and of their correlations with body type may be under more influence from culture than from some underlying genetic basis. There is then, at this point, little scientific support for a biological connection between shapes of parts of the body, or the body as a whole, and expressions of personality and temperament characteristics.

Obviously, a more detailed, point-by-point examination could be carried out comparing the findings of medical dermatoglyphics with the claims of palmistry. It seems clear enough to me, however, that the scientifically established connections between palmar and digital features and other aspects of individual biology offer no positive evidence in support of the relationships advocated by the palmists. This, of course, does not preclude the possibility of some sort of connection beyond the bounds of this type of examination or even of our current knowledge. The probability, though, seems remote at this point.

For instance, Lyall Watson, to cite the only "scientific" example of which I am aware, proposes in *Supernature* that an intimate connection between internal physical and mental conditions and the lines of the palm makes sense, since the nervous system, sense organs, and skin are all derived from the same embryonic layer, and since so many diseases and mental states are known to have some manifestations or effects on the condition of a person's skin. The inside and outside of the body, in other words, are in constant contact and interplay throughout life. As evidence for this, with regard to palms, Watson claims that palmar creases "break down" at the moment of death, when signals from the brain, which have maintained them, cease. A check with the Connecticut State Medical Examiner's Office indicated that this claim was unfounded; I was told that the crease lines remain after death. At any rate, although there are certainly numerous intimate connections between parts of the body, I think Watson fails to provide any reason to believe that any of these manifest themselves in specific ways in the lines of the palms. The occult art of palmistry and the scientific study of the meaning and cause of dermatoglyphic features must remain, for now at least, two quite distinct categories of knowledge.

This does not mean, however, that there may not be some connection between the two in historical perspective. It remains to be explained just where the idea for palmistry came from in the first place. There are certainly other parts of the body that could be "read" for purposes of divination. It strikes me as possible that the concept may have originated as a result of observations of medical phenomena like those described above.

That this is at all plausible was made clear to me during my own research on dermatoglyphics. I was studying processes of microevolution among the Hutterian Brethren of Canada, using fingerprints as genetic markers. The Hutterites are extremely knowledgeable with regard to anything related to agriculture, but are fairly ignorant of many things outside that realm; in other words, I had no reason to suspect that they knew anything about dermatoglyphics other than the fact that "the government" took fingerprints. In the course of my work, one Hutterite man came to me and inquired about the "funny" line in his palm; no one else in the colony, he noted, had one. It turned out to be a simian line.* I am fairly certain that his observation of it was not prompted by anything other than curiosity. It would seem perfectly reasonable then that early literate or even preliterate human groups could have noted correlations between unusual palm and finger features and physical or, especially, mental aberrations (observations that would have been much more obvious than that of my Hutterite subject's) and could then have developed broader ideas about the connections between these features and things like mental state and even future events.

To be sure, despite the lack of scientific support, and perhaps especially *because* of the clear distinctions between these two spheres of knowledge, those who "believe" in palmistry will continue to do so. After all, there is a certain comfort in having access to knowledge that is hidden from view or "hiding" in the future. And there is a certain discomfort in the ever-changing, never-absolute world of science. That my palm reveals my innermost being and my future life, that ancient astronauts built the amazing pyramids, that the mind can fix broken watches and locate archaeological sites—all these are intriguing and exciting and somehow comforting in their simplicity and absoluteness. It is an understandable emotion. Thus it is up to those of us involved in research like the above not only to examine (and often debunk) such paranormal phenomena but also to communicate the excitement and intrigue of *our* "occult" ("hidden" or "concealed") knowledge: the fascinating mysteries of black holes in space, of the extinction of the dinosaurs, of invisible particles smaller than atoms, and, yes, even of fingerprints and genes.

References

Alter, Milton. 1966. "Dermatoglyphic Analysis as a Diagnostic Tool." *Medicine 46*, no. 1.

Cummins, Harold, and Charles Midlo. 1961. *Fingerprints, Palms and Soles: An Introduction to Dermatoglyphics*. New York: Dover.

de Givry, Grillot. 1958. *Pictorial Anthology of Witchcraft, Magic, and Alchemy*. Chicago and New York: University Books.

Encyclopedia of Occult Sciences. 1939. New York: Robert M. McBride.

Gettings, Fred. 1979. *Palmistry Made Easy*. No. Hollywood, Calif.: Wilshire

---

*An ethical, methodological note: I chose not to tell him about the correlation between simian lines and genetic disorders. It was obviously not a factor for him, and an explanation complete enough not to cause him undue concern might not have been possible.

Book Co.

Mulvihill, John J., and David W. Smith. 1969. "The Genesis of Dermatoglyphics." *Journal of Pediatrics 75:* 579-89.

Park, Michael. 1969. "Physique and Temperament in Nilotics and Eskimos." Unpublished manuscript.

――――. 1979. "Dermatoglyphics as a Tool for Population Studies: An Example." Unpublished doctoral dissertation, Department of Anthropology, Indiana University, Bloomington.

Sheldon, W. H. 1942. *The Varieties of Temperament: A Psychology of Constitutional Differences.* New York and London: Harper and Brothers.

Watson, Lyall. 1973. *Supernature.* Garden City, N.Y.: Anchor/Doubleday.

Psychic Flim-Flam: An Example

21

Prediction After the Fact: Lessons of the Tamara Rand Hoax

Kendrick Frazier and James Randi

The Tamara Rand psychic prediction hoax has faded from memory, but the lessons the incident provides for the public and the news and entertainment media should not be forgotten. Nor should it be overlooked that CSICOP had a hand in helping to expose the hoax and to get the TV networks to issue retractions. Rand's transgressions in trying to put herself across as a psychic and to capitalize on a tragic national event varied from those of other self-proclaimed "psychics" only in blatancy and degree. For those who have followed the tactics of such persons, the only surprises were the swiftness and boldness with which her plan was carried out and its equally quick collapse and public exposure.

On the morning of April 2, 1981, you will remember—four days after the assassination attempt on President Reagan in Washington—the NBC-TV "Today" show and ABC-TV's "Good Morning America" joined the Cable News Network in broadcasting a tape that was claimed to have been made on January 6, 1981, by Rand, a noted Los Angeles "psychic" who makes a handsome living by advising movie stars whether or not to sign contracts. On that videotape, viewers saw and heard Rand predict that Reagan would be shot in the chest area by a sandy-haired young man from a wealthy family. The assailant, said Rand, would have the initials, "J. H.," first name possibly Jack, last name something like "Humley." She foresaw a "hail of bullets" as well. All this was to take place during the last week of March or the first week of April.

It was an impressive apparent prediction, the kind of thing that can turn a local "psychic" into a world-class celebrity, with the fame and fees that go with that exalted status. Yet to veteran psychic watchers the prediction was too precise. This wasn't a case of after-the-fact reinterpretation of a set of vague and ambiguous statements, the usual stock-in-trade of the breed.

The wire services had already helped broadcast Rand's miracle prediction worldwide, but Paul Simon, an Associated Press reporter in Los Angeles, contacted CSICOP chairman Paul Kurtz in Buffalo. Simon was skeptical about Rand's claims. Kurtz asked whether it was certain the videotape had been prepared before the assassination attempt as claimed and urged that Simon check the time sequence. Simon called Kurtz back several times that day to report on his investigation. He said he had contacted station personnel at WTBS in Atlanta, Georgia, which Rand claimed had broadcast the tape Saturday night, March 28, two days before the predicted event. They denied that such a tape had been aired. He then called KTNV in Las Vegas, where the tape was said to have been prepared, and technicians there told him that they had taped it on Tuesday night, March 31, more than 24 hours *after* the shooting of Reagan.

Simon said that Arthur Lord, bureau chief in Los Angeles for NBC, admitted that he had accepted the word of the producer of the TV tape and of Tamara Rand and that he had been bamboozled. Unfortunately, at the same time, NBC-TV officials in Los Angeles were telling the press that they "stood by every word" of the broadcast. That evening Kurtz sent telegrams on behalf of CSICOP to the "Today" show (NBC) and to "Good Morning America" (ABC) requesting that they retract the "prediction" story since all the evidence pointed toward a hoax.

The next morning, April 3, early viewers of the NBC-TV "Today" show were startled to hear host Jane Pauley admit that the Rand affair "may be a hoax," and ABC-TV briefly mentioned that they were "looking into the matter." Amazingly, even though Simon's story and other wire service and newspaper articles were now labeling the incident a hoax, the ABC-TV "Telenews" program, broadcast that evening on ABC affiliates across the country, told the Tamara Rand story again as if it were true. Kurtz sent another telegram of protest, and there was a subsequent retraction. The following Monday, NBC devoted a substantial segment of the "Today" show to the hoax.

Fortunately, most of the network programs did eventually acknowledge that the affair had been a hoax and that they had fallen for it.

The *coup de grâce* arrived when Dick Maurice, host of the faked TV show, who had maintained the prediction was genuine, finally admitted that it was an outright hoax. In a front-page column in the *Los Angeles Sun* on April 5, Maurice proclaimed:

"I am sorry.

"I have committed a terrible wrong. I have committed the cardinal sin of a columnist. I have perpetrated a hoax on the public and feel very much ashamed.

"My interview with Tamara Rand in which she predicted the assassination attempt on President Ronald Regan is a lie. Ed Quinn's (vice president and general manager of KTNV-TV) statement about the actual taping taking place on March 31 is the truth."

Maurice said he had gone along with the plan as a favor to Rand, a friend whose career as a psychic they both were sure would be advanced by the publicity about such a spectacularly successful "prediction." Maurice resigned all of his columnist, television, and radio-show efforts. (The newspaper publisher didn't accept the resignation, saying he didn't believe in kicking a man who was down.) In late May, in an appearance on Tom Snyder's "Tomorrow Coast to Coast" show (dealing with journalistic malfeasance), Maurice said that as a result of the incident he had "lost everything," including his house, and had for a time seriously contemplated suicide.

Rand, despite it all, gained worldwide publicity in a game in which notoriety is nearly as good as deserved fame. There is little doubt she will bounce back and continue in the psychic-prediction business and will be as strong as ever, now a pitied victim of those dreadful skeptics, her hoax forgiven as an isolated and temporary aberration from *real* "psychic" predicting.

The most important question is how such a thing could happen in the first place. Her claimed prediction was newsworthy, but why was it so readily aired nationally, before any independent checks could be made into its authenticity? A large part of the answer lies in the widespread credulous acceptance by much of the public and the media of the possibility of psychic wonders. Without that gullible mind-set, it is difficult to see how Rand's claim could have failed to raise all sorts of red flags to the gate-keepers of the media. None of it would be possible if much of the press and public didn't believe such claims and predictions were possible.

The radio and TV talk-shows, the supermarket tabloids, and even the legitimate news media have credulously disseminated the claims of legions of psychics so often that it is easy for everyone to believe there are such things as real psychics. Yet all the evidence points to this fact: There are no such people as psychics; there are only people who claim to be psychic. And, we might add, the other necessary ingredient in the equation is an army of people willing to believe in psychic miracles.

The Tamara Rand hoax came shortly before the case of the award of a Pulitzer Prize to a young *Washington Post* reporter for a sensational article that turned out to have been fabricated. That incident unleashed an admirable round of intense soul-searching among responsible reporters and editors. Important journalistic issues were raised and freshly aired, among them the role of editors in striving to ensure the truth and accuracy of what their publications print. The public saw honest journalists wrestling with complex questions of responsibility and ethics.

Unfortunately, psychic claims seem to be treated entirely differently. Certainly the community of "psychics" has no ethical standards to discuss or uphold since their business is based on deception in the first place. TV and radio talk-shows conveniently classify themselves as entertainment rather than news media, thus neatly skirting most ethical responsibilities to strive for the truth. In a revealing statement on the "Tomorrow" show,

Maurice said he had often had as many as three psychics a week appearing on his TV talk-show. Until radio, television, and newspapers openly challenge the claims of "psychics"—or, if not that, decide not to give them a ready forum—the abuses of the truth and the exploitation of the public's interest in, and gullibility over, "psychic" miracles will continue.

Part II
Evaluating Fringe Science

Astrology

22

Scientific Tests of Astrology
Do Not Support Its Claims

Paul Kurtz and Andrew Fraknoi

An estimated 1,200 newspapers in North America carry astrology columns. While many editors chuckle and tell you that no one takes these columns seriously, the evidence does not bear this out. For example, a June 1984 Gallup poll showed that 55% of American teen-agers (ages 13-18) believe that astrology works. Continuous exposure to the ideas of astrology in newspapers contributes to that credulity.

Astrologers assert that astrology has a successful record stretching back 4,000 years and that this record speaks for itself. Yet dozens of scientific tests of astrological columns, charts, and horoscopes clearly contradict this claim.

The present formulation of astrology was largely codified by Ptolemy in the second century A.D. The basic premise is that the position of the heavenly bodies at the time and place of an individual's birth influences or is correlated with his or her personality, physical characteristics, health, profession, and future destiny. Classical astrology regarded the earth as the center of the universe, with the planets, stars, sun, and moon orbiting around it. The heavenly bodies were originally considered divine and possessing "magical" characteristics. Thus Mars, thought to be the color red, represented the god of war and signified courage and aggression. Venus was soft and white and was the goddess of love and beauty.

What does science have to say about astrology? First, modern astronomy has negated its key principle: that the earth is the center of our solar system. We now know that the planets circle the sun, that our solar system is on the outskirts of a galaxy, which itself is only a part of an expanding universe that contains millions of galaxies. Moreover, new planets (Uranus, Neptune, Pluto) have been discovered that were unknown to ancient astrologers. It is interesting that the presumed astrological influences of the planets did not lead astrologers to discover them long before astronomers did.

Second, we now know that a person's personality and physical characteristics are determined by his or her genetic endowment inherited from both parents and by later environmental influences. Several decades of planetary exploration have confirmed that there is no appreciable physical influence on the earth from planetary bodies. Indeed, the obstetrician hovering over the infant during delivery exerts a much greater gravitational pull than the nearest planet.

Third, there have been exhaustive tests of astrological claims to see if they have any validity. Astrologers predict that individuals born under certain signs are more likely to be personality types that become politicians or scientists. Thus you would expect the birth dates of these two groups to cluster in those signs. John McGervey, a physicist at Case Western Reserve University, looked up the birth dates of 16,634 scientists listed in *American Men of Science* and 6,475 politicians listed in *Who's Who in American Politics* and found the distributions of these signs were as random as for the public at large.

Are some signs relatively more compatible or incompatible with each other, as astrologers maintain? Professor Bernard Silverman, a psychologist at Michigan State, obtained the records of 2,978 couples who married and 478 couples who divorced in Michigan in 1967 and 1968. He found no correlation with astrologers' predictions. Those born under "compatible" signs married—and divorced—just as often as those born under "incompatible" signs.

In order to look for trends favoring astrological signs ruled by Mars (courage and aggression) as opposed to signs ruled by Venus (love and beauty), James Barth and James Bennett at George Washington University examined the horoscopes of men who re-enlisted in the Marine Corps between 1962 and 1970. No such correlation was found.

What about the often-heard claim of famous astrologers that they have made countless correct predictions over the years? Astronomers Roger Culver and Philip Ianna examined 3,011 specific predictions by well-known astrologers and astrological organizations. The results indicated that only 10% of these predictions were realized. The public reads the predictions in newspapers and magazines; the fact that 90% of these predictions never come true is not publicized.

Newspaper charts and horoscopes deal primarily with the sun signs rather than with other so-called planetary influences. Even astrologers admit that the sun-sign astrology featured in newspaper columns has little reliable basis for prediction of the day's events. Incidentally, very few astrology columns agree on what is supposed to occur.

Why then do so many people believe that astrology works? Careful inspection of astrological predictions in a typical newspaper column shows that the statements are so general and vague that they can apply to anyone.

The results of one experiment show why these statements sometimes seem to work. C. R. Snyder, a psychologist at the University of Kansas, and his colleagues drew up a personality description that incorporated the characteristics they found most people believed they possessed. They showed this description to three groups of people, each of whom was asked to rate, on a scale of 1 to 5, how well they were described by it. The individuals in the first group were told it was a universal personality sketch, and the average rating was 3.2. Individuals in the second group were asked for the month in which they were born and were then told the statement was a horoscope for their signs. On the average, they rated it 3.76. The individuals in the third group were asked for the day on which he or she was born and were told that the description was his or her *personal* horoscope. This group rated the same description an average of 4.38. Apparently those who want to believe will do so!

We respectfully ask that newspapers let their readers know that astrology columns should be read only for their entertainment value and that they have no reliable basis in scientific fact.

23

Alternative Explanations in Science: The Extroversion-Introversion Astrological Effect

Ivan W. Kelly and Don H. Saklofske

Recent years have shown a significant growth of interest in purportedly paranormal phenomena. It often appears that many individuals will believe almost anything that wears the mantle of science. Too many of us forget that scientific truth is acquired by the rigorous testing of ideas and by the elimination of alternative explanations. It is premature to entertain a paranormal hypothesis unless all plausible "normal" alternative hypotheses have been ruled out. Presentations of ideas to the scientific marketplace where they can be confronted by informed critics should precede general publication. In the paranormal field this does not always occur, and so the public is often misled by a presentation of disputable data in support of paranormal claims.

An example of the importance of ruling out alternative explanations can be found in recent psychological literature. Jeff Mayo, a prominent astrologer, approached Hans Eysenck, the renowned British personality theorist, to collaborate with him on a test of astrology. Astrologers contend that there is a relationship between the personality of human beings and configurations of the constellations and planets. Most prominent in astrological theory is the zodiac, or sun sign, of an individual. This refers to the time of year during which an individual is born. There are twelve zodiac signs, and they can be classified in various ways. As well as its individual characteristics, each sign is said to be either positive or negative: positive signs describe a tendency to be extroverted; negative signs, a tendency to be introverted. Mayo and Eysenck were to examine the popular astrological claim that individuals possessing an extroverted personality are born more often than would be expected by chance under the odd-numbered or positive signs (Aries, Aquarius, Gemini, Leo, Libra, and Sagittarius). Introverted individuals, by this claim, tend to be born more often than would be expected by chance under the even-numbered or negative signs (Cancer, Capricorn, Pisces, Scorpio, Taurus, and Virgo). If there was some validity to this assertion, it would be reflected in the scores of personality tests. The Eysenck Personality Inventory (EPI) was chosen as the vehicle to examine this claim.

A total of 2,324 adults (917 males and 1,407 females) completed the test. The results of the study provided prima facie support for both hypotheses. All the positive zodiac sign groups had higher than average extroversion scores and all the negative zodiac sign groups had higher than average introversion scores. Although the differences between the groups were marginal, they were statistically significant (that is, they could not be accounted for by chance).

Even before the study was published, astrologers had heard about the results and trumpeted their triumph to the media. Sydney Omarr, the noted astrologer, boasted about the results in early 1977 in his daily column, which appeared in hundreds of newspapers across North America. The Canadian magazine *MacLean's* also gave prominent coverage to the findings. Eysenck was featured on the NBC television special "In Search of . . . Astrology," where these results were further disseminated to the public.

The study was subsequently published in the *Journal of Social Psychology* under the title "An Empirical Study of the Relation Between Astrological Factors and Personality" (Mayo, White, and Eysenck 1978). The article created considerable controversy in psychological literature. There followed in rapid succession two studies that supported the findings (Smithers and Cooper 1978; Jackson 1979) and then two failures to replicate (Veno and Pamment 1979; Jackson and Fiebert 1980). More recently, Saklofske, Kelly, and McKerracher (1981) examined the responses of 241 New Zealand university students (average age 20.9 years) to the Eysenck Personality Questionnaire (EPQ) in relation to odd and even zodiac signs.[1] There were 165 females and 76 males in the sample. Regarding the main hypothesis examined, there were found to be no significant differences in extroversion between odd and even zodiacal signs for the total sample or for males and females considered separately. Although females scored higher than males on the neuroticism scale irrespective of zodiac sign, once again the scores showed no significant differences between odd and even signs for either sex or for the total sample. On the psychoticism scale, males scored higher than females. However, when psychoticism relative to odd and even zodiac signs was considered there were no significant differences for males, females, or the total sample. In order to ensure that subjects were responding truthfully to the questionnaire items, lie scale items were utilized. There were no sex differences on this scale in response to the items; and when odd and even zodiac signs were considered, again there were no differences noted for females, males, or the total sample. [2]

There were some other puzzles. As Kelly (1979-80) asked, Why did other studies that had examined possible zodiac sign and personality interactions using other personality tests not give significant findings on scales (such as sociability) that are related to the introversion-extroversion dimension of the EPI? This question was soon answered.

Although the results of the Mayo-White-Eysenck study were construed as supporting the astrological hypotheses, both psychology and chronobiology could produce alternative hypotheses for the data. Kelly (1979, 1979-80) pointed out that an obvious alternative explanation would be a seasonal effect operating on a critical perinatal development stage or phase. The chronobiologist Hans Wendt (1978) suggested a related explanation in terms of an imprinting or learning phenomenon that may be contingent upon certain triphasic biological rhythms.[3] Another possibility would be that very early critical stages (nearer conception, perhaps) are also susceptible to some sort of electromagnetic field vector.

Finally, one could consider the systematic self-attributions people make because of their personal notions of "astrology." Individuals aware of their zodiac sign may answer test questions in light of what they know about the characteristics associated with their sign.[4,5] Pawlik and Buse (1979) pursued the latter hypothesis and used controls for their subjects' own beliefs of what their zodiac signs were supposed to mean. They found that the so-called astrological effect in their Hamburg students could be statistically ascribed to their self-attributions. Subsequent research by Eysenck and Nias (Nias 1979, Eysenck 1979) replicated the findings of Pawlik and Buse. They concurred that the entire astrological effect was due to the subjects' expectations and familiarity with the characteristics associated with their zodiac signs. Eysenck and Nias tested over 1,000 children, who showed no sign of the effect. In a second study, adults (Salvation Army students) were selected specifically to exclude any with knowledge of astrology, and they too showed no sign of the effect. The effect appeared in the Mayo-White-Eysenck research because even the group classified as lacking knowledge of astrology probably knew about zodiac signs (and this, after all, is all that is necessary). In fact, in the Mayo-White-Eysenck study subjects had been classified on the basis of their answers to the question: "Do you know how to interpret an astrological chart?" Such a chart involves a relatively advanced level of knowledge.

It is also significant that, of the several studies relating EPI scores to zodiac signs, the one most perfectly in accord with astrological claims (Mayo-White-Eysenck) contained the highest proportion of people inclined toward astrology.

The Mayo-White-Eysenck "astrological effect" is an instructive example that underscores the need to eliminate all "normal" alternative hypotheses before embracing a paranormal one. In this case, what at first appeared to be supportive of a paranormal hypothesis was, with subsequent research, destined to have a nonastrological explanation after all and, indeed, to count *against* the astrological hypothesis.

Update

Since this article was first published, Eysenck and Nias (*Astrology: Science or Superstition?* London, England: Maurice Temple Smith, 1982) described their research in more detail and also found that children (who are less likely than adults to know about astrology) did not even begin to match the results of the Mayo study. Some other studies that were subsequently published and gave *no support* to the original study are: James Russell and Graham F. Wagstaff, "Extroversion, neuroticism, and time of birth," *Bulletin of the British Psychological Society*, 22:27-31, 1983; and Don Saklofske, Ivan Kelly, and D. W. McKerracher, "An empirical study of personality and astrological factors," *Journal of Psychology*, 11:275-280, 1982.

In addition, further discussion of the importance of self-attribution in astrological research can be found in Geoffrey Dean, Ivan Kelly, James Rotton, and Don Saklofske, "The Guardian astrology study: a critique and re-analysis," *Skeptical Inquirer*, IX:327-338, 1985.

References

Dean, M. 1980. *The Astrology Game.* Don Mills, Ontario: Nelson Foster & Scott.
Delaney, J., and H. Woodyard 1974. "Effects of Reading an Astrological Description on Responding to a Personality Inventory." *Psychological Reports* 34:1214.
Eysenck, H. 1979. "Astrology—Science or Superstition?" *Encounter*, December, pp. 85-90.
Jackson, M. P. 1979. "Extroversion, Neuroticism, and Date of Birth: A Southern Hemisphere Study." *Journal of Psychology* 101:197-98.
Jackson, M., and M. S. Fiebert 1980. "Introversion-Extroversion and Astrology." *Journal of Psychology* 105:155-56.
Kelly, I. W. 1979. "Astrology and Science: A Critical Examination." *Psychological Reports* 44:1231-40.
———1979-80. "Studies of Astrology and Personality." *Psychology* 16:25-32.
Mayo, J., O. White, and H. J. Eysenck 1978. "An Empirical Study of the Relation Between Astrological Factors and Personality." *Journal of Social Psychology* 105:229-36.
Nias, D. K. B. 1979. Personal communication, December.
Pawlik, K., and L. Buse 1979. "Selbst-Attribvierung als differentiell-psychologische Moderator variable: Nachprufung und Erklarung von Eysencks Astrolgi-Personlichkeit-Korrelationen." *Zeitschrift fur Sozialpsychologie* 10:54-69.
Saklofske, D. H., I. W. Kelly, and D. W. McKerracher 1981. "Astrology and Personality: Yet Another Failure to Replicate." Unpublished manuscript, University of Saskatchewan, Saskatoon.
Smithers, A. G., and H. J. Cooper 1978. "Personality and Season of Birth." *Journal of Social Psychology* 105:237-41.
Veno, A., and P. Pamment 1979. "Astrological Factors and Personality: A Southern Hemisphere Replication." *Journal of Psychology* 101:73-77.

Wendt, H. 1978. "Season of Birth, Introversion, and Astrology: A Chronological Alternative." *Journal of Social Psychology* 105:243-47.

Notes

1. The Eysenck Personality Questionnaire (EPQ) is the latest version of the personality measures developed by the Eysencks. It includes the scales of the EPI and an additional scale labeled "psychoticism" or "tough-mindedness."

2. One astrologer, Malcolm Dean (1980, pp. 260, 263) has gone so far as to state that the "world-views" of the experimenters are responsible for the negative results in most zodiac sign-personality studies. This would seem to be taking ad hockery and special pleading to unparalleled heights. It should be mentioned here that the data on which this study was based were not gathered with the testing of astrology in mind. The data were collected to examine achievement in New Zealand students.

3. Amazingly, Dean (1980, p. 260) argued that Wendt's hypothesis was, in fact, congruent with astrological claims! He stated:

> [Wendt] suggested that the Mayo/Eysenck results were simply due to biological rhythms which many people experience with three crests each year. Other biological rhythms have been discovered with this pattern, he noted: Circulatory functions tend to peak in March, July, and November ... But, his paper provided astrologers with a curious correlation with signs which they had overlooked. Blood is a fluid, and the heart is related in astrological tradition to the Sun (hence circulation). And by a strange "coincidence" in March, July, and November, the Sun moves from a water sign to a Fire sign!

But surely Wendt's hypothesis shows the astrological juggling to be superfluous. This is like someone arguing that a knock on your door proves the presence of a spirit. When you check and find a person there our spiritist argues that there is a person and a spirit knocking at the same time. But our initial reasoning, namely, that the person was responsible for the knocking, is more than adequate. If we have a hypothesis that is consistent with psychological theory why entertain an exotic one?

4. Mayo-White-Eysenck were not completely unaware of this alternative possibility. In their paper they wrote:

> It is clear that the results are in good agreement with the hypotheses outlined at the beginning of this paper: Is there any alternative explanation to that suggested by astrology? The only alternative which occurred to us is that some of the Ss might have known about the connection between personality and zodiacal birth sign traditionally posited by astrologers, and that consequently they might have altered their question-naire responses in agreement. There are two reasons for disbelieving this hypothesis. In the first instance, the relation between the water signs and emotion is much more widely known than that between extroversion-introversion and the odd-even num-bered birth signs; yet the former hypothesis was much less strongly supported. Our alternative hypothesis would have led us to expect the opposite. In the second place, about one-third of the Ss had some knowledge of astrological principles, two-thirds did not. An analysis of the scores obtained from these two groups did not show any significant differences, suggesting that knowledge of astrological principles was not a causal factor. We cannot in the nature of things rule this alternative hypothesis out completely, but it does not seem to us to account for the facts. It is a weakness of the study that only Ss are included who requested astrological predictions from the senior author—i.e., who were not a random sample of the population. It is difficult to see how any selection along these lines could have produced the results obtained,

however. and we do not believe that this constitutes a serious weakness of the experiment. (p. 234)

The later research by Eysenck and Nias showed it to be the best hypothsesis after all.

5. There was a precedent in the psychological literature for this possiblity. Delaney and Woodyard (1974) found that a prior reading of horoscopes influenced the self-concept of individuals. as measured by a personality test. even when the subjects were told that "the [personality] questionnaire was 'concerned with Your personality *not* that personality which was astrologically predicted'" (p. 1274). They did not determine. however. how enduring the influence of the horoscope reading was.

24

The Alleged Lunar Effect

George O. Abell

The Lunar Effect: Biological Tides and Human Emotions. By Arnold L. Lieber, M.D. Anchor Press/Doubleday, New York, 1978. 168 pp. $7.95.

The cycle of phases of the moon, and full moon in particular, are associated with a plethora of folklore, ranging from lycanthropy (werewolfism) to the frequency of human births. Certainly there are real effects of the moon—tides and eclipses, for example—and it is entirely possible that some lunar influences exist that have not yet been recognized by the scientific community.

One person who believes this to be the case is Arnold Lieber, a Miami psychiatrist who has recently described his ideas in *The Lunar Effect,* prepared in collaboration with writer Jerome Agel, who is listed as "producer."

Lieber, noting what he regards to be especially anomalous activity of many mental patients during the time of full moon, theorizes that the moon's gravitational pull produces "biological tides" in man and other organisms. The human body is 80 percent water, he reminds us, so that erratic behavior may result from tidal action on bodily fluids analogous to the action of the moon on the oceans of the earth. If so, he argues, the effect should be most pronounced at the time of full and new moon, when ocean tides are highest.

To test his theory, Lieber and an associate, Carolyn Sherin, investigated the incidence of violent crime. In a sample of 1,892 homicides over a 15-year period in Dade County (Miami), they found a peak in the number occurring on the day of full moon and another peak two days after new moon. In a second sample of 2,008 homicides over a 13-year period in Cuyahoga County (Cleveland) they found one peak three days after full moon and another two days after new moon. Lieber suspected that the time displacement between the peaks in the two samples might be due to a phase lag associated with latitude.

Other investigators, in particular Alex Pokorny, professor of psychiatry at Baylor College of Medicine, have been unable to replicate Lieber's finding of a peak in the number of homicides at full or new moon in investigations of several independent samples. Lieber argues, however, that the studies of Pokorny and others consider only the time of death, while his own statistics are based on the time of initial injury. To verify this point, Lieber reexamined the homicides in Dade and Cuyahoga counties but used time of death rather than assault, and he combined these data with a much larger sample of 10,000 homicides in New York City, which listed only death times. He found these death times to have no correlation with lunar phase.

Encouraged by his findings, in late summer 1973 Lieber predicted that "an increase in accidents of all kinds" and "an increase in the number of homicides" would occur in January and February of 1974, when the "Earth, the Moon, and the Sun would be in a straight line . . . with the Moon at perigee unusually close to Earth" (pages 41-42). He reports, "The murder toll for the first three weeks of the new year was three times higher than for all of January 1973," and he describes several violent events over the world that occurred during "this time of cosmic coincidence." He points out, further, that deaths in Dade County, psychiatric emergency-room visits, and admissions to the psychiatric institute at Jackson Memorial Hospital were all unusually high during the first three months of 1974.

Needless to say, astrologers are delighted with the appearance of Lieber's book, not because it confirms predictions of traditional astrology, but because it argues for cosmic influences, which, at least to astrologers, makes their doctrine seem more credible. And on the face of it, Lieber's mass of evidence and statistics seems very impressive indeed. It is, therefore, worth a second look.

Lieber believes the key to lunar influences is found in the tides the moon produces on biological organisms. Tides are gravitational phenomena, and gravitation is an attraction between all material bodies. The attraction is strongest for bodies close together, but drops off rapidly the greater their separation. Now the reason the moon produces tides is that, as astronomical bodies go, it is relatively close to the earth, so that its pull on the side of the earth nearest it is slightly stronger than on the side of the earth farthest from it. The result is that the moon tries to distort the earth, stretching it into a "football" shape, with the long dimension pointing toward the moon. The solid earth actually does distort in this way, but only by about 20 centimeters. (At the same time, the earth similarly distorts the moon.)

This small tidal distortion of the rigid earth is not great enough to bring the forces on it into equilibrium; consequently, the water on the surface tries to flow in such a way as to pile up both under the moon and on the side of the earth far from the moon. The earth is rotating with respect to the moon, however, so that the directions of the tide-raising forces on the oceans reverse twice each day—far too rapidly for the water to have time to flow significantly over the surface of the earth. But these forces, switching back and forth as they do, set the oceans sloshing back and forth in their basins, so that at most places along the shores the tides do "come in and go out" twice each day.

The heights of the tides produced by the moon have nothing to do with its phase (whether it is new, full, or whatever), but they do depend on the moon's distance. Its orbit is eccentric so its distance varies, and once each month (actually 27.55 days) it is 10 percent closer to us than it is two weeks later. At those monthly times of the moon's nearest approach (perigee) the tides produced are more than 30 percent higher than when the moon is at its farthest (apogee).

The sun also produces tides, but the sun is 400 times as far away as the moon is, so even though its total gravitational pull on the earth is more than 100 times that of the moon, the *difference* of its pull on different sides of the earth—that is, its tidal force—has less than half the effect of the moon. Still, when the sun, earth, and moon are in a line, as they are every two weeks (at either full or new moon), their tidal forces combine and produce higher than average tides (spring tides), while when they are at right angles to each other in the sky (first or third quarter) their tides partly cancel each other (neap tides). Thus it is not that the moon's gravitational pull is stronger at new and full moon (as Lieber states), but that its tidal force joins that of the sun. This is why Lieber attaches significance to the full and new moon, although the difference between spring and neap tides is not dramatic; in fact, it is comparable to the effect of the moon's changing distance from the earth.

Even so, the effect of the moon on the oceans is important only because the tidal force acts over the 8,000-mile diameter of the earth. What effect has the moon on a man or woman? The earth's pull on a person is called his *weight*. On the average the moon's gravitational pull amounts to only three parts in a million of one's weight; for a 200-pound man, this is about 0.01 ounce. But the moon's pull on you isn't even relevant, because it also pulls on the earth. What might matter is the difference between your weight in the presence of the moon's gravitational effect and what it would be if there were no moon. At the most, that difference amounts to only 0.01 gram, or about 0.0003 ounces, less than the effect of a mosquito on your shoulder.

But even the moon's effect on your weight is not what *really* counts. If there is anything to the idea of biological tides, it is due to the difference of the moon's pull on different parts of your body, which is pretty small compared to the 8,000-mile diameter of the earth. It works out that the effect on your blood (or any fluid in your body whose flow or circulation could be distorted by the moon) is about one part in 3×10^{13} (or 30 trillion) of the weight of that fluid. A magazine in your hand exerts a tidal force on your body that is tens of thousands of times stronger than that of the moon.

In short, we would not expect tides by the moon to have a significant effect on human behavior, and would, I should think, be very surprised to find a lunar influence of the sort Lieber proposes. On the other hand, if he presents convincing evidence for a lunar effect, we should certainly take a strong interest, for such an effect, if real, could signal new science. So how convincing is Lieber's case?

First, I find it strange that Lieber implies (p. 16) that the effect is obvious and that nearly everyone is aware that, for example, people in mental institutions act strangely at full moon, but then (p. 18) says that it is a small effect that requires large samples to reveal. Still, what about those large samples?

Lieber does not give the actual numbers of homicides in his Dade County and Cuyahoga County samples, either by calendar date or by date in the lunar cycle; he only shows a graph of the data. Therefore, I looked up his and Sherin's original paper (*American Journal of Psychiatry,* 129:69-74, 1972) and read the numbers off their plots as well as I could. A simple X^2 test showed that the distribution of Dade County homicides through the lunar cycle did not differ from what one would expect by chance at least 7 percent of the time, and that the actual day-to-day fluctuations are quite in line with chance. Although there was a peak at full moon, I would judge the peak as typical of random noise. And the second peak in the lunar cycle was not at new moon, as predicted, but two days later.

Now, Lieber (in his and Sherin's paper) points out that if one asks the probability of obtaining peaks within one (or two) days of the times of the observed peaks, that probability is low. But this is a fallacious way to test a matter statistically. Every random distribution has random fluctuations. One cannot simply pick out the high noise peaks and ask what is the probability of finding them exactly where they occurred; it is already *known* that there are peaks at those places. It is like drawing a card from a deck, noting that it is four of hearts, arguing that the chance was only one in 52 of drawing that particular card; although true, it is exactly the same as the probability of drawing any other card named in advance, and of course does not indicate anything remarkable about the deck or drawing. Lieber could have made the case seem all the more remarkable by noting that there were 1,892 homicides in Dade County during the 480 million seconds (15 years) of the period sampled, and that the chance of a homicide occurring during the particular second that, say, homicide number 87 occurred is only 1,892 divided by 480 million, or about one in 200,000. Of course each homicide had to occur during *some* second. One cannot test the chance occurrence of an event already known to have occurred; what he is doing is calculating the probability of the identical thing happening again in a given random trial. In short, I find the Dade homicide data typical of noise, and unconvincing.

I am also not convinced by Lieber's explanation of why Pokorny failed to replicate his results. According to Pokorny, 85 percent of homicide victims die within one hour of injury, but Lieber contends that the 15 percent who die later would destroy the subtle correlation he finds. On the other hand in the New York City sample of 10,000 homicides one would expect that surely the 85 percent who die at once would show some correlation with full or new moon if the lunar effect were real.

How about the Cuyahoga sample? There, too, the day-to-day fluctuations of numbers of homicides within the lunar cycle are within what is expected by chance. The actual distribution of events is somewhat unusual; it differs from what is expected by chance more often than occurs only 3 percent of the time. Still, that is not overly surprising. And more to the point, the Cuyahoga sample has a very different distribution from that of Dade County. There are three peaks in the Cuyahoga County sample, the second highest being near third quarter moon, and the third highest is the one two days after new moon. Moreover, the peak near full moon lagged three days behind the peak in the Dade County sample, while the one near new moon had no lag at all. Lieber seems to be changing the rules for what constitutes agreement.

Now every year tens of thousands of reports of experiments and observations are published. Some are sound, but many are based on poor experimental techniques, on improper controls, on biased or selected data, and sometimes they are even fudged. It is not an uncommon practice for those with novel theories to comb the scientific literature for results that seem to support their theses. There is an old adage, "If ye seek hard enough, ye shall probably find." Among the many thousands of accounts to choose from, it is indeed not rare to find two or three that will serve a cause.

Lieber found what he needed in the studies of the metabolic activity of hamsters by Northwestern University biologist Frank Brown. Brown is himself somewhat of a maverick. He has been studying rhythms of various kinds in living organisms. Brown holds that these rhythms are externally stimulated—say by the sun or moon—whereas most of his colleagues are of the opinion that, while natural selection may have favored evolution of rhythms that are in step with environmental cycles, the actual timing is controlled by internal clocks in the organisms. An example is the familiar jet lag that most of us feel after an air trip from Europe to the United States; for several days our internal clocks keep waking us in the wee hours until we manage to get those clocks reset to our workaday cycle.

The activity Brown measured in his hamsters roughly matched, in lunar-phase cycle, the Cuyahoga County fluctuations in homicide rate. I measured the correlation between the two and found the coefficient to be a significant 0.5; but remember that the hamster study was selected because it did resemble the Cuyahoga murder data. It hardly provides convincing proof of the totally unexpected theory of biological tides.

Lieber's attempt to make a case out of violent activity in Miami in early 1974 involves a misunderstanding of elementary astronomy. When new or full moon occurs while the moon is also at its nearest (perigee), the spring tides are especially high. The moon was indeed near perigee at the full moons occuring on January 8, February 6, and March 8, 1974, and the corresponding spring tides were large ones. However, these high tides occur only for a few days; the moon was actually at its farthest (apogee) during the spring tides corresponding to new moon on January 23, February 22, and March 23, so those tides were below average for spring tides. During a 27.55-day period the moon goes through all of its possible distances.

Lieber reports (pp. 42-43) that "all hell broke loose, especially during the first two weeks of January, with the moon rocking only 217,000 miles from Earth." (Actually, during that period the moon's distance from the center of earth varied from 221,564 miles to 246,085 miles.) Lieber then describes nine murders that occurred in that period in Miami—but without giving the dates, and he mentions some isolated acts of violence elsewhere in the world.

To support this theory, Lieber would have to demonstrate that violence in all parts of the world was significantly increased, not just during a two- or three-week period but within a day or two of those full moons when the moon was near perigee. Meanwhile, violence should have been lower than typical during the intervening new moons, when the moon was at apogee.

In Chapters 3 through 8 of *The Lunar Effect* Lieber describes a host of phenomena allegedly or possibly associated with the moon, including Harry Rounds's study of blood factors in cockroaches (blood in cockroaches?), the human menstrual cycle, birthrates, geophysical effects, "wolfish" tendencies in man, and even disappearances in the Bermuda Triangle.

I am highly skeptical of even the human menstrual cycle having anything to do with the moon, despite widespread opinion to the contrary (and on page 52 of *The Lunar Effect*). The moon's cycle of phases is 29.53 days, while the human female menstrual cycle averages 28 days (although it varies among women and from time to time with individual women); this is hardly even a good coincidence! The corresponding estrous cycles of some other mammals are 28 days for opossums, 11 days for cows and mares, 24 to 26 days for macaque monkeys, 37 days for chimpanzees, and only 5 days for rats and mice. One could argue, I suppose, that the human female, being more intelligent and perhaps more aware of her environment, adapted to a cycle close to that of the moon, while lower animals did not. But then the 28-day period for the opossum must be a coincidence, and if it is a coincidence for opossums, why not for humans?

Lieber accuses science of having prejudice against a lunar effect. I cannot imagine why such a prejudice should exist; scientists I know would be overjoyed to find a significant new correlation and to achieve some professional acclaim for their discovery. I think most of us are quite open-minded about the possibility of undiscovered lunar influences—especially psychological ones. But scientists do tend to be prejudiced against faulty statistical treatments, teleological reasoning, biased selection of data, and the adoption of farfetched arguments at the expense of far simpler ones to explain phenomena.

My impression is that Arnold Lieber is sincere in his attempt to investigate what seem to him to be interesting effects, and I do not regard *The Lunar Effect* as pseudoscience in the usual sense. But it is certainly very bad science.

25

The Moon Is Acquitted
of Murder in Cleveland

N. Sanduleak

In his book *The Lunar Effect* (see the review by Abell in Chapter 25), Miami psychiatrist Arnold Lieber (1978) proposes that the moon is able to adversely affect the mental and emotional stability of humans by means of raising physiologically disruptive "biological tides" in our bodies akin to the tides it raises in the earth's oceans. Because the human body is substantially composed of water, this analogy (which smacks of sympathetic magic) can seem highly plausible to the general public, which is unaware of the infinitesimal tidal action the moon has on an object as small as a human.

The "biological tides" theory was invoked by Lieber to explain the results of a study by Lieber and Sherin (1972) that purportedly showed that the incidence of successful homicidal assaults in two localities (Miami and Cleveland) was significantly higher at or near the times of *both* the new and full moon, when the combined lunar and solar action is near a maximum. This study has often been cited in the media as providing scientific proof of what, we are told, every policeman, bartender, and emergency-room attendant knows to be true—that people tend to act crazy and become more violent when the moon is full. Note, however, that this popular notion makes no mention of a similar rampage at the time of new moon, which would be required by the proposed tidal mechanism.

Other studies, e.g., Pokorny (1964), Pokorny and Jachimczyk (1974), and Lester (1979), found no correlation between frequency of homicides and lunar phase. Lieber has challenged these studies as being improperly conducted, in that they used the conventionally recorded time of death of the victim rather than the actual time of the assault per se, which presumably would be more closely synchronized with the malevolent lunar influence. As Abell noted, this should hardly matter if there is a strong correlation with lunar phase, since Pokorny found that 85 percent of homicide victims die within one hour after being attacked. Lieber also claims that a comparison of the Miami and Cleveland results indicates that the exact timing of peak homicidal periods is a function of geographical location.

Thus, according to Lieber's criteria, the only valid method for replicating his results would involve another study in Miami or Cleveland also using the time of assault. I have made such a replication study for Cuyahoga County (greater Cleveland). The results are the subject of this report.

The data were again graciously provided by the statistical department of the Cuyahoga County Coroner's Office. The Lieber-Sherin sample involved 2,008 homicides during the 13-year period from 1958 to 1970. The present study begins where they ended and includes 136 lunar synodic periods between January 1, 1971, and December 31, 1981. Our much larger sample (3,370 homicides) over a two-year-shorter interval sadly reflects the increased level of violence during more recent times. As was done in the earlier study, only those cases (96 percent) were included where the date if not the hour of the assault was well-established. The method for assigning the lunar-phase day for each calendar day followed exactly that used by Abell and Greenspan (1979) in their investigation of a

TABLE 1

Homicidal Assaults in Cuyahoga County
January 1, 1971, to December 31, 1981

Phase Day	Assaults	Phase Day	Assaults
1	119	16	100
2	120	17	124
3	122	18	102
4	111	19	117
5	121	20	126
6	106	21	118
7	119	22	119
8	103	23	108
9	126	24	108
10	135	25	108
11	104	26	116
12	105	27	94
13	104	28	104
14	125	29	118
15 (Full)	121	30	67 (127)*

Total 3370

*Normalized value based on the use of 64 synodic periods of 29 days duration and 72 synodic periods of 30 days duration giving a mean synodic month of 29.5294.

correlation between birthrate and lunar phase. By definition, lunar-phase day 15 coincides with the day of the full moon. Table I gives the observed numbers of homicidal assaults as a function of the lunar-phase day. In Figure 1, our results are compared with those given on page 47 of *The Lunar Effect,* where, by the way, the mean is incorrectly shown to be 70 rather than the actual value of near 67.

The standard chi-square test applied to our sample gives values (X^2 = 25.09, *df* = 29, *p* = 0.67) that indicate with a high degree of probability that the day-to-day fluctuations in homicidal assaults are random in nature and are not correlated with lunar phase. The "homicidal peaks" found by Lieber and Sherin to lag several days behind new and full moon (note that a third peak near third-quarter moon was ignored by them because it did not fit the tidal theory) are therefore most likely to be nothing more than statistical artifacts. Rotton, Kelly, and Frey (1983) were able to obtain the raw data used by Lieber and Sherin for their Dade County (Miami) study and concluded from an independent analysis that no lunar phase relationship could be demonstrated. Thus the conclusions drawn by Lieber and Sherin from both their Cleveland and their Miami

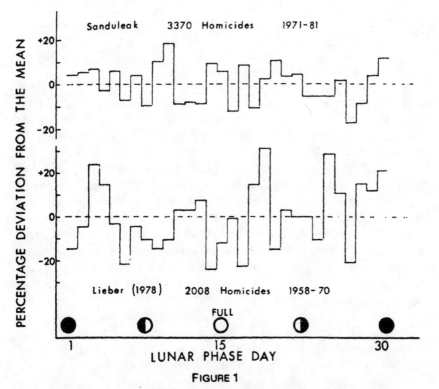

FIGURE 1

Comparison of the two studies of the Cuyahoga County homicidal assault rate versus lunar phase.

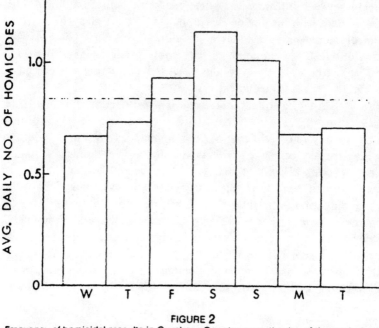

FIGURE 2
Frequency of homicidal assaults in Cuyahoga County versus the day of the week showing the "weekend effect."

samples have now been shown to be invalid.

One very well known correlation did emerge from our study. As is
shown in Figure 2, "Saturday night special" is a justified nickname for a
small handgun. This marked increase in homicidal attacks on weekends
(undoubtedly related to increased alcoholic intake) is not even mentioned
by Lieber. Note that for intervals of about three months the full and new
moon can become temporarily synchronized with weekends; that is, they
will occur on Friday through Sunday. Thus, in correlation studies covering
only a few months, it is possible to statistically confuse this very real
"weekend effect" with a putative lunar effect. In very long term studies,
such as this one, this is not a problem, since full and new moons will occur
with essentially equal representation on each day of the week. Our data
also show the well-known seasonal effects, in that slightly above average
numbers of assaults took place during July and August (short tempers in
hot, muggy weather) and during December and January (increased hostility
due to cabin fever?).

Although Lieber is quick to suggest a "biological tides" theory, it is
obvious from his discussion that he does not fully appreciate the nature of
the lunar and solar tidal action. The magnitude of a tidal force varies as
the inverse cube of the distance of the tide-raising body. Since the orbit of
the moon about the earth is slightly elliptical, its distance varies by as

much as 14 percent, which causes a variation of about 60 percent in its tidal action. The point of closest approach to the earth (perigee) and the point of greatest distance (apogee) occur at differing phases from one month to another. Thus it is possible for the tidal action of the moon to be actually *greater* at the first- and third-quarter phases (with the moon at perigee) than it is at full or new moon (with the moon at apogee). It is the addition of the solar tidal component (roughly half as large as the lunar component) that causes the *combined* lunar and solar tidal action to be maximized near the new and full moon phases. Thus, given the proposed tidal mechanism, Lieber's book should have properly been entitled "The Luni-Solar Effect."

Since the variation in lunar phase is not an exact measure of the variation in luni-solar tidal action, it was necessary to extend and complete this study by providing the results shown in Table 2. Here the frequency of homicidal assaults (same data as used in Table 1) is given as a function of the Tidal Index. The Tidal Index was devised by the author as a numerical value that is approximately proportional to the magnitude of the luni-solar tidal force. It derives from a simplified model that takes into account the relative positioning of the sun, earth, and moon and the varying earth-moon and earth-sun distances but assumes that all three bodies move in a common plane. A BASIC language program was written to carry out these lengthy computations on a personal computer. The Tidal Index ranges from a minimum value of 0.97 at the smallest possible neap tide (e.g., June 30, 1971) to a maximum of 1.72 at the greatest possible spring tide (e.g., January 8, 1974). A test of the data in Table 2 ($X^2 = 0.50$, $df = 2$, $p = 0.78$) gives no indication of a relationship between assaults and the Tidal Index.

In conclusion, this study found no evidence that the frequency of timing of homicidal attacks in Cuyahoga County, Ohio, during 1971-1981 was related in any way to the phases of the moon or the action of luni-solar tidal forces. Nationwide data for both homicide and suicide likewise

---

TABLE 2

The Number of Homicidal Assaults as a
Function of the Luni-Solar Tidal Index

Range in Tidal Index	Number of days	Number of Assaults Observed	Expected
0.97-1.19 (LOW)	764	640	640.79
1.20-1.49 (MID)	2460	2048	2063.26
1.50-1.72 (HIGH)	794	682	665.95
Totals	4018	3370	3370.00

show no relationship to lunar phase (Lester, 1979; MacMahon, 1983). Indeed, a survey of the literature by Campbell and Beets (1978) found that no conclusive statistical evidence existed for the reality of any kind of lunar effect on human behavior. More recently, Rotton and Kelly (1984) applied the meta-analysis technique to 39 lunar-effect studies in the literature and drew the same conclusion.

How then does one account for the anecdotal testimony so readily provided by police, bartenders, and maternity-ward nurses? I would propose the following speculations. On a particularly busy or memorably stressful day, an emergency-room attendant (or any person dealing with the public) turns to a colleague and says: "Wow, things are really going wild around here today. There must be a full moon." Of course no one takes the trouble to determine or record the actual lunar phase, but everyone within earshot of the that oft-uttered remark is likely to subsequently recall that there *was* a full moon on that unusually eventful day. Numerous repetitions of this scenario over the years might help convince someone that they have personal experience of the moon's ability to influence human behavior. I seriously doubt, however, that even the most ardent proponents of a lunar effect could specify the current phase of the moon when tested on the spur of the moment. I have tested audiences and found that only a very small percentage could. Another likelihood involves people selectively remembering occasions when a full moon did indeed coincide with unusual activity (probably on a holiday or weekend) while they disregard all those times when nothing much happens at full moon or when total mayhem reigns near the quarter phases.

Clearly, it is these psychological factors and the role of the media in the propagation of this now remarkably pervasive and often deep-seated delusion that constitute the "lunar effect" most deserving of further investigation.

Acknowledgment

I would like to thank Mrs. Liz Tidwell of the Coroner's Office and Dr. Steve Shore for their assistance in the compilation of the data used in this study.

References

Abell, G. O. 1979. SKEPTICAL INQUIRER, 3 (no. 3):68-73.
Abell, G. O., and B. Greenspan. 1979. "The Moon and the Maternity Ward." SKEPTICAL INQUIRER, 3 (no. 4):17-25.
Campbell, D. E., and J. L. Beets. 1978. "Lunacy and the Moon." *Psychological Bulletin*, 85:1123-1129.
Lester, D. 1979. "Temporal Variation in Suicide and Homicide." *American Journal of Epidemiology*, 109:517-520.

Lieber, A. L. 1978. *The Lunar Effect*. New York: Dell.

Lieber, A. L., and C. R. Sherin. 1972. "Homicides and the Lunar Cycle: Toward a Theory of Lunar Influence on Human Emotional Disturbance." *American Journal of Psychiatry*, 129:101-106.

MacMahon, K. 1983. "Short-Term Temporal Cycles in the Frequency of Suicide, United States, 1972-1978." *American Journal of Epidemiology*, 117:744-750.

Pokorny, A. D. 1964. "Moon Phases, Suicide and Homicide." *American Journal of Psychiatry*, 121:66-67.

Pokorny, A. D., and J. Jachimczyk. 1974. "The Questionable Relationship Between Homicides and the Lunar Cycle." *American Journal of Psychiatry*, 131:827-829.

Rotton, J., I. W. Kelly, and J. Frey. 1983. "Geophysical Variables and Behavior: X. Detecting Lunar Periodicities: Something Old, New, Borrowed, and True." *Psychological Reports*, 52:111-116.

Rotton, J., and I. W. Kelly. 1984. "Much Ado About the Full Moon: A Meta-Analysis of the Lunar-Lunacy Research." *Psychological Bulletin* (in press).

UFOs

26

The Claim of a Government UFO Coverup

Philip J. Klass

Clear Intent: The Government Coverup of the UFO Experience. By Lawrence Fawcett and Barry J. Greenwood. Prentice-Hall, Englewood Cliffs, N.J., 1984. 264 pp. $8.95.

The theory that the U.S. government has been withholding "the truth" about UFOs has sustained the UFO-faithful during their nearly 40 years of wandering in the barren desert of UFOlogy. Leaders of the UFO movement hope that the new book *Clear Intent* will provide sorely needed sustenance to rejuvenate the movement from its present comatose condition.

For example, Walter Andrus, international director of the Mutual UFO Network (MUFON) predicted a year before the book's publication that it would be "the vehicle that will force the Pentagon and our government intelligence agencies to reveal why they have conducted a 'Cosmic Watergate,' or coverup, with respect to their involvement with UFOs."

More recently, this leader of the nation's largest UFO group said: "I feel confident that the Freedom of Information Act documents published [in the book] will significantly stir the American public's interest in the UFO phenomenon and the 'Cosmic Watergate' that has been so prevalent since the 1947-1948 era." In this reviewer's opinion, the first of Andrus's predictions will not occur and the second is not likely.

If Andrus and the book's author are correct, every American president since Harry Truman, despite their sharply divergent views on many issues, have been able to agree on only one thing—that "the truth" about UFOs must be withheld from the public. Even Richard Nixon seemingly was able to sustain this UFO secret for his six years in office despite his inability to contain the Watergate scandal for more than a few months.

Having myself carefully studied most of the documents quoted in *Clear Intent* (as reported in my own book *UFOs: The Public Deceived*) and additional documents released under FOIA that authors Fawcett and Greenwood chose to omit or censor, I reach quite the opposite conclusion.

Even the authors admit that they are perplexed by some actions of government agencies that argue against a coverup.

For nearly three decades after UFOs were "discovered" in 1947, the U.S. Air Force was the victim of UFOlogists' charges of coverup. In the late 1940s and early 1950s, there was some secrecy, for reasons explained by Capt. Edward J. Ruppelt in a briefing of top officials of the Air Defense Command in early 1953, when Ruppelt directed the USAF's Project Blue Book UFO investigations. In Ruppelt's briefing, later declassified and published, he said: "1 ne required security classification for admittance to this briefing is Secret. The reason for this is that in some instances we may get into a discussion of classified equipment, classified locations or classified projects during the question and answer period that follows this briefing.

"When the [UFO] project was first started, it was classified Top Secret. This is probably the reason for the rumors that the Air Force has Top Secret information on this subject; it does not. The only reason for the original classification was that when the project was first started the people on the project did not know what they were dealing with and, therefore, unknowingly [sic] put on this high classification," Ruppelt told top Air Defense Command officials. Not surprisingly, Fawcett and Greenwood make no mention of Ruppelt's once-classified briefing.

More than 20 years later, in the mid-1970s, Ruppelt's statements were confirmed when the USAF, after having closed down the project in 1969, made public all of its Project Blue Book files in the National Archives. UFOlogists poured over the many tens of thousands of pages of file material looking for a "smoking gun." When they could not find any, they did not admit they had been wrong. Instead they simply changed targets and charged that it was the Central Intelligence Agency that was involved in a UFO coverup. There was some slight basis for their suspicions.

In mid-1952, the CIA had become interested in UFOs, and there were some within the agency who were eager to launch a major investigation. But, in early 1953, the CIA had secretly convened a panel of top scientists, headed by Dr. H. P. Robertson, to consider the USAF's most impressive UFO cases. The Robertson Panel concluded that all of these UFO reports were explainable in prosaic terms and showed no evidence of advanced foreign technology, either Soviet or extraterrestrial, and so the CIA promptly abandoned plans for its own UFO investigation. Word of the Robertson Panel had leaked to UFOlogists in 1958, and in 1966 the highlights of its "Secret" report were even published in *Saturday Review*.

If Congress had not passed the Freedom of Information Act, UFOlogists could never have known with certainty that the CIA had never launched a major UFO effort and that its post-1953 interest was so scant that it did not assign even one person to monitor UFO reports on a full-time basis. When attorney Peter Gersten, with a longtime personal interest in UFOs, volunteered to obtain once-classified UFO material through FOIA, UFOlogists expected they would strike gold. Largely through Gersten's efforts, approximately 3,000 pages of material were located and released from CIA files and those of the Defense Intelligence Agency, the National Security Agency, and the Federal Bureau of Investigation in the late 1970s. It turns out to be "fool's gold."

Most of the material dates back more than a quarter of a century—a key point that most UFOlogists, including the authors of *Clear Intent*, gloss over or ignore. Included are newspaper clippings of foreign UFO sightings, submitted by overseas embassies, as well as magazine articles. There are more than 50 pages of correspondence with a single UFOlogist who is certain the CIA itself is building UFOs to brainwash the American public. (Even today, this UFOlogist reportedly clings to that view.)

The authors withhold from their readers considerable information that would damage their coverup hypothesis. For example, when Gersten filed his original FOIA request in 1978, he asked the CIA to release files on a number of specific UFO incidents. To Gersten's amazement, the CIA responded by asking him to change his request to ask that the agency release *all* of its UFO file material, which he promptly did. Curious behavior if the CIA was trying to withhold UFO information. But it made good sense if the CIA had nothing to hide and was anxious to spare itself the need to make repeated searches of its dusty files in response to a never-ending stream of FOIA requests on each of many thousands of UFO incidents.

If the alleged coverup is accorded the highest priority in the nation, how could intelligence agencies such as CIA, NSA, and DIA be so stupid as to release material that the authors claim to be so incriminating? Why not simply destroy it, or deny its existence? After all, the authors claim that NSA alone shreds more than 80,000 pounds of paper every day. The authors are even more puzzled by actions that seem to them to indicate that there are persons within these agencies who are eager to help UFOlogists penetrate the alleged coverup.

For example, in 1968, an NSA employee whose name is withheld because of the Privacy Act, and who obviously was a believer in UFOs, gratuitously wrote and circulated a paper on the subject that found its way into the agency's files. In response to Gersten's FOIA request, NSA released his paper with a disclaimer that it had no official status and simply reflected the 1968 views of one of its employees. Fawcett and Greenwood are perplexed that NSA would release a paper supportive of UFOs and note that "it would have been easy for the NSA to release something which argued strongly against UFOs," i.e., to have prepared an anti-UFO paper and released that. This prompts the authors to suspect that "people within the NSA want the public to know what is happening." The alternative explanation, that there is no coverup, is not considered by the authors.

In another instance, UFOlogists sought a report on a UFO incident that allegedly had occurred in a forest near a USAF base in England. In response to an FOIA request, a USAF organization was able to borrow and supply a copy of a brief memo on the incident from the Royal Air Force when it found it had no copy in its own files. This beyond-the-call-of-duty initiative prompts the authors to ask: "Could there be a segment within the Air Force that wants us to have the facts on UFOs? It's a distinct possibility."

"Another striking [CIA] memo," according to the authors, dated October 1, 1958, reports that an unnamed (i.e., name censored for privacy) unemployed civilian who had been experimenting with new types of photographic film claimed to have made photos of UFOs on a number of occasions. The writer of the memo (name censored) indicated he himself would like to see the photos but indicated he did not want to get involved if there was no CIA interest. Fawcett and Greenwood claim that it is not known whether the CIA ever analyzed these UFO photos "because no further file material on this affair has been released but it certainly belies claims by the CIA at the time of not being interested in UFOs."

Had the authors been more diligent they would have found another released CIA memo, signed by Philip G. Strong and dated October 29, 1958, in reply to the October 1 memo. The reply was addressed to the director of the Photographic Intelligence Center, who then was Arthur Lundahl, a man with a longtime interest in UFOs. Strong's reply said that his office of scientific intelligence "has an interest in keeping track of UFOs, however the overall community responsibility for investigating UFOs . . . is vested in the Department of the Air Force."

The next significant documents the authors find in the CIA files are dated nearly 20 years later—a series of memoranda exchanged between unnamed CIA employees, involving an unsolicited report submitted by an unnamed (censored) scientist outside the agency. As a result, an unnamed (censored) CIA employee made inquiries to determine if the CIA, or any other government agency, would be interested in examining the unsolicited report. In a memo dated April 26, 1976, he wrote that "it does not seem that the government has any formal program in progress for the identification/solution of the UFO phenomena." Based on this and equally "incriminating" documents, Fawcett and Greenwood claim to see an "unambiguous pattern of continuing interest and monitoring of the UFO phenomenon by the CIA."

It is hardly surprising that the authors are confident that there is even more incriminating evidence in the 156 pages of UFO-related material that NSA has refused to release and in the 57 pages that the CIA has declined to make public. The unreleased NSA documents cover a period of 21 years, from 1958 to 1979, or *an average of less than eight pages per year.* More important is the nature of NSA's mission and of these documents, according to the agency's stated rationale for not releasing them, which has been upheld by the U.S. Supreme Court in response to Gersten's appeal.

One of NSA's principle missions is to eavesdrop on the radio communications of the Communists and other potential enemies and to try to decode their encrypted messages. In some instances NSA's "listening posts" are located in "neutral" countries whose leaders are sympathetic to the West but would be embarrassed if such cooperation, past or present, were made known. Additionally, NSA would not want to reveal that it had been able to decode messages in the 1958-1979 period, for that would reveal that these codes—modified versions of which may still be in use by Soviet Bloc countries—had been compromised.

One can only speculate as to what these NSA-intercepted Soviet Bloc UFO-related messages might involve. Possibly some were messages exchanged between Moscow and the Soviet embassy in Washington, asking about the results of USAF investigations into UFO reports, while others might involve UFO reports by Communist Bloc military pilots. The significant fact is that if any of these NSA-intercepted/decoded messages reveal that the USSR has discovered "the truth" about UFOs, it seems strange that the Soviet Union would cooperate with its arch-enemy to maintain a coverup.

What of the 57 pages of material still withheld by the CIA? Is it possible that so large an agency would generate *an average of only two sheets of paper per year* on so important a subject if the agency were truly keenly interested and actively investigating UFOs? The insignificant number of documents itself belies the claim of Fawcett and Greenwood, and others.

We know from released CIA documents that in 1952 the agency's analysts were puzzled by the fact that there had been UFO reports from many countries—but not one (at that time) from the Soviet Bloc countries. This had prompted concern that perhaps UFOs were a secret Soviet reconnaissance vehicle or part of a psychological warfare effort. Certainly the agency would have asked its secret agents behind the Iron Curtain to investigate, and their reports undoubtedly are included in the unreleased CIA material. While some of the agents may no longer be alive, their families may still reside in the USSR and need to be protected.

After reading all of the released documents, Fawcett and Greenwood reach a startling conclusion: The U.S. government "probably [does] not have a definite 'answer' to the UFO problem, but they monitor the phenomenon in pursuit of an answer." Thus *the "Cosmic Watergate" exists to coverup the government's ignorance, not its knowledge, as other UFOlogists claim using the very same evidence!*

If the government is too stupid to recognize the implications of evidence in its own files, Fawcett and Greenwood are not so handicapped. They are certain that "UFOs are a real, material, physical phenomenon . . .

display intelligence of a very high order . . . [and] that these life forms are here for a purpose. A 'clear intent' has been demonstrated numerous times."

The authors conclude that "UFOs have overflown U.S. military and other government facilities since World War II. . . . This activity has extended to other nations. . . . A definite threat to U.S. national security is considered as fact within the highest levels of government. This scenario also applies to foreign governments." The authors make a brief, obscure admission that many UFO reports turn out to have prosaic explanations, but every UFO reported in the vicinity of a military base seems to them to be an "authentic" one. Military officials who are responsible for protecting this nation from any potential military threat, whether terrestrial or other, seem oblivious to the threat perceived by the authors.

Or has the threat finally been recognized? Is it possible that President Reagan's now-famous "Star Wars" speech of March 23, 1983, was not really intended to explore the use of spaceborne lasers and similar Buck Rogers concepts as a possible defense against Soviet ballistic missiles? Perhaps that was simply a cover, and the multi-billion-dollar Star Wars effort is intended to defend against UFOs. It is surprising that Fawcett and Greenwood did not raise this possibility even though their book went to press after Reagan's speech.

Perhaps this wild speculation will be added if this book goes into subsequent printings, which seems likely, because it will be popular among UFO buffs. It will be popular because it reads more like the books of Donald Keyhoe, who popularized UFOlogy in the 1950s and early 1960s, and all but ignores claims of crashed saucers and UFO-abductions that have dominated recent vintage books and challenged the credulity of even hard-core UFO-believers.

The most accurate and unchallengeable statement to be found in *Clear Intent* is the authors' acknowledgment to "the FBI, CIA, NSA, DIA, State Department, Air Force and many other federal agencies for providing valuable documentation, without which this book could not have been possible." (If the book's royalties were distributed on the basis of content, probably 80 percent should go to the U.S. government.)

If there is indeed a "Cosmic Watergate," it seems surprising that it never occurred to any one of those assigned to direct the coverup through these many decades that all this information that UFOlogists find so incriminating should be shredded or otherwise destroyed, especially after FOIA became the law of the land. But, who knows, perhaps all of the persons responsible for maintaining the coverup secretly wanted the public to have the facts. To borrow a phrase from the authors, "It's a distinct possibility." But quite unlikely.

27

An Eye-Opening Double Encounter

Bruce Martin

As a graduate of Northwestern University, I retain a deep respect for its faculty. Some years ago I learned that a retired Northwestern professor and director of its Dearborn Observatory, J. Allen Hynek, had endorsed several purported UFO sightings. That a former Northwestern astronomy professor could accept the sightings as UFOs made me think that there might be some validity to such reports.

A reported UFO sighting in Charlottesville, Virginia, two years ago has been strongly supported by Hynek. I have lived in the area for the past 25 years and thus have had an opportunity both to follow this UFO claim and to observe Hynek's handling of it. My experience in this eye-opening double encounter may interest those who have also wondered about Hynek's role in UFOlogy.

On Friday, April 2, 1982, the Charlottesville afternoon paper, the *Daily Progress*, displayed a front-page story about a UFO sighting by Nannette Morrison on the previous Tuesday, March 30. The article featured a picture of Morrison pointing to the sky where reportedly the UFO hovered. Morrison observed the UFO at about 11:30 P.M. while returning home. She claimed that the object paced her car and then hovered overhead. Even though she turned off her car radio, Morrison could not hear anything. Based on the description in the newspaper account of the streets traveled and allowing time for Morrison to go inside to get her mother, the whole event would have lasted a few minutes.

In response to this article, on April 14, 1982, the *Daily Progress* carried a letter to the editor by a Charlottesville resident, Jimmy Smith, stating that he and his brothers had seen a brightly lighted object at about the same time on the same night and identified it as a passenger jet, possibly a 737. According to the local airport authorities, a 737 did land about 11:45 P.M. In his letter Smith stated that they could not hear the engines until after the plane passed over. In a thoughtful final two paragraphs Smith also mentioned that while driving he had noticed several times that planes looked as though they were pacing his car. At other times the planes just appeared to hang in the air and not move. (Depending upon their angle of approach, this has also been my experience: planes sometimes appear to move very slowly or to be motionless, and little or no noise is heard until they have passed.)

A year and a half later, on November 19, 1983, during the "Conference on the Psychic" at the local library, Nannette Morrison delivered a talk billed as "Psychic Investigator will relate the psychic phenomena to the UFO experience." She gave an expanded version of the newspaper account, adding a second sighting, on April 1, 1982. Her talk also described encounters with extraterrestrials by others, such as the "Andreasson affair" and the abduction of Betty and Barney Hill. Morrison added that in abductions extraterrestrials are especially interested in human reproductive organs and have been known to insert a sharp needle into the human navel. Morrison also described six manifestations of UFOs and psychics.

To me, however, the most startling part of Morrison's talk was her use of Hynek's name to legitimize her claims. She said Hynek had talked to her for more than an hour by phone; he had asked if she had seen figures or had mental telepathy with the UFO occupants. She also stated that Hynek had said that only certain people are chosen to see UFOs.

I found the views attributed to Hynek astonishing. Therefore, in late November 1983, I wrote Hynek a letter enclosing copies of both the original Morrison article from the front page of the local paper and the Jimmy Smith letter. I also cited the conference and told Hynek that Morrison spoke of him frequently. I specifically asked Hynek his opinion of abduction claims and if it is his view that only certain people are chosen to see UFOs. Hynek never replied.

Meanwhile, in the November/December 1983 issue of Hynek's *International UFO Reporter*, there appeared a three-page article written by Hynek entitled "A Remarkable Double Encounter," in which both Morrison sightings are discussed. The account is essentially that given by Morrison at the November 1983 conference. Since the second encounter, which took place two days after the first, occurred a half-hour after Hynek had finished an hour-long interview with Morrison, he opened by stating that this is the closest he had come to a CE-1 event. In his version the first encounter lasted 15 to 20 minutes. The second encounter two nights later, on April 1, 1982, also at about 11:30 P.M., lasted only 2 minutes. Morrison was inside her house and felt drawn to the window, where she again saw a bright, hovering UFO. She again called her mother, who also saw it, whereupon it zipped away at incredible speed. In his analysis of the pair of encounters, Hynek declares that they argue strongly against a natural event.

To my surprise there appeared in the March/April 1984 issue of the *International UFO Reporter* a one-page comment entitled, "Double

Encounter Questioned," which cites me for calling Hynek's attention to the Jimmy Smith letter (mistakenly assigned by Hynek as April 2, rather than April 14, 1982). Early sections of the Jimmy Smith letter are quoted but not the final two paragraphs referred to above. Hynek begins the comment by stating that, though he feels the evidence for the encounters is "solid," he never "swept under the carpet logical counter views." He resists (as he says) jumping to the conclusion that just because the time of the sighting and the plane's flight were similar the sighting has been solved. He reports Morrison's reply, "that she does not live anywhere near the airport but some 25 miles to the southeast." She also repeated that the UFO paced her car turn for turn for about 20 blocks and that it did not bank or change direction. Since jets are not capable of hovering soundlessly or of reversing direction without turning, and because it is unlikely that a plane would fly so low so far from the airport, Hynek concludes that Morrison's reply is convincing.

Inspection of the U.S. Geological Survey map shows, however, that Morrison's home is not southeast but mainly south and slightly west of the airport. Moreover, her home is but 5.3 miles from the airport runway, which points in the direction of her house. Even by automobile the distance to the airport is only 8 miles. Morrison's estimate of 25 miles would take one far past the airport and one-quarter of the way to suburban Washington. These contradictions of Morrison's report are critical because they account for the low altitude of the plane. Morrison's house lies under an airport landing pattern. As one of her neighbors said to me, "We see that UFO every night." Evidently the parts of the Morrison UFO sightings that are not imagination may be attributed to jets that happened to come in for a landing at a certain angle as well as the difficulty even experienced observers have in estimating distances to lighted objects in the night sky. Whether Morrison passes as an acute observer of objects in the night sky is not, however, the main point of this article.

As a leading investigator of UFO phenomena, former astronomy professor J. Allen Hynek brings to the subject the prestige of studied, objective scientific judgment. In this case of a purported double encounter, which he himself describes as remarkable, what are Hynek's standards of judgment? No trip to Charlottesville is reported, only phone calls to the witness. Evidently he made no check even of regularly scheduled commercial flights, which would have suggested that a plane was in the area at the time of the sightings. A check of maps that ought to be available to Hynek would have revealed that Morrison lives on the north side of town, only 5.3 miles from an airport runway that is aligned with her house. The experiences of neighbors whose homes also lie under the landing pattern are not solicited. In a university town with 16,000 students and an area population approaching 100,000, he does not ask why no one else reported spotting either of the two UFOs. Instead, Hynek rejects the reasonable interpretation of Jimmy Smith in favor of a presumed extraterrestrial craft encounter. It is difficult to detect what if any objective standards Hynek applied to this case.

Finally, what does this case reveal about the present critical mentality of J. Allen Hynek, former astronomy professor? It is discouraging to see someone who has practiced as a scientist, and presumably applied an objective, analytical attitude to his own professional work, take up a subject so much in the public mind as UFOs and apply what seem like no standards whatever. Hynek appears to be but another sad case in which the mania for the occult has overwhelmed critical judgment and submerged the practice of meticulous checking and rechecking of evidence that every reliable scientist observes.

28

Hypnosis and UFO Abductions

Philip J. Klass

Nearly two hundred persons now claim to have been abducted by "UFO-nauts," taken aboard a flying saucer, typically for a physical examination, and then safely released, and the number of such tales is mushrooming. If extraterrestrials are indeed visiting the earth, their curiosity is understandable; but one might expect that they would carry a few earthlings back to their native planet for a more useful dissection, rather than simply repeat the superficial examinations reported.

If only one of the alleged abductees had managed to bring back a single souvenir, the UFO question would be resolved incontrovertibly. But since this has not occurred, UFO proponents rely on regressive hypnosis as their principal tool to substantiate the tale of abduction, and almost invariably it seems to confirm the account. The key question is whether hypnosis is really an effective "lie-detector."

James A. Harder, one of the principal practitioners of hypnosis for UFO incidents, claims that "it is impossible to lie under hypnosis." Harder, director of research for APRO, one of the nation's oldest UFO organizations, is a professor of civil engineering. On the strength of Harder's use of hypnosis, he has endorsed the alleged UFO abductions of Charles Hickson and Calvin Parker in Pascagoula, Mississippi, in 1973 and Travis Walton in Arizona in 1975. (My own investigations indicate both incidents are hoaxes.)

The first known use of regressive hypnosis in connection with a reported UFO abduction occurred in 1963-64. The case involved Betty Hill and her late husband, Barney, who were treated by Dr. Ben Simon, a respected Boston psychiatrist. Simon achieved fame in psychiatric circles during World War II when he used hypnosis to successfully treat military personnel suffering battle-induced psychoses at the U.S. Army's Mason General Hospital, where he was chief of neuropsychiatry.

When the full story of the Hill case was made public in the fall of 1966, in two widely read articles in *Look* magazine and in a book entitled *Interrupted Journey*, leaders of the UFO movement generally were *not* impressed with the tale. Possibly this was because the story was too incredible by the movement's then current standards. Simon sharply disagreed with the conclusions of the book's author, and in a brief introduction he wrote for the book he cautioned that hypnosis is not a "magical and royal road to the Truth."

Shortly after the *Look* articles were published I interviewed Dr. Simon, and he told me he was certain that the alleged UFO abduction was fantasy, not fact, although he was equally certain that the Hills had seen a bright light in the night sky that they had found frightening. To demonstrate the basis for his conclusions, Simon played some tapes for me of Barney reliving under hypnosis his experience of looking at the bright light. I could hear the terror in Barney's voice, and Dr. Simon told me he had never before had a patient become so agitated under hypnosis.

But when we listened to Betty and Barney reliving the alleged experience of being taken aboard a flying saucer by strange-looking creatures, their voices were relaxed and casual, as if they were describing a visit to a neighborhood shopping center. Simon cited numerous other reasons for his conclusion that the tale of abduction was fantasy. Much later, when he appeared on NBC-TV's "Today" show on October 20, 1975, prior to that network's two-hour pseudo-documentary on the Hill incident, Simon was asked whether he believed a UFO abduction had really occurred. He responded that "the abduction did *not* happen," and he characterized it as "fantasy."

When so experienced a practitioner of hypnosis as Simon rejects the idea that simply because a tale of UFO abduction is told under hypnosis it must be true, one might expect that far less experienced practitioners would be cautious in its use for this purpose. But this has not deterred R. Leo Sprinkle, an APRO consultant, who is the leading practitioner of hypnosis in UFO-abduction cases. (Sprinkle, a psychologist, is director of counseling and testing at the University of Wyoming.) In a paper presented to the American Psychological Association in Toronto, on August 28, 1978, Sprinkle reported that he had used hypnosis on 25 persons and had "obtained information . . . which supports their claims of 'abduction' experiences." He said he was inclined "to accept, tentatively, the claims of UFO abductions as 'real.' " He added: "I do not know if these 'abductees' have experienced physical abduction, or whether they have experienced 'out of the body' events [an alleged psychic phenomenon]. In either case, the experiences seem 'real' to the 'abductee.' "

The crucial issue is not whether the tale "seems real" to the subject but *whether the alleged abduction actually occurred.* On January 23, 1977, I wrote to Sprinkle: "To your knowledge, has anyone conducted controlled experiments to evaluate the effectiveness of regressive hypnosis in determining whether the subject is intentionally trying to perpetrate a hoax or a falsehood?" I added that if such experiments had not been conducted it seemed to me that this should be done "before UFO investigators invest any more time in its use as a means of trying to sort out reality from nonreality."

Sprinkle replied promptly, saying that when he first began to experiment with the use of hypnosis on abduction claimants in the mid-1960s he was unconvinced that an abduction had really occurred. But with further use of hypnosis, Sprinkle said, he "began to recognize that there was no way for me to 'know' whether the UFO witness did or did not experience an abduction." Then Sprinkle made a remarkably candid admission: "Now I have persuaded (conned?) [*sic*] myself that ... the apparent abduction experiences are 'real'; at least they are real in the minds of these persons ... I accept these experiences as 'real,' although I'm not in a position to determine the 'level of reality.'" In my reply, I sought clarification of "level of reality." Sprinkle tried to explain that there are *many* "levels of reality," a concept my simple mind was unable to grasp.

In May 1977, the UFO movement's growing reliance on hypnosis to support tales of UFO abductions was shaken by a paper published by Alvin H. Lawson, a professor of English at California State University, Long Beach. The paper was entitled: "What Can We Learn from Hypnosis of Imaginary 'Abductees'?" Lawson had a long-standing interest in UFOs and had offered courses in UFO literature. He reported on an experiment in which imaginary UFO abductions were induced hypnotically in a group of subjects who were then questioned about their experience. The hypnosis was administered by William C. McCall, an M.D. with clinical experience in its use.

Not only were the subjects able to improvise answers about what had happened to them aboard the imaginary flying saucer, Lawson reported, but their stories "showed no substantive differences" from tales in the UFO literature by persons who claimed to have actually experienced an abduction. This prompted Lawson to observe: "The implications of the study for future hypnotic regression of Close Encounter cases, and for abduction cases now deemed of the highest credibility, are unclear at this time."

The results of this experiment, and Lawson's conclusions, were attacked sharply by Harder in the September 1977 issue of the *APRO Bulletin*. Harder said that, while Lawson admitted that the experiment did not prove that all UFO abduction reports were imaginary, "his paper may well lead naive readers to think that there is a strong case that they all are [imaginary]." Harder criticized Lawson's paper for pointing out the similarities between the stories told by "real" and "imaginary" abductees, while failing to point out "a very important difference." That difference, Harder claimed, "was that the 'real' abductees were convinced that their experiences were real whereas the 'imaginary' abductees were not."

Lawson offered a revised paper on the same experiment at the August 28, 1978, meeting of the American Psychological Association. He stuck by his original report that there were "no substantive differences" in the accounts given by "real" and "imaginary" abductees. But he added that "despite the many similarities, there are crucial differences — such as alleged physical effects and multiple witnesses — which argue that UFO abductions are separate and distinct from imaginary and hallucinatory experiences." However, Lawson also warned that "one should be cautious about the results from hypnotic regression in UFO case investigations . . . A witness can lie, or believe his own lies, and thus invalidate any investigation. A more common result may be that hypnotized witnesses subtly confuse their own fantasies with reality — without either the witness or the hypnotist being aware of what is happening." It is clear that Lawson is much more knowledgeable about the limitations and pitfalls of hypnosis than Harder, who so often employs this technique.

The obvious lessons to be drawn from the Lawson/McCall experiments and papers have been largely ignored by the leaders of the UFO movement. For example, J. Allen Hynek, scientific director of the Center for UFO Studies, has strongly endorsed the tale of Mrs. Betty Andreasson, told under hypnosis, who claims not only to have gone aboard a flying saucer but also to have flown to its native planet. In the foreword to a book about the Andreasson case, Hynek wrote: "In the past, I frankly would not have touched an invitation to write the foreword for a book treating 'contactees,' abduction, mental telepathy, mystical symbolism, and physical contact and examination by 'aliens.' But across the years I have learned to broaden my view of the entire UFO phenomenon. Those who still hold that the entire subject of UFOs is nonsense will be sorely challenged if they have the courage to take an honest look at the present book."

Martin T. Orne, past president of the International Society of Hypnosis and director of the Institute of Pennsylvania Hospital's unit for experimental psychiatry, is an internationally recognized authority on hypnosis. In a paper published in the October 1979 issue of the *International Journal of Clinical and Experimental Hypnosis*, entitled "The Use and Misuse of Hypnosis in Court," Orne completely demolishes the basic premises upon which Harder, Sprinkle, and other UFOlogists have operated in using hypnosis in an effort to substantiate tales of UFO abductions. While Orne does not discuss the misuse of hypnosis in UFO cases, except for one oblique reference, it is obvious that his warnings and recommended safeguards apply to UFOlogy as well as to forensic use.

Orne notes that the courts "have recognized that hypnotic testimony is not reliable as a means of ascertaining the truth," and he says this view "is supported by scientific data." He cites experiments showing that "it is possible for an individual to feign hypnosis and deceive even highly experienced hypnotists ... Further, *it is possible for even deeply hypnotized subjects to willfully lie*" (emphasis added). This flatly contradicts Harder's self-serving claim.

Orne warns: "We should keep in mind that psychologists and psychiatrists are not particularly adept at recognizing deception." (Surely this also applies to a professor of civil engineering.) "We generally arrange the social context of treatment so that it is not in the patient's interest to lie to us ... As a result, the average hotel credit manager is considerably more adept at recognizing deception than we are." Orne acknowledges that "military psychiatrists and other health professionals who are required to make dispositional judgments on a daily basis do become adept at recognizing manipulation and deception." But Orne says that relatively few "who are experienced in the use of hypnosis have had this type of background. Consequently, they have little experience or concern about being deceived or used."

Orne cautions: "Hypnotic suggestions to relive a past event, particularly when accompanied by questions about specific details, puts pressure on the subject to provide information for which few, if any, actual memories are available. This situation may jog the subject's memory and produce some increased recall, *but it will also cause him to fill in details that are plausible but consist of memories or fantasies from other times*" (emphasis added). He adds: "It is extremely difficult to know which aspects of hypnotically aided recall are historically accurate and which aspects have been confabulated."

"There is no way, however, by which anyone — even a psychologist or psychiatrist with extensive training in the field of hypnosis — can for any particular piece of information determine whether it is an actual memory versus a confabulation *unless* there is independent verification," Orne states. He cites experiments by others that show that "free narrative recall will produce the highest percentage of accurate information but the lowest amount of detail. Conversely, the more an eyewitness is questioned about details, the more details will be obtained — but with *a marked decrease in accuracy*" (emphasis added). (Examination of transcripts of hypnosis sessions with "abductees" reveals that great pressure was applied for details rather than allowing the subject to use free narrative.)

Orne's paper suggests that the use of hypnosis by pro-UFO investigators can generate what he calls "pseudo-memories," which may enable a subject to tell a convincing story later when not under hypnosis. Such "pseudo-memories can and often do become incorporated into the individual's memory store as though they had actually happened ... If a witness is hypnotized and has factual information casually gleaned from newspapers or inadvertent comments made during prior interrogation or in discussion with others ... many of these bits of knowledge will become incorporated and form the basis of any pseudo-memories that develop."

One of Orne's warnings is especially appropriate for hypnotic interrogations conducted by Sprinkle and Harder, both of whom lean strongly to the hypothesis that the earth is being visited by extraterrestrial craft. Orne writes: "Furthermore, if the hypnotist has beliefs about what actually occurred, it is exceedingly difficult for him to prevent himself from inadvertently guiding the subject's recall so that he [the subject] will eventually 'remember' what he, the hypnotist, believes actually happened."

(During my own investigation into the Travis Walton "abduction" case, I talked with Jean Rosenbaum, a Durango, Colorado, psychiatrist who was brought into the case and was in Scottsdale, Arizona, when Harder used hypnosis to interrogate Walton to probe for details of the incident. Rosenbaum told me that Harder's "interviewing techniques are very interesting in that all of his questions are loaded.")

Orne notes that "the more frequently the subject [describes] the event, the more firmly established the pseudo-memory will tend to become. In [conducting] the experimental demonstration, we are dealing with an essentially trivial memory about which the subject has no inherent motivations [to be untruthful]. Nevertheless the memory is created by a leading question, which, however, on casual observation, seems innocuous." Orne warns: "Hypnosis has not resulted in accurate memories but rather has served to produce *consistent memories*" (emphasis added).

After discussing these potential pitfalls, Orne proposes four important procedural safeguards. One of these is that hypnosis "should be carried out by a psychiatrist or psychologist with special training in its use." (In one instance, when Sprinkle had to return home after a hypnosis session with an "abductee" and UFO-writer Jerome Clark wanted to continue the investigation, according to Clark's subsequent article in the August 1976 *UFO Report*, Sprinkle urged Clark to "conduct hypnotic regression" himself. "Since I had never performed hypnosis before, I was dubious about the prospect, but Sprinkle had taught me the methods and said he could see no reason why Sandy [the subject] and I could not work together. So three weeks later ... Sandy and I got together in an effort to continue the interview ... Sandy fell quickly into a hypnotic trance and was able to reply quickly and easily to my questions," Clark wrote.)

Orne cautions that the hypnotist "should not be informed about the facts of the case verbally; rather he should receive a written memorandum outlining whatever facts he is to know, carefully avoiding any other communication which might affect his opinion ... It is extremely undesirable to have the individual conducting the hypnotic sessions have any involvement in the investigation of the case." (Based on this safeguard, Sprinkle and Harder would be disqualified in their UFO investigations.)

Another important safeguard recommended by Orne is that "all contact of the psychiatrist or psychologist with the individual to be hypnotized should be *videotaped from the moment they meet until the entire interaction is completed* [emphasis added]. The casual comments which are passed before or after hypnosis are every bit as important to get on tape as the hypnotic session itself. (It is possible to give suggestions prior to the induction of hypnosis that will act as post-hypnotic suggestions.)"

Additionally, Orne says: "No one other than the psychiatrist or psychologist and the individual to be hypnotized should be present in the room before and during the hypnotic session. This is important because it is all too easy for observers to inadvertently communicate to the subject what they expect, what they are startled by, or what they are disappointed by." (This recommended safeguard almost invariably is violated in UFO investigations.)

Orne also recommends that tape recordings of prior interrogations be made "because the interactions which have preceded the hypnotic session may well have a profound effect on the sessions themselves." Orne cautions that a subject may unwittingly have been given cues to certain information "which might then be reported for apparently the first time by the witness during hypnosis."

After I read Orne's paper, it was obvious that it should have been of great importance to UFOlogists who make use of hypnosis, and so I wrote to Sprinkle on March 24, 1980, seeking his reactions. He replied on April 7, saying: "Dr. Orne is an acknowledged authority on the use of hypnosis ... However, I am sure that he would agree with the principle that 'Science' is based upon accumulated evidence of many observations, as well as upon the views of authorities." This was a curious response inasmuch as Orne had cited 37 different scientific papers and court cases to support his conclusions and recommendations.

Sprinkle said that the safeguards recommended by Orne "seem most appropriate for the forensic uses of hypnosis in court," but he questioned whether they were also applicable to UFO investigations because there is no "crime" and no "criminal" or "victim." (Sprinkle takes a curious view of "UFO abductions." He does not consider "abductees" to be "victims" even when they claim to have been taken aboard a flying saucer against their will and subjected to physical examinations. Sprinkle explained his benign views: "They do not seem to perceive themselves as 'kidnapped.' In fact, they see themselves as citizens of a 'higher civilization.'" In response to my subsequent questions, Sprinkle told me that if a member of his immediate family were kidnapped by an earthling, he *would* report it to the Federal Bureau of Investigation. But if a member of his family were kidnapped by an extraterrestrial, he would report it "to the world," *not to the FBI.*)

In subsequent correspondence, Sprinkle indicated that he had begun to use video-taping of hypnotic sessions, when the subject was willing, but other than this he indicated no plans to introduce the rigorous safeguards recommended by Orne. Sprinkle explained: "In my opinion, there are three general 'models' of UFO investigation involving hypnotic procedures. One is the 'forensic model,' as indicated by Dr. Orne; another is the 'psychotherapy model,' indicated by Dr. Simon in his work with Betty and Barney Hill; and a third model is one of a combination of the forensic/ therapeutic models, which I call the 'educational model.' "

"The educational model is based upon the view that two goals are important in UFO investigation: as much information as possible should be obtained, but not at the expense of the dignity of the individual ... In the legal model, the search for 'truth' is intense ... In the therapy model, the truth is less important than the personality structure and the welfare of the individual ... The educational model is one which suggests that the individual's growth is important, but sharing that information with others can assist the individual, and others, in their educational development," Sprinkle wrote. This "educational model," he admitted, "may be considered a 'loose' model by proponents of either the therapy or legal models, because [of] the manner in which 'truth' is being explored, sought, and shared."

Asked to respond to the crucial question of whether he believed that hypnosis was of any value in determining that a UFO-abduction had actually occurred, Sprinkle replied: "I believe that the use of hypnotic techniques is helpful to UFO abductees and contactees in exploring their memories of their experiences and that it is helpful to them in assisting them to come to terms with the abductions which have occurred — in this very reality!"

29
Hypnosis Gives Rise to Fantasy and Is Not a Truth Serum

Ernest R. Hilgard

The use of hypnotic recall as evidence in UFO abduction cases is an abuse of hypnosis. It is an abuse, first, because of the role that fantasy plays for all hypnotically responsive subjects and, second, because abundant evidence exists that fabrication can take place under hypnosis.

For example, under hypnosis I implanted in a subject a false memory of an experience connected with a bank robbery that never occurred, and the person found the experience so vivid that he was able to select from a series of photographs a picture of the man he thought had robbed the bank.

At another time, I deliberately assigned two concurrent — though spatially very different — life experiences to the same person and regressed him at separate times to *that date*. He gave very accurate accounts of both experiences, so that a believer in reincarnation, reviewing the two accounts, would have suspected that the man had really lived the two assigned lives.

These particular examples have not been published, but many similar accounts have been. For example, it has been shown experimentally that, while acting the part of a spy, a subject can hold a "cover" when posing as a citizen of another nation and in an occupation not his own. Under hypnosis, the person does not give himself away (Orne 1971).

The role of fantasy in hypnosis has been amply documented by Josephine R. Hilgard (1979).

References

Hilgard, J. R. 1979. *Personality and Hypnosis: A Study of Imaginative Involvement*, 2nd ed. Chicago: University of Chicago Press.

Orne, M. T. 1971. "The Potential Uses of Hypnosis in Interrogation." In A. D. Biderman and H. Zimmer, eds., *The Manipulation of Human Behavior*. New York: Wiley.

Fringe Archaeology

30

Deciphering Ancient America

Marshall McKusick

A retired professor of marine biology from Harvard, Barry Fell, attracted an immense lay following with his best-selling book, *America B.C.: Ancient Settlers in the New World* (1976). The book created a sensation because the author combined Harvard credentials with very unorthodox conclusions: he claimed that pre-Columbian America had been colonized by Celts, Phoenicians, Egyptians, Libyans, and other ancient peoples. Its commercial success may be judged by the fact that the American Booksellers Association presented it to the White House library as one of the 250 best books published in the United States during 1973-1977.

Has the reading public both here and abroad been victimized by archaeological fiction posing as legitimate research? Some professional archaeologists have claimed that Fell is a deluded scholar whose statements represent compounded errors. Nevertheless, *America B.C.* obviously offered the reading public something it wanted, perhaps the hope that American prehistory was an ethnic record of European and Mediterranean ancestors. Others enjoyed the vigorous style with which Fell attacked traditional archaeological research. The conflict between the professionals and Fell has continued with the publication of his most recent book, *Saga America* (1980), which introduced Vikings and other adventuresome peoples as an explanation for the past.

In the midst of claims and counterclaims Fell has been able to convince a large number of followers that American prehistory is a record of Old World civilization. Among the claims that Fell makes are:

1. Egypto-Libyans left stone memorials, such as the Long Island Tablet and the Davenport Tablet from Iowa, or had been the source of the pictographic writing on these stones, taught the Canadian Micmac pictographic writing, left written messages in the greater Southwest and elsewhere, and explored Polynesia. Later, the Greeks and Libyans mingled and were responsible for the Greco-Libyan place names found in New

England. To assist in overseas Pacific voyages, Libyan-Americans established a naval academy in Nevada and schools in Colorado and California, and they mapped Hawaii.

2. Celto-Phoenicians established an international fur trade between the hemispheres, shipping lumber, silver, copper, gold, furs, and other commodities to Europe and the Mediterranean in exchange for manufactured goods. The Celts of Spain taught the Phoenicians an Irish form of ogham alphabet, which was used for American inscriptions written in both Celtic and Phoenician. Iberian Celts established commercial banks in the west.

3. Other inscriptions and coins found in America demonstrate that many nations influenced pre-Columbian America, including Greeks, Romans, Jews, Islamic Arabs, North African Christians, and Byzantines.

4. Scandinavian Vikings, later joined by Irish and other Celts, traveled extensively in North America by circumnavigating Florida and exploring the Mississippi and western tributaries. The Norse left written inscriptions in the Southwest, Oklahoma, and elsewhere, as well as building Newport Tower in Rhode Island.

Scripts and Languages

Successful decipherment of a text depends upon the correct identification of the alphabet or the other characters used. Where these are phonetic in value, as in alphabetic or syllabic scripts, the pronunciation of the word identifies the language. While this is a simplification of the complexities involved in a true decipherment of a previously unread script, it does represent a fundamental necessity. If the script is incorrectly identified, there is no chance of a successful decipherment. On these grounds the attacks by scholars on Fell's identifications of scripts lie at the heart of their rejection of his work. Scholarly challenges have been made of Fell's identification of ogham, Egyptian, Libyan, Punic, runic, and petroglyphic characters.

Fell contends that ogham (or *ogam)* script was used in New England and elsewhere in North America by ancient Celts and, in addition, that the Phoenicians learned to use it from these Celts, writing their Punic language without vowels. However, two British archaeologists familiar with Celtic ogham have demonstrated that these scripts have not been correctly identified in America (Ross and Reynolds 1978). Irish ogham developed after the fourth century as a system using groups of linear marks to represent 15 consonants and groups of dots for 5 vowels. Although this script is innovative and separate from traditional European alphabetic characters, the phonetic values of the signs are derived from written Latin. The distribution of ogham was restricted to one part of Ireland and adjacent colonies in Wales and Scotland and was largely limited to writing funerary inscriptions on stone. Because of its Latinized origin, it is impossible for ogham to have been known and used by Spanish Phoenicians or American Phoenicians 1,000 years before it was invented in Ireland. The two British archaeologists visited Vermont where the alleged ancient ogham was reported. They found that some of these marks were of comparatively recent origin. At the Crow site in Vermont, Ross and Reynolds (1978) positively identified the marks; they were not ancient writing but were marks left by some Yankee farmer who ran his plow across a field boulder. The archaeologists identified the plow as a single share, the Gloustershire type, and the plow marks paralleled the stone fence row. Other so-called ogham marks were identified as erosional grooves and natural striations. At the New Hampshire site named Mystery Hill, Fell's reported appearance of a Roman numeral "thirty" on the Beltane Stone was alleged to be part of a Celtic calendar. More recently Cazeau and Scott (1979) demonstrated that some of the markings on this stone were caused by granite veins that were misidentified as an intentional inscription.

Fell's ogham hypothesis proved to be very attractive to the unwary antiquarians who reported linear markings in New England and throughout North America. As one example, Fell read three parallel lines as ogham script "BL" and, supposing it to be a Phoenician word, added the dipthong "AA," which he then identified as a dedication to the Canaanite god Baal. Another instance of self-deception is Fell's decipherment of some lines on the Bourne Stone from Masachusetts, marks allegedly "Phoenician-ogham" and translated as a report of the annexation of the territory into the Carthaginian empire by General Hanno—literally: "A Proclamation of annexation. Do not deface. By this Hanno takes possession" (Fell 1976, p. 160). No Phoenicians are known to have been west of Morocco, and no valid evidence exists for the speculation that they reached Massachusetts or that they wrote in an Irish script.

Where Phoenician-ogham cannot be made to serve the purposes of Fell's flights of fancy, other linguistic inventions are forthcoming. One of these is a rendering of the Grave Creek stone as the Phoenician "Punic" language written in "Iberic script." According to Fell (1976, p. 158), this small tablet was found in 1838 in a stream bed in central West Virginia and reads: "The memorial of Teth. This tile (His) brother caused to be made." Because Fell either is ignorant of the relevant archaeological literature or rejects it without proper citations, a lay reader will be deceived by this interpretation. The Grave Creek tablet was thoroughly discredited as a hoax perpetrated to add an attraction to the public museum associated with the 1830s excavations of the Grave Creek mounds. Testimony from the workmen, obtained in later years, showed this artifact had no clear provenience; and the short inscription actually represents an assortment of letters from different alphabets. All of this information has been available since the evaluation by Thomas (1894).

Scripts that Fell identified as Libyan, hieroglyphic Egyptian, and hieratic Egyptian have fared no better than his decipherments of plow-mark ogham. The Davenport Tablet, which Fell identifies as a trilingual inscription in Egyptian, Libyan, and Punic, is thoroughly documented as a hoax made by members of the Davenport Academy of Science in the fall of 1876. This hoax became part of a broader conspiracy to deceive a local German clergyman who irritated Yankee academy members by his mound-digging activities. In the 1880s Smithsonian staff members exposed various Davenport frauds, and this affair became more thoroughly documented in the 1960s when confessions and other written testimony appeared (McKusick 1979a). Fell still defends his theory of Egyptians in Iowa (1980, p. 109), although Jonas Greenfield, professor of Semitic languages at Hebrew University, informs me that the scripts are not Egyptian, Punic, or Libyan. The Davenport Tablet is marked up with obvious Greek letters, musical cleff signs, ampersands, and a mixture of letters never associated with each other in the Old World. Cyrus Thomas, one of the Smithsonian archaeologists of the 1880s, suggested that the forgers copied this hetero-geneous collection from the alphabets illustrated in *Webster's Unabridged Dictionary,* 1872 edition.

The Long Island Tablet found in the 1880s is a bilingual record written, according to Fell (1976, p. 270), in Egyptian and Libyan char-acters. Anthropologists, beginning with Brinton in 1893, identify the marks as typical aboriginal petroglyphs, including a bow and arrow, a man and a canoe, a possible wigwam, and those resembling a bird, a deer, a fish, an eel, and a bear's paw. No mention or citation of the anthropological interpretation is made by Fell, who transformed the canoe into a Libyan galley and translated the inscription to read: "This ship is a vessel from the Egyptian Dominions."

The transformation of aboriginal petroglyphs into Old World scripts occurs throughout *Saga America,* where crudely redrawn sketches from standard anthropological studies are variously labeled and deciphered as Arabic, Libyan, Chinese, and even Scandinavian. These anthropological studies were based upon petroglyphs from California and adjacent states in the Great Basin and Southwest, and even from British Columbia, locations that lead Fell to speculate about far-flung ancient conquests in western North America by Celts, Vikings, and various Mediterranean nations. No anthropologist familiar with aboriginal petroglyphs accepts any of Fell's identifications.

Runic letters of Scandinavia continued to be used for short inscriptions long after the more efficient and complete Latin alphabet came into use in medieval times. While some runic inscriptions have been found in the Greenland Norse settlements, and farther north, where one was found on a rock cairn, not one genuine runestone has been reported from North America. For years specialists in medieval Scandinavian studies have written about the famous American runic forgeries, such as the Kensington and Spirit Pond stones, and the misidentified "runes" from Oklahoma and elsewhere (Wallace 1971). Fell, in *Saga America*, accepts various types of false evidence as genuine. Some of these so-called runes are simply not Scandinavian letters, and among examples one can point to the misidentification of markings on Newport Tower, which Fell and others claim to be Scandinavian. Another group of errors is the identification of pictographs as Norse-inspired pictorial art, although no such drawings are found in Scandinavia. Finally, Fell has reported the Pelham Stone from New Hamphsire to be an attempt by a Norse trader to contact the Indians by means of petroglyphs, a supposition unsupported by any known evidence. With such inventions, it is hardly surprising that he reports Welsh-ogham from Oklahoma and traces joint ventures in America by Celts and Norsemen.

Languages from the Scripts

Specialists in Amerindian languages have failed to find any trace of Old World grammar or vocabulary having a pre-Columbian origin (Goddard and Fitzhugh 1978). However, Fell ignores such conclusions and projects major contacts expressed in language relationships. Among the linkages he reports are so-called "interface tongues," which are represented as Old World and New World mixtures exemplified by Libyan-Zuni, Iberic-Pima, Basque and Celtic-Algonkian, and Greco-Libyan-Algonkian. Micmac-Algonkian "hieroglyphic writing" is said to represent the adoption of a writing style rather than a language transferral.

Since the scripts have been misidentified by Fell, much of his discussion about Celtic, Welsh, Iberic-Punic, and other languages is illusionary. Nevertheless, not all of his conclusions can be rejected on these grounds alone. Not all of his claims deal strictly with phonetic scripts from abroad — for example, he identifies the modern spelling of New England place names as concealing more ancient Celtic root words. In this case, a Celtic specialist has shown that not one of the numerous examples mentioned by Fell represents a genuine pre-Columbian loan word of Celtic origin (Nicolaisen 1979). There is an aura of uninformed speculation surrounding such linkages as Libyan-Zuni and Libyan-Polynesian, but so far no professional linguist has published an analysis of such claims. Since the grammar and morphology of these three languages reflect totally different origins from the perspective of historical linguistics, the comparison must be considered fallacious, pending the application of orthodox linguistic methodology. The task may not be easy in one sense, because a linguist will have a great deal of difficulty with the "Libyan" part of the equation. Fell has scrambled together various scripts from North Africa, invented phonetics and vocabulary, and ignored grammar and has seemingly constructed the Libyan language without reference to known linguistic relationips from the Mediterranean area. It appears that "Libyan," which he traces to locations throughout the world, is an illusionary and phantom language.

The Archaeology of Deception

The history of successful decipherments is littered with footnotes about unsuccessful attempts that for a time seemed somewhat plausible. From eighteenth century onward there were many attempts made to "read" Egyptian hieroglyphics and other scripts, and these efforts by well-intended antiquarians were frequently published. Fell's numerous claims represent a culmination of amateurish speculations. He has ventured into American prehistory and attempted to rewrite the past by ignoring professional archaeological studies. His evidence has rested upon phony linguistics. However, in contrast to Fell's claims of numerous foreign scripts, archaeologists have yet to accept a single valid case of a pre-Columbian inscription in the New World (McKusick 1979b). Fell's well-illustrated books appear to present archaeological evidence, until one recognizes that most of the artifacts either are aboriginal petroglyphs or have some other explanation, such as plow-marked ogham, natural striations, or misidentified lines of recent origin.

It may be added that two other lines of inquiry also appear in his writings. Various Old World coins have been found in the New World, which he accepts as supporting proof of his claims of massive pre-Columbian trade and cultural contacts. A study of this coin evidence shows that some of the coins, such as "Hebrew shekels," were recently minted souvenirs, and not one other example has a solid archaeological context suggesting antiquity (Epstein 1980). Furthermore, the archaeological structures illustrated and described by Fell, such as those from New England, are not Celtic and Phoenician shrines. Those that have been studied in Vermont have a known Yankee origin and include stone cellars once put to a variety of uses, such as storing turnips for the sheep industry (Neudorfer 1979). Supporting excavations have been made in Massachusetts, where stone structures of various kinds are related to former Yankee settlements (Cole 1980). Because the Mystery Hill site in New Hampshire was extensively disturbed and rebuilt by a former owner, some of the structures are now difficult to identify; but there is nothing that suggests either a Phoenician or a Celtic origin and, in particular, Fell's "sacrificial altar" is a Yankee lye stone (Swauger 1980).

The challenges presented by Fell and his imitators are difficult to meet because traditional linguistics and archaeology lack the drama of the strange, mysterious, and the unexplained, which is the stock-in-trade of the popularizers. Until a greater effort is made to reach the public with more factual accounts of prehistory, the mythology about lost races and ancient nations will persist as a substitute for scholarship.

References

Cazeau, C. J., and C. D. Scott 1979. *Exploring the Unknown.* New York: Plenum Press.

Cole, J. R. 1980. "Enigmatic Stone Structures in Western Massachusetts." *Current Anthropology* 21:269-70.

Epstein, J. 1980. "Pre-Columbian Old World Coins in America." *Current Anthropology* 21:1-20.

Fell, Barry 1976. *America B.C.* New York: Times Books.

_____ 1980. *Saga America.* New York: Times Books.

Goddard, I., and W. W. Fitzhugh 1978. "Barry Fell Reexamined." *Biblical Archeologist* 41:85-8.

McKusick, M. 1979a. "The Davenport Stone: A Hoax Unraveled," *Early Man* 1:9-12.

_____ 1979b. "Canaanites in America?" *Biblical Archeologist* 42:137-40.

Neudorfer, G. 1979. "Vermont's Stone Chambers, Their Myth and Their History." *Vermont History* 47:79-147.

Nicolaisen, W. 1979. "Celtic Place-Names in America B.C." *Vermont History* 47:148-160.

Ross, P., and A. Reynolds 1978. "Ancient Vermont." *Antiquity* 53:100-107.

Swauger, J. 1980. "Petroglyphs, Tar Burner Rocks and Lye Leaching Stones." *Pennsylvania Archaeologist* (in press).

Thomas, C. 1894. "Report on Mound Explorations." *12th Annual Report, Bureau of American Ethnology.* Washington, D.C.

Wallace, B. 1971. "The Points Involved," In *The Quest for America,* ed. Geoffrey Ashe. London: Pall Mall.

31

American Disingenuous:
Goodman's 'American Genesis' —
A New Chapter in 'Cult' Archaeology

Kenneth L. Feder

Prepare to enter the world of anthropologist Jeffrey Goodman, Ph.D. We are about to examine his latest work, *American Genesis,* and are about to enter the "twilight zone" of psychic archaeology, bizarre interpretations, and misrepresentation of others' research.

"Scientist Stuns Anthropological World," read the headline of the *Chicago Tribune* article reprinted in my local newspaper, the *Hartford Courant,* on April 20, 1981. The piece concerned the "revolutionary" thesis proposed by Dr. Jeffrey Goodman that the crucible of human evolution was not located in Africa, as virtually all anthropologists and human paleontologists have claimed, but was, instead, in North America, specifically California. According to the article, Goodman went on to claim that human beings did not, as again virtually all archaeologists believe, enter the New World from the Old via the Bering land-bridge at a relatively recent geological date but, rather, appeared *first* in California at a very ancient date and populated the rest of the world from there. Thus all other human groups — Africans, Australians, Europeans, and Asians — can be traced to American Indian roots that were far more ancient than anyone had previously believed. According to the article, the scientific world was "stunned" by Goodman's hypotheses, and his ideas had "some anthropologists waving their shovels."

I must say that I was not "waving my shovel," nor was I the least bit stunned. I had run into Goodman before, albeit indirectly, when I was writing a highly critical, skeptical piece on "psychic archaeology" (Feder 1980). Goodman's previous book was, in fact, titled *Psychic Archaeology: Time Machine to the Past* (1977) (see Cole 1978b), and in it he discussed the use of "psychically" derived information in finding and interpreting the Flagstaff site, a major piece of evidence also used in *American Genesis.* *

---

* The source of Goodman's Ph.D. is an interesting side issue. In *Psychic Archaeology* (p. 97), he mentions his master's degree in anthropology from the University of Arizona, one of the top schools in this field in the country. He further discusses his acceptance into

In any event, having carefully examined his latest work, I am happy to report that Goodman has not let us down. It is a disingenuous fantasy that completely misrepresents the prehistory of the New and Old Worlds and implies that all those (particularly archaeologists) who disagree with him are being racist. This 200-page book is now in a second paperback printing, and demands professional response.

American Genesis makes the following claims:

1. An understanding of the American Indian has been hindered by racism and prejudice. As Goodman says (1981, p. 7), and as most would agree, "The American Indians have been one of the world's most misunderstood, maligned and persecuted races."

2. Indians were certainly in the New World at the very latest 12,000 years ago. Their presence is reflected by the remains of a culture known as "Paleo-Indian." These people, whose artifacts have been found throughout the United States, hunted the large game animals present in North America at the end of the latest period of glaciation.

3. There is a body of data indicating that Indians may have been present in the New World earlier than 12,000 years ago and possibly as long as 30,000 years ago.

Almost all researchers would support statements 1 and 2, and the majority of archaeologists agree with point 3. From here, however, virtually all archaeologists would part company with Goodman.

4. There is very good evidence that people were in the New World 70,000 years ago and perhaps as long as 500,000 years ago. (Only a very few professional prehistorians would agree here.)

5. Skeletal evidence indicates that more than 70,000 and possibly up to one-half million years ago, the people present in the New World were *fully modern* in appearance, predating the first appearance in the Old World of folks who look just like us by at least 35,000 and as much as 465,000 years.

6. These earliest humans lived in California – Goodman, in fact, characterizes California as a "Garden of Eden" (1981, p. 4) – and spread out to populate the rest of the world from this home base. Individual members of early California hunting bands explored as far west as

---

their Ph.D. program, where he decided to "concentrate on developing the empirical practicality of psychic archaeology" (p. 98). On the cover of *American Genesis*, Goodman's name is followed by the title "Ph.D." I made what seemed to be the inevitable inference that one of the finest anthropology programs around had awarded a Ph.D. in psychic archaeology. Curious, and unable to find Goodman's dissertation in *Dissertation Abstracts*, I contacted the Anthropology Department at the University of Arizona. I was informed in no uncertain terms that they had *not* awarded Goodman a Ph.D., but that a large number of people have made that assumption based upon the information Goodman provides in his books. The University of Arizona Anthropology Department made it their business to track down the actual source of Goodman's doctoral degree – California Western University, a school not listed in the *Guide to Departments of Anthropology*, the official list of anthropology programs published by the American Anthropological Association.

France and returned home to tell people of their New (our Old) World.

7. There is abundant archaeological evidence to indicate that America, along with being the womb of physical evolution, was also the cradle of cultural innovation. Artifacts, including bifacial projectile points, bone tools, ceramics, basketry, and stone carvings, together with other data, indicate that most if not all of the great inventions and innovations of prehistory appeared first in the New World, often in contexts of twice the antiquity of the previously presumed oldest examples from the Old World.

Thus Goodman is attempting to turn prehistory on its head. The accumulated discoveries of researchers in the prehistory of the Old and New Worlds are seen as being almost completely invalid. A scenario of the slow evolution of humanity from a series of African, Asian, and European hominids is viewed as incorrect. The movement of humans into the New World from Asia only after attaining fully modern "sapienhood" in the Old World is to be similarly discarded. In other words, Goodman would have us discard just about all of our constructs of human physical and cultural evolution.

Could archaeologists be so wrong? Can we have been so completely mistaken in our interpretations of the prehistoric record? Has Goodman provided a bold new synthesis for archaeology that will inevitably lead to a revolution in the science of the past?

To answer these questions, we must examine some of Goodman's specific claims. We will begin, as Goodman does, with the Bering landbridge. Archaeologists know that the ancestors of American Indians made their way into the New World via a land connection with the Old World that was made available during periods of glaciation. Goodman rejects this.

He claims that the land bridge was not rich in animal life and therefore conditions would not have been conducive to human migration. Goodman claims that the land bridge had terrible weather and was so harsh an environment that even as hardy a creature as the woolly rhinoceros did not make it across. So, because the woolly rhino did not make it, nothing or nobody could have. The fact that almost all of Canadian fauna is of Asiatic origin is not dealt with by Goodman. He also neglects to mention that, as hardy as the woolly rhino may have been, all evidence indicates that it was adapted to forests and forest steppe areas, which were not present on the land bridge. In fact, remains of the woolly rhino are quite rare in northeastern Asia for this reason, so its lack in the New World is not surprising (Flerow 1967). If the land bridge could not have supported animal populations, someone should tell this to the bison, snow sheep, muskox, moose, elk, brown bear, ermine, weasel, wolverine, wolf, red fox, lynx, arctic hare, lemmings, voles, and so on, who all made the trip from Asia to the New World on that very land-bridge.

The Map Goodman Omits in American Genesis: In *American Genesis*, Goodman supplies a map of the world replete with arrows indicating hypothetical migration routes of the original human beings outward from his "Paleo-Indian California Garden" (like in Eden). However, he provides no map and no arrows indicating where these "Paleo-Indians" came from. The map above remedies this by indicating the true source of American Indian populations according to Goodman's previous work—i.e., the so-called lost continents of Atlantis and Lemuria!

In terms of archaeological data related to the Bering land-bridge route Goodman (1981, p. 42) states, for example: "Dr. K. R. Fladmark of Simon Fraser University notes that the known distribution of early archaeological sites in the New World *does not* match that expected from an initial population from the Bering route." [Emphasis added.]

This seems very clearly to imply that Fladmark, a noted Canadian archaeologist, rejects a land-bridge migration route and thus adds support to Goodman's hypothesis. In fact, this is a gross misrepresentation of Fladmark's research. To quote from Fladmark's article (1979, p. 55) referenced by Goodman: "The intent of this paper is to examine and compare the feasibility of late Pleistocene coastal and interior routes for man entering southern North America *from* Beringia." [Emphasis added.]

Fladmark is obviously not questioning the feasibility of the land-bridge route; he is merely questioning where people went after they used that route. Goodman admits this later on, but his initial misrepresentation is likely to make more of an impression on the reader than his later clarification.

Goodman provides further "damning" evidence against the Bering land-bridge route. Referring to American Indian mythology regarding their own origins, Goodman (p. 20) states: "Conspicuously missing in all the known myths are any stories that bear the slightest resemblance to the notion of a Bering route; none seem to describe an arduous journey from Asia across the ice and snow of the North." Expecting Indians to "remember" the Bering land-bridge through myth is like expecting a modern Parisian to remember painting Pleistocene animals on cave walls. This argument is merely ludicrous.

Finally, Goodman claims that geological and meteorological evidence indicates that, even if people could have crossed the land bridge, ice and bad weather would have blocked their way to the south. But this notion of an insurmountable barrier of ice blocking the migration of humans south from Alaska into the continental United States is not supported by available evidence.

The funniest aspect of Goodman's argument here is that three chapters later he proposes the idea that American Indians migrated and, in fact, made regular round-trip journeys to Asia, and ultimately to Europe, via the Bering land-bridge (Goodman 1981, pp. 120–21).

Next, Goodman discusses particular sites in the New World in support of his hypothesis. He lists a series of sites, all purported to be much older than the oldest accepted Paleo-Indian sites (1981, p. 69). It should be noted that none of these sites, though very controversial, relate in any way to Goodman's more extreme claims of human physical evolution in California.

For example, Goodman refers to the Calico Hills site in California as one supported by the late Louis Leakey and estimated to be 100,000 to 500,000 years old. The "site" has been extremely controversial since its discovery, and whether the artifacts are the result of human or geological activity is still uncertain (see Haynes 1973). However, even if Calico turns out to be a valid site, it does not support Goodman's hypothesis that modern humanity *evolved* there, though Goodman (1981, p. 98) consistently implies that this is indeed what Louis Leakey thought. In an interview (Yates 1981), Goodman said: "Dr. Leakey was convinced that modern man originated here in Southern California." Leakey of course never said this in print; and, if he did believe it, we may ask what on earth he, Mary, and their son Richard were doing excavating in Africa when it was "all happening" in California?

Goodman claims that skeletal evidence, primarily from California, shows that not only are there very ancient sites in the New World but that fully modern *Homo sapiens sapiens* were present here at least 70,000 years ago. This is astounding, because the earliest evidence of physically modern humans in the Old World, where everyone else thinks they evolved, points to only about 35,000 to 40,000 years ago. Goodman bases his claim on five prehistoric skeletons from California, all of which have been dated by the amino acid racemization dating technique to between 44,000 and 70,000 B.P. (Before Present). The skulls of these skeletons are completely modern in appearance and, in fact, look just like those of contemporary California Indians. If the dates are valid, this would suggest tremendous genetic stability, to say the least. Recent work with the racemization technique, however, indicates that temperature fluctuations, the rate of soil deposition, and even the kinds of plants growing in the soil near the skeletal material can seriously affect the dates. Two of the skeletons so crucial to Goodman's argument were recently dated with a uranium series technique and turned out to be between 8,000 and 11,000 years old (Bischoff and Rosenbauer 1981). Thus Goodman's most serious claim is based only on very questionable dates on a handful of skeletons. Should we ignore the entire fossil hominid record of the Old World because of five very questionable dates on these California materials?

Goodman goes on to claim that these early New World humans invented most of what are now considered to be the hallmarks of humanity and civilization. Only later, the argument goes, did they bring these things to the rest of the world. Goodman (1981, p. 178) states: "Our debt to the Paleo-Indians could include the first domestication of plants, the first domestication of animals, the first practice of freeze-drying food, pottery, the calendar, astronomy, and the applied understanding of the physics behind electro-magnetics and Einstein's gravity waves." Where they found the time to make spears and kill woolly mammoths while doing all of this, we are not told. Those who are familiar with the paranormal literature may notice a similarity in the extent of the claimed precocity of American Indian culture and that usually ascribed to the residents of the "lost continent" of Atlantis. This similarity is not coincidental, as we shall see.

Can Goodman trace the diffusion of these revolutionary ideas from California to the rest of the world? Is there any evidence to back up his claims? Since stone tools are among the most common kinds of artifacts found by archaeologists, we can start here. Can we trace the movement of stone-tool types from their invention in the New World into the Old World during prehistory, thus supporting Goodman's concept of migration from east to west?

Goodman claims that the leaf-shaped Solutrean spearpoints of Ice Age Europe were based on the fluted or channeled point of the New World. Thus Paleo-Indians traveled from California to Europe, bringing with them their distinctive toolkit and introducing bifacial projectile points into the Old World. This ignores a couple of minor items, such as the fact that the most distinctive aspect of fluted points – their fluting – is entirely absent from the Old World material. There is also the problem of some of the European points being about 5,000 years *older* than the earliest American ones. Goodman disagrees here and, basing his hypothesis on exactly one point from one site, claims that in the New World fluting is not 12,000 years old, but 38,000 years old. The site is in Lewisville, Texas. There, in a purported hearth whose C-14 date is 38,000 B.P., a fluted point was recovered. The date perturbed archaeologists when the site was excavated in the 1950s, and people have tried to explain it in terms of a provenience problem or even fraud. Clearly the site is out of line with the hundreds of other dates available for the fluted-point technology in the New World. What does Goodman say about all this?

Goodman says the fluted point from Lewisville is obsidian. It is not. Goodman says (1981, p. 74) subsequent finds made by Dennis Stanford of the Smithsonian Institution resulted in "stone tools and debris which support the original evidence." According to Stanford (1981, p. 91), *no* subsequent stone tools were found at Lewisville by him or anyone else. Stanford, who has built a career on his quest for finding pre-Paleo sites and who has excavated at Lewisville, states that all of the hearth material recovered at the site used in radiocarbon dating was contaminated with local, naturally occurring lignitic coal and that the dates could be off by as much as 27,500 years (Stanford 1981, p. 91). Lewisville may be a genuine Paleo site after all – but it is probably around 10,500 years old, an expectable date. It would be hasty indeed to rewrite world prehistory on the basis of these data.

According to Goodman, basketry is also older in the New World than in the Old. This hypothesis is based on evidence from only one site. Goodman (1981, p. 85), describing the archaeological material recovered from Meadowcroft Rockshelter in southwestern Pennsylvania, writes: "And from a slightly deeper level came a radiocarbon date of 20,000 years ago for what is believed to be a basketry fragment." Later on (p. 123), when discussing the source of rope wicks used in oil lamps in Europe during the Upper Paleolithic, he states: "The rope wick would take advantage of the same technical skill used in the making of baskets which appeared at Meadowcroft in Pennsylvania over 20,000 years ago."

So, what was "believed to be a basketry fragment" just 38 pages earlier, now becomes definite "baskets." In fact, the excavators of Meadowcroft have been very careful to label the *single* artifact in question a "carbonized fragment of cut bark-like material/possibly basketry fragment" (Adovasio et al. 1979), which is dated to 17,650 B.P. ±2,400 years. In other words, a "fragment of cut bark-like material" with a widely ranging carbon date becomes definite 20,000-year-old baskets in the New World for Goodman.

This brings us to a discussion of the Flagstaff site. This site, excavated by Goodman, is a central piece of evidence in his argument. Inex-
plicably, we are not told in *American Genesis* that the supposed site was
"discovered" by psychic means and similarly interpreted. Goodman
claims an age of between 100,000 and 170,000 years for this Arizona site.
Somehow, based on a number of extremely questionable artifacts, Goodman wishes us to believe that his Flagstaff site shows that the Hopi
Indian creation-myth is literally true and should supplant the theory of
evolution. Goodman's *pièce de résistance,* however, is his "Flagstaff
stone" (1981, p. 173): "a flat stone, a piece of hard volcanic ash approximately four inches by six inches in size which had a number of straight
lines on both of its sides. It looked like an engraving, it had to be an
engraving, . . . it was as if a Paleo-Indian had left his irrefutable signa-
ture here for us." Goodman (p. 173) goes on to claim, "I believe that here
in one artifact alone there is evidence of fully modern man's earlier presence and earlier sophistication in the New World than in the Old World."
Thus we have, just outside of Flagstaff, Arizona, a prototype of the cave
art of the European Upper Paleolithic at least three and one-half times
the age!

Goodman assures us that experts in Paleolithic art have analyzed the
Flagstaff stone and were impressed. Alexander Marshack of Harvard's
Peabody Museum examined the stone and supposedly said that the
markings looked intentional and were quite similar to those from Cro-
Magnon sites in Europe (Goodman 1981, p. 174). However, as Goodman
states (p. 175): "The highly weathered and now very soft surface of the
stone was apparently damaged in cleaning and thus many of the lines had
been stripped of bits of information which would have let Marshack
make a more conclusive determination. Marshack said that if we could
resolve these problems then the 'Flagstaff stone' would be one of the
most important artifacts ever found in the entire world."

Goodman neglects to tell us *exactly* what Marshack wrote to him
after he examined the stone. In Dennis Stanford's review (1981, p. 92) of
American Genesis, we see *exactly* what Marshack told Goodman. Marshack said that he believed the grooves on the rock were intentional,
however, "Every groove without exception had been deepened and
straightened, reworked after it was dug out of the ground . . . thus the
stone cannot be used as evidence that early man engraved it."

This litany of incorrect data, misrepresentation of others' research
or statements, and the apparent mishandling of a possibly important
artifact barely scratches the surface of Goodman's own particular brand
of pseudoscience. I have not mentioned his claim of having evidence of
the Paleo-Indian domestication of corn 80,000 years ago (1981, p. 179),
based on a few questionable pollen grains from Mexico, and the domestication of horses inferred from representations of pregnant mares in the

cave paintings of Europe (1981, p. 180). How about Paleo-Indians inventing aspirin, insulin, and birth control pills more than 10,000 years ago? (Goodman 1981, pp. 178–79)

What is Goodman leading up to? If people did not evolve in the Old World, and if modern humans arrived fully developed in California with an astoundingly sophisticated technology 500,000 years ago, where, pray tell, did they come from? Goodman never really says, though he poses a significant question (1981, p. 91): "Was modern man's world debut the result of slow development or the result of a quantum leap inspired from some outside source?" (Emphasis added.)

What could be this "outside source"? Goodman doesn't tell us in American Genesis, just as he does not tell us that the Flagstaff site was supposedly found through psychic powers. However, in Goodman's earlier book, Psychic Archaeology: Time Machine to the Past (1977), we find that the psychic who found the Flagstaff site, and who was allegedly accurate in his predictions of what was to be found there, provided the answer to the ultimate origin of American Indians: They came to the New World 500,000 years ago from the now lost continents of Atlantis and Lemuria (Goodman 1977, p. 88). The sites in the New World that are so much older than those of the Old World represent outposts or colonies from Atlantis. Some of the great inventions Goodman ascribes to American Indians in American Genesis are listed by his psychic as coming directly from Atlantis (Goodman 1977, p. 92).

Goodman has merely written a new chapter in the saga of what Cole (1980) calls "cult archaeology," in which the field is treated not as a scientific enterprise of discovery and explanation but as a foundation for pseudoscientific belief. Within the "field" of cult archaeology we may include, as Cole does, the ancient-astronaut "theory," the search for Noah's Ark, Atlantisology, inscription mania, psychic archaeology and, now, "American Genesis."

Goodman's arguments depend on psychic archaeology, rejection of human evolution, abandonment of well-supported cultural chronologies, ignoring genetics and epidemiology, and the existence of the lost continent of Atlantis. To call this credulity is an understatement.

So we have come full circle; a "rebel" scholar who writes a book to present a new, revolutionary hypothesis concerning the origins of the American Indians is simply revising an explanation first suggested by the Spanish author Lopez de Gomara in 1522 (Huddleston 1967, p. 24). It is not surprising that Goodman does not desire to spread this around.

There is a final irony here. Goodman presents himself as a great defender of the significance of Indian prehistory. Only he fully appreciates just how impressive the accomplishments of American Indians really were (1981, p. 197): "Today, in the teeth of the facts, many archaeologists still believe that every prehistoric invention of consequence, was made in the Old World instead of the New." So we all put down the accomplishments of American Indians, but Goodman is a beacon of knowledge and understanding.

Goodman should read the works of the originator of the Atlantis-Indian Connection, Lopez de Gomara, who characterized Indians as "stupid, wild, insensate asses" who went around naked, were liars, ingrates, and cannibals, and engaged in public sexual intercourse with animals (Huddleston 1967, p. 24).

Goodman does no favor to American Indians, whose genuine past shows them to be the equal of any group of people in the world, by concocting an outrageous and disingenuous fantasy. Now that Goodman has set all of us archaeologists straight, he will be going on to solve the problems of evolution in his next book, *The Genesis Mystery: The Sudden Appearance of Modern Man*. So, to return to the very beginning of this paper: just as we did with television's "Twilight Zone," we can tune in again for another excursion into fantasy.

Acknowledgments

The final manuscript of this paper has greatly benefited from the suggestions, comments, and assistance of Dr. Michael A. Park of the Department of Anthropology at Central Connecticut State College and Dr. John R. Cole of the Department of Anthropology at the University of Northern Iowa. Both are committed to the scientific unraveling of the true mysteries of our species and active participants in opposition to the created "mysteries" of pseudoscience.

I also thank Mrs. Harriet B. Martin, secretary to the Department of Anthropology at the University of Arizona, for her help in finding out where Dr. Goodman received his Ph.D.

Thanks are also due, as always, to Mrs. Ann Ruddock, our departmental secretary, for unraveling the mysteries of my penmanship and producing the final manuscript.

References

Adovasio, J. M., J. D. Gunn, J. Donahue, R. Stuckenrath, J. Guilday and K. Lord. 1979-80. Meadowcroft Rockshelter—Retrospect 1977: Part I. *North American Archaeologist* 1(1):3-44.

Austin, Janice. 1976. A test of Birdsell's hypothesis on New World migration. Paper presented at the annual meeting of the Society for California Archaeology.

Bischoff, James L., and Robert J. Rosenbauer. 1981. Uranium series dating of human skeletal remains from the Del Mar and Sunnyvale sites, California. *Science* 213:1003-1005.

Cole, John R. 1978a. Anthropology beyond the fringe. *Skeptical Inquirer* 2(2): 62-71.

———. 1978b. Review of J. Goodman, *Psychic Archaeology: Time Machine to the Past. Skeptical Inquirer* 2(2):105-108.

————. 1980. Cult archaeology and unscientific method and theory. In *Advances in Archaeological Methods and Theory*, vol. 3, edited by Michael B. Schiffer. New York: Academic Press.

Fagan, Brian. 1981. American genesis? *Early Man Review* 3(4):24–26.

Feder, Kenneth. 1980. Psychic archaeology: The anatomy of irrationalist prehistoric studies. *Skeptical Inquirer* 4(4):32–43.

Fladmark, K. R. 1979. Routes: Alternative migration corridors for early man in North America. *American Antiquity* 44:55–69.

Flerow, C. C. 1967. On the origin of the mammalian fauna of Canada. In *The Bering Land Bridge*, edited by D. M. Hopkins. Stanford: Stanford University Press.

Goodman, Jeffrey. 1977. *Psychic archaeology: Time machine to the past.* New York: Berkley.

————. 1981. *American Genesis.* New York: Berkley.

Haynes, Vance. 1973. The Calico Site: Artifacts or geofacts? *Science* 181:305–309.

Huddleston, Lee. 1967. *Origins of the American Indians: European concepts 1492–1729.* Austin: University of Texas Press.

Reeves, B. O. K. 1973. The nature and age of the contact between the Laurentide and Cordilleran icesheets in the western interior of North America. *Arctic and Alpine Research* 5:1–16.

Stanford, Dennis. 1981. Who's on first? *Science 81.* June:91-92.

Turner, Christy G. II. 1981. A review of *American Genesis: The American Indian and the Origins of Modern Man. Archaeology* 35:72-74.

Yates, Ronald. 1981. Scientist stuns anthropological world. *Hartford Courant,* April 20. (Article originally appeared in the *Chicago Tribune*.)

32

The Nazca Drawings Revisited: Creation of a Full-Sized Duplicate

Joe Nickell

Called "Riddles in the Sand" (*Discover* 1982) they are the famous Nazca lines and giant ground drawings etched across 30 miles of gravel-covered desert near Peru's southern coast.

The huge sketch-pad came to public prominence in Erich von Däniken's *Chariots of the Gods?*—a book that consistently underestimates the abilities of ancient "primitive" peoples and assigns many of their works to visiting extraterrestrials. Von Däniken (1970) argues that the Nazca lines and figures could have been "built according to instructions from an aircraft." He adds: "Classical archaeology does not admit that the pre-Inca peoples could have had a perfect surveying technique. And the theory that aircraft could have existed in antiquity is sheer humbug to them."

Von Däniken does not consider it humbug, and he obviously envisions flying saucers hovering above and beaming down instructions for the markings to awed primitives in their native tongue. He views the large drawings as "signals" (von Däniken 1970) and the longer and wider of the lines as "landing strips" (von Däniken 1972). But would extraterrestrials create signals for themselves in the shape of spiders and monkeys? And would such "signals" be less than 80 feet long (like some of the smaller Nazca figures)?

As to the "landing strip" notion, Maria Reiche, the German-born mathematician who for years has mapped and attempted to preserve the markings, has a ready rejoinder. Noting that the imagined runways are clear of stones and that the underlying ground is quite soft, she says, "I'm afraid the spacemen would have gotten stuck" (McIntyre 1975).

It is difficult to take von Däniken seriously, especially since his "theory" is not his own and it originated in jest. Wrote Paul Kosok (1947), the first to study the markings: "When first viewed from the air, [the lines] were nicknamed prehistoric landing fields and jokingly compared with the so-called canals on Mars." Moreover, one cropped photo exhibited by von Däniken (1970), showing an odd configuration "very reminiscent of the aircraft parking areas in a modern airport," is actually of the knee joint of one of the bird figures (Woodman 1977). (See Figure 1.) The spacecraft

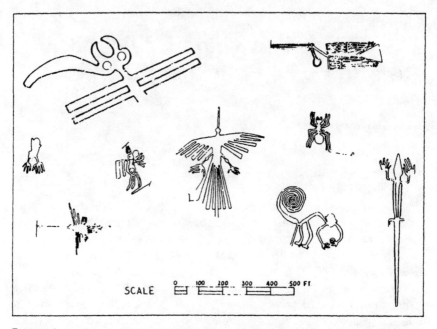

FIGURE 1. Etched upon the Nazca plains in Peru are giant drawings like these. Their large size has fueled misguided speculation that they were drawn with the aid of "ancient astronauts" or by sophisticated surveying techniques, the secrets of which are lost.

that parked there would be tiny indeed.

Closer to earth, but still merely a flight of fancy, in my opinion, is the notion of Jim Woodman (1977) and some of his colleagues from the International Explorers Society that the ancient Nazcas constructed hot-air balloons for "ceremonial flights," from which they could "appreciate the great ground drawings on the *pampas.*" If one believes that the theory is also inflated with hot air, one must at least give Woodman credit for the strength of his convictions. Using cloth, rope, and reeds, Woodman and his associates actually made a balloon and gondola similar to those the Nazcas might have made had they actually done so. Woodman and British balloonist Julian Nott then risked their lives in a 300-foot-high fly-over of the Nazca plain. Their balloon was descending rapidly and after they had thrown off more and more sacks of ballast they jumped clear of their craft some ten feet above the *pampas.* Free of the balloonists' weight, the balloon shot skyward and soared almost out of sight, only to finally crash and drag briefly across the ground.

The Nazca markings are indeed a mystery, although we do know who produced them—von Däniken notwithstanding. Conceding that Nazca pottery is found in association with the lines, von Däniken (1970) writes: "But it is surely oversimplifying things to attribute the geometrically arranged lines to the Nazca culture for that reason alone."

No knowledgeable person does. The striking similarity of the stylized figures to those of known Nazca art has been clearly demonstrated (Isbell 1978; 1980). In addition to this iconographic evidence must be added that from carbon-14 analysis: Wooden stakes mark the termination of some of the long lines and one of these was dated to A.D. 525 (±80). This is consistent with the presence of the Nazca Indians who flourished in the area from 200 B.C. to about A.D. 600. Their graves and the ruins of their settlements lie near the drawings.

The questions of who and when aside, the mystery of *why* the markings were made remains, although several hypotheses have been proffered. One is that they represent some form of offerings to the Indian gods (McIntyre 1975). Another is that they form a giant astronomical calendar or "star chart." Writing in *Scientific American*, William H. Isbell (1978) states:

> As Reiche has pointed out for many years, certain of the Pampa Colorada lines mark the position of the sun at the summer and winter solstices and certain other lines also appear to have calendrical significance. A computerized analysis of line orientation conducted by Hawkins, although it failed to demonstrate that a majority of the lines have astronomical significance, showed that twice as many of them were oriented with respect to annual solar and lunar extremes than would be expected on the basis of chance.

Isbell himself suggests that an important function of the markings was economic and "related to the drafting of community labor for public works," although at best that is only a partial explanation.

Still another suggestion (first mentioned by Kosok) comes from art historian Alan Sawyer (McIntyre 1975): "Most figures are composed of a single line that never crosses itself, perhaps the path of a ritual maze. If so, when the Nazcas walked the line, they could have felt they were absorbing the essence of whatever the drawing symbolized." Sawyer is correct in observing that most of the figures are drawn with a continuous, uninterrupted line. But there *are* exceptions, and it is possible that the continuous-line technique is related to the method of producing the figures, as we shall discuss presently.

In any case, these are only some of the hypotheses; whatever meaning(s) we ascribe to the Nazca lines and drawings must be considered in light of other giant ground-markings elsewhere. Even putting aside the Japanese and European ones—e.g., the White Horse of Uffington, England, which is known from as early as the twelfth century (Welfare and Fairley 1980)—we are left with numerous ground drawings in both North and South America.

In South America giant effigies are found in other locales in Peru, for example, and in Chile, in the Atacama Desert (Welfare and Fairley 1980). Interestingly, the plan of the Incan city of Cuzco was laid out in the shape of a puma, and its inhabitants were known as "members of the body of the puma" (Isbell 1978; 1980).

Turning to North America, there is the Great Serpent Mound in Ohio and giant effigies in the American Southwest. In 1978, with the aid of an Indian guide, I was able to view the ground drawings near Blythe, California, in the Mojave Desert. Like the Nazca figures, the Blythe effigies are large and give the impression they were meant to be viewed from the air. Also in common with the Nazca figures, they were formed by clearing away the surface gravel to expose the lighter-colored soil. However, although they are thought to date from a much later period (Setzler 1952), none of the Blythe figures match the size of the largest Nazca drawings; and the human figures and horselike creatures are much cruder in form, typically having solid-area bodies and sticklike appendages—quite unlike the continuous-line drawings of Nazca (yet somewhat similar to some of the Chilean effigies). Moreover, absent from the Blythe site are the "ruler-straight" lines that may or may not have calendrical significance.

In short, there are similarities and dissimilarities between the Nazca and other ground drawings that complicate our attempts to explain them. Certainly the Blythe and other effigies have no attendant von Dänikenesque "runways"; neither do their crude forms suggest they were drawn with the aid of hovering spacecraft. And there is nothing whatever to warrant the assumption that they were made to be viewed by select native balloonists on aerial sorties.

It seemed to me that a study of how the lines were planned and executed might shed some light on the ancient riddle. English explorer and film-maker Tony Morrison has demonstrated that, by using a series of ranging poles, straight lines could be constructed over many miles (Welfare and Fairley 1980). (The long lines "veer from a straight line by only a few yards every mile," reports Time [1974].) In fact, along some lines, the remains of posts have been found at roughly one-mile intervals (McIntyre 1975).

By far the most work on the problem of Nazca engineering methods has been done by Maria Reiche (1976). She explains that Nazca artists prepared preliminary drawings on small six-foot-square plots. These plots are still visible near many of the larger figures. The preliminary drawing was then broken down into its component parts for enlargement. Straight lines, she observed, could be made by stretching a rope between two stakes. Circles could easily be scribed by means of a rope anchored to a rock or stake, and more complex curves could be drawn by linking appropriate arcs. As proof, she reports that there are indeed stones or holes at points that are centers for arcs.

But Reiche does not detail the specific means for positioning the stakes that apparently served as the centers for arcs or the end points of straight lines. In her book she wrote, "Ancient Peruvians must have had instruments and equipment which we ignore and which together with ancient knowledge were buried and hidden from the eyes of the conquerors as the one treasure which was not to be surrendered." Be that as it may, Isbell (1978) states: "Maria Reiche, using scale models, has made major advances toward demonstrating how Nazca ground art was produced. Although more research needs to be done, the prehistoric engineering skills are no longer completely unknown."

Isbell himself suggests that the Nazcas used a grid system adapted from their weaving experience, a loom "establishing a natural grid within which a figure is placed." All that would be necessary, he observes, would be to simply enlarge the grid to produce the large drawings.

However, as one who has used the grid system countless times (in reproducing large trademarks and pictorials on billboards—summer work during my high school and college years), I am convinced the grid system was not employed. To mention only one reason, a characteristic of the grid method is that errors or distortions are largely confined to individual squares. Thus, the "condor" drawing in Figure 1—with its askew wings, mismatched feet, and other asymmetrical features—seems not to have been reproduced by means of a grid.

Other, even less likely possibilities would be the plotting of points by a traverse surveying technique (such as is used today to plot a boundary of land) or by triangulation. Having some experience with both of these, I note that such methods depend on the accurate measurement of angles, and there appears to be no evidence that the Nazcas had such a capability.

I decided to attempt to reproduce one of the larger Nazca figures—the 440-foot-long condor in the center of Figure 1—using a means I thought the Nazcas might actually have employed. I was joined in the project by two of my cousins, John May and Sid Haney. The method we chose was quite simple: We would establish a center line and locate points on the drawing by plotting their coordinates. That is, on the small drawing we would measure along the center line from one end (the bird's beak) to a point on the line directly opposite the point to be plotted (say a wing tip). Then we would measure the distance from the center line to the desired point. A given number of units on the small drawing would require the same number of units—larger units—on the large drawing.

For this larger unit we used one gleaned by Maria Reiche from her study of the Nazca drawings and approximately equivalent to 12.68 inches. For measuring on the ground, we prepared ropes marked off with paint into these Nazca "feet," with a knot tied at each ten-"foot" interval for a total length of 100 units. To aid in accuracy in plotting on the ground, we decided to employ a "T" made of two slender strips of wood. With this we could ensure that each measurement made from the center line would be at approximate right-angles to the line.

My father, J. Wendell Nickell, took charge of logistics—including obtaining permission to use a suitable giant "drawing board" (a landfill area in West Liberty, Kentucky, owned by Dr. C. C. Smith, to whom we are grateful) and securing the services of a pilot for the subsequent aerial photography. Since we could not mark the lines by clearing gravel to expose lighter-colored earth, as the Nazcas did, we planned to simply mark them with white lime, as one marks a playing field. With the addition of my young cousin, Jim Mathis, and my 11-year-old nephew, Conrad Nickell, our work crew of Indians was complete.

On the morning of August 7, 1982, the six of us assembled at the site and immediately began by laying out the center line. Some nine hours, one meal, and much ice-water later, we had plotted and staked the last of 165 points and had connected them with twine.

Here, I think, we differed slightly from the Nazcas, for I seriously doubt they expended just over a mile of string (the total distance traversed by the outline). I rather suspect that they made their furrows (or at least preliminary scratched lines) as they progressed in plotting the various points. We could not do this, since rain threatened and would certainly obliterate our lines of powdered lime. But we did find it helpful (though not

FIGURE 2.   A duplication of the giant "condor" drawing made full size and utilizing only sticks and cord such as the Nazcas might have employed. The experimental drawing—possibly the world's largest art reproduction— is viewed here from just under 1,000 feet.

essential) to connect our points in sequence, to prevent possible confusion with stakes sometimes clustered rather closely together. (Otherwise we would have needed only a single long length of cord, to be used for the final marking of each straight line.)

The rains did come, and while no harm was done to our staked-out condor, large puddles (then more rain and still more puddles) prevented our completing our project for about a week. Finally, the ground had dried, the weather forecast was good, and the pilot was on standby. My father and I then spent much of one day marking the lines, finishing just in time to see the airplane circling.

Jerry Mays, a skilled local pilot, then took John and me up in his Cessna for a preliminary look and the taking of photographs, which John accomplished at just under 1,000 feet.

Our work was a success. In fact the results were so accurate that we are convinced we could have easily produced a more symmetrical figure by this method. Thus it would seem—unless they employed an even simpler method of making the enlargement—that the Nazcas plotted considerably fewer points. That, coupled with mere visual estimation of right angles and less careful measurement (distances might simply be stepped off), could account for the imperfections we observed. Also, an entire small area, such as a foot, could have been done completely freehand. (Our own freehand work was minimal: We produced the circle of the head by scribing it with a rope. All other curves were *marked* freehand; of course we *had* plotted the numerous points that served as a guide, although we bypassed stakes slightly in attempting to draw smooth curves.)

It is frequently asserted that the Nazca drawings are recognizable only from the air. That is not quite true, certainly not of the smaller figures, such as the effigy of a fish, which is only 80 feet long (Reiche 1976). Neither is it true of some drawings—attributed to the Nazcas' predecessors—that are found on hill slopes (McIntyre 1975; Isbell 1978, 1980). Here, seemingly, is a clue to how the Nazcas could have been confident of the accuracy of their method of enlargement. Once a technique was found to be successful for producing large drawings on slopes, where they could actually be viewed from the ground, the same technique could be expected to consistently yield good results—wherever figures were drawn and whatever their size.

Moreover, even the large drawings can be appreciated to some extent from the ground. With our condor, we were able to see whole portions—such as body and head, leg and foot, the entire fan of the tail—and thus had determined the figure was reasonably accurate even before our fly-over. We felt that an observer would be able to recognize it as a bird.

To test this possibility, my father took wildlife biologist Harold Barber to the site. Although Barber knew nothing of our project, and Nazca was deliberately not mentioned, on viewing the figure he recognized the drawing as one of the Nazca birds. That he was familiar with the Nazca ground drawings was unfortunate for our experiment (and rain prevented another); but the salient point is that he was able to identify the figure as a bird rather than as a spider, fish, monkey, or some other figure. In fact, when he was later shown pictures of several Nazca bird drawings, he immediately and correctly identified ours as the condor.

In summary, we do know that it was the Nazcas who produced the drawings. While their large size does suggest the possibility that they were meant to be viewed from above, as by the Indian gods, the figures can be recognized, at least to some extent, from the ground. The drawings could have been produced by a simple method requiring only materials available to South American Indians centuries ago. The Nazcas probably used a simplified form of this method, with perhaps a significant amount of the work being done freehand. There is no evidence that extraterrestrials were involved; but, if they were, one can only conclude that they seem to have used sticks and cord just as the Indians did.

Acknowledgments

In addition to those mentioned in the text, the author also wishes to thank his mother, Ella T. Nickell; Robert H. van Outer and the University of Kentucky Photographic Services; Carl Burton; and the May Grocery Co. and the Blair Wholesale Grocery Co., both of West Liberty, Kentucky.

References

Discover. 1982. "Riddles in the Sand" (June): 50-57.

Isbell, William H. 1978. "The Prehistoric Ground Drawings of Peru." *Scientific American* 239 (October): 140-53.

_____. 1980. "Solving the Mystery of Nazca." *Fate* (October): 36-48.

Kosok, Paul. 1947. "The Markings of Nazca" (written in collaboration with Maria Reiche). *Natural History* 56: 200-38.

McIntyre, Loren. 1975. "Mystery of the Ancient Nazca Lines." *National Geographic* (May): 716-28.

Reiche, Maria. 1976. *Mystery on the Desert* (1968), rev. ed. Stuttgart: Privately printed.

Setzler, Frank M. 1952. "Seeking the Secrets of the Giants." *National Geographic* 102: 393-404.

Time. 1974. "Mystery on the Mesa" (March 25).

Von Däniken, Erich. 1970. *Chariots of the Gods?* New York: G. P. Putnam.

_____. 1972. *Gods from Outer Space.* New York: Bantam Books.

Welfare, Simon, and Fairley, John. 1980. *Arthur C. Clarke's Mysterious World.* New York: A & W Publishers.

Woodman, Jim. 1977. *Nazca: Journey to the Sun.* New York: Pocket Books.

Creationism and Shroud Science

33
Science and the Mountain Peak

Isaac Asimov

Some scientists are making their peace with theology. If we listen to them, they will tell us that science has only managed to find out, with a great deal of pain, suffering, storm, and strife, exactly what theologians knew all along.

A case in point is Robert Jastrow, an authentic professor of astronomy who has written a book called *God and the Astronomers*. In it and in recent articles such as "Have Astronomers Found God?" *(Reader's Digest,* July 1980) he explains that astronomers have discovered that the Universe began very suddenly and catastrophically in what is called a big bang and that they're upset about it.

The theologians, however, Jastrow says, are happy about it, because the Bible says that the Universe began very suddenly when god said, *Let there be light!*

Or, to put it in Jastrow's very own words: "For the scientist who has lived by his faith in the power of reason, the story ends like a bad dream. He has scaled the mountains of ignorance; he is about to conquer the highest peak; as he pulls himself over the final rock, he is greeted by a band of theologians who have been sitting there for centuries."

If I can read the English language, Jastrow is saying that astronomers were sure, to begin with, that the Bible was all wrong; that if the Bible said the Universe had a beginning, astronomers were sure the Universe had no beginning; that when they began to discover that the Universe did have a beginning, they were so unhappy at the Bible being right that they grew all downcast about their own discoveries.

Nothing in Common

Furthermore, if I can continue to read the English language, Jastrow is implying that since the Bible has all the answers—after all, the theologians have been sitting on the mountain peak for centuries—it has been a waste of time, money, and effort for astronomers to have been peering through their little spyglasses all this time.

Perhaps Jastrow, abandoning his "faith in the power of reason" (assuming he ever had it), will now abandon his science and pore over the Bible until he finds out what a quasar is, and whether the Universe is open or closed, and where black holes might exist—questions astronomers are working with now. Why should he waste his time in observatories?

But I don't think Jastrow will, because I don't really think he believes that all the answers are in the Bible—or that he takes his own book very seriously.

In the first place, any real comparison between what the Bible says and what the astronomer thinks shows us instantly that the two have virtually nothing in common. And here are some real comparisons:

1. The Bible says that the Earth was created at the same time as the Universe was *(In the beginning, god created the heaven and the earth)*, with the whole process taking six days. In fact, whereas the Earth was created at the very beginning of creation, the Sun, Moon, and stars were not created until the fourth day.

The astronomer, on the other hand, thinks the Universe was created 15 billion years ago and the Earth (together with the Sun and the Moon) was not created until a little less than five billion years ago. In other words, for ten billion years the Universe existed, full of stars, but without the Earth (or the Sun or the Moon).

2. The Bible says that in the six days of creation, the whole job was finished *(Thus the heavens and the earth were finished, and all the host of them. And on the seventh day god ended his work which he had made)*.

The astronomer, on the other hand, thinks stars were being formed all through the 15 billion years since the Universe was created. In fact, stars are still being formed now, and planets and satellites along with them; and stars will continue being formed for billions of years to come.

3. The Bible says that human beings were created on the sixth day of creation, so that the Earth was empty of human intelligence for five days only.

The biologist, on the other hand, thinks (and the astronomer does not disagree) the earliest beings that were even vaguely human didn't appear on the Earth until well over 4½ billion years after its creation.

4. The Bible doesn't say when the creation took place, but the most popular view among the theologians on that mountain peak is that creation took place in 4004 B.C.

As I've said, the astronomer thinks creation took place 15 billion years ago.

5. The Bible says the Universe was created through the word of god.

The astronomer, on the other hand, thinks the Universe was created through the operation of the blind, unchanging laws of nature—the same laws that are in operation today.

(Notice, by the way, that in these comparisons I say, "The Bible says ..." but "The astronomer thinks . . ." That is because theologians are always certain in their conclusions and scientists are always tentative in theirs. That, too, is an important distinction.)

Theologians on Their Backs

These are enormous differences, and it would be a very unusual astronomer who could imagine finding any theologians on *his* mountain peak. Where are the theologians who said that creation took place 15 billion years ago? That the Earth was formed ten billion years later? That human beings appeared 4½ billion years later still?

Some theologians may be willing to believe this *now*, but that would only be because scientists showed them the mountain peak and carried them up there.

So what the devil is Jastrow talking about? Where is the similarity between the book of Genesis and astronomical conclusions?

One thing. One thing only.

The Bible says the Universe had a beginning. The astronomer thinks the Universe had a beginning.

That's all.

But even this similarity is not significant, because it represents a *conclusion*, and conclusions are cheap. *Anyone* can reach a conclusion— the theologian, the astronomer, the shoeshine boy down the street.

Anyone can reach a conclusion in any way—by guessing it, by experiencing a gut feeling about it, by dreaming it, by copying it, by tossing a coin over it.

And no matter who reaches a conclusion, and no matter how he manages to do it, he may be right, provided there are a sharply limited number of possible conclusions. If eight horses are running a race, you might bet on a particular horse because the jockey is wearing your favorite colors or because the horse looks like your Aunt Hortense—and you may win just the same.

If two men are boxing for the championship and you toss a coin to pick your bet, you have one chance in two of being right—even if the fight is rigged.

How does this apply to the astronomical and theological view of the Universe? Well, we're dealing with something in which there are a sharply limited number of conclusions—more than a two-man prizefight, but fewer than an eight-animal horserace. There are, after all, just three things that might be happening to the Universe in the long run:

A. The Universe may be unchanging, on the whole, and therefore have neither a beginning nor an end—like a fountain, which, although individual water drops rise and fall, maintains its overall shape indefinitely.

B. The Universe may be changing progressively; that is, in one direction only, and may therefore have a distinct beginning and a different end—like a person, who is born, grows older (never younger), and eventually dies.

C. The Universe may be changing cyclicly, back and forth, and therefore have an end that is at the beginning, so that the process starts over endlessly—like the seasons, which progress from spring, through summer, fall, and winter, but then return to spring again, so that the process starts over.

A Myth Shared by All Mythmakers

If theologians or astronomers, or the shoeshine boy down the street, choose from one of these three by picking one of three different cards in a hat, they will each have a one-in-three chance of being right.

It is not, however, by sheer chance that a decision is usually come to in this case. All our experience tells us that various familiar objects have a beginning and an end. A loaf of bread is baked and finally is eaten; a suit of armor is fashioned and finally rusts; a human being is born and finally dies. For that reason, alternative B seems intuitively the likely situation with respect to the Earth and the Universe.

It is not surprising, then, that people generally think there must have been a beginning to the Universe, and that even theologians think so. It is not only our Biblical theologians who think so, almost all mythmakers did. The world had a beginning in the Greek and Norse mythologies and, for that matter, in the Babylonian mythology from which the exiled Jews borrowed (with modifications) their own Genesis tale.

It is not only our theologians who are sitting on Jastrow's mountain peak, but a whole melange of primitive bards and medicine men. Now, it seems, the astronomers also suspect that alternative B is correct, that the Universe had a beginning, and they are sitting on the same mountain peak.

But conclusions don't matter. The mountain peak means nothing, since you can get there by guessing. *What counts is the route one takes to the mountain peak.*

Theologians, mythmakers, legend-constructors, dreamers—all of them, *but not* scientists—derive their conclusions from intuition, or by whatever words you use to mean intuition—divine revelation, transcendental meditation, sudden enlightenment, dreams, inspiration. The words don't matter; they all mean that the conclusion is born from within one person.

But how can *you* check out a conclusion that comes from inside some other person? The results of one person's intuition cannot force another to believe. In other words, they are not compelling.

Oh, yes, the charismatic dreamer can sweep people along on a tide of emotionalism. He can convert them into armies and send them out to kill—or, as we have recently seen, to commit suicide. What's *that* got to do with the truth?

These hot-blooded attempts to find truth by intuition mean nothing to the intellectual history of humanity, except insofar as they have barred the way to an increase in our knowledge of the Universe and have succeeded in keeping us longer in the mire of ignorance.

Science—of which astronomy is one branch—is the one human endeavor that does not rely on intuition. Intuition pops up here and there on the road science travels, but the final decision on which branch to follow at each of an infinite number of intersections is based on careful observation and measurement of natural phenomena and deliberately arranged experiments. Deductions and inductions are made from those observations and measurements, according to the established and universally accepted rules of reason. What's more, everything is done in the open, and nothing can be accepted unless and until those observations and measurements are repeated—independently.

Even then the acceptance of a particular "truth" is never more than tentative. It is always subject to change, since further, better, and more extensive observations and measurements may be made, and more subtly reasoned inductions and deductions can lead to more elegant and useful conclusions.

The result is that, despite controversy in the preliminary stages (and the controversy can be acrimonious, emotional, ill-advised, or all three—for scientists are human beings, too) a consensus is eventually reached. Because arguments from reason *are* compelling.

What's more, scientists accept defeat. They may do so with poor grace, but they accept it. There are examples of this all through the history of science.

What counts, then, is *not* that astronomers are currently of the opinion that the Universe is changing progressively and therefore had a beginning. What counts is the long chain of careful investigation that led to the observation of red-shifts in galactic spectra (the lengthening of light-waves emitted by galaxies, which shows those galaxies are moving away from us) that supports that opinion.

Questions No Theologian Can Answer

What counts is *not* that astronomers are currently of the opinion that there was once a big bang, in which an enormously concentrated "cosmic egg" that contained all the matter there is exploded with unimaginably catastrophic intensity to form the Universe. What counts is the long chain of investigation that led to the observation of the isotropic radio wave background (shortwave radio waves that reach Earth faintly, and equally, from all directions) that supports that opinion.

So when the astronomer climbs the mountain, it is irrelevant whether theologians are sitting at the peak or not, if they have not *climbed* the mountain.

As a matter of fact, the mountain peak is no mountain peak; it is merely another crossroad. The astronomer will continue to climb. Jastrow seems to think the search has come to an end and there is nothing more for astronomers to find. There occasionally have been scientists who thought the search was all over. They are frequently quoted today, because scientists like a good laugh.

What was the cosmic egg and how did it come to explode at a particular moment in time? How did it form? Was there something before the big bang? Will the results of the explosion make themselves felt forever, or will the exploding fragments at some time begin to come together again? Will the cosmic egg form again and will there be another big bang? Is it alternative C that is the true explanation of the Universe?—these are only some of the infinite number of questions that those astronomers who are not convinced it is all over are interested in. In their search they may eventually reach new and better conclusions, find new and higher mountain peaks, and no doubt find on each peak guessers and dreamers who have been sitting there for ages and will continue to sit there. And the scientists will pass by on a road that, it seems possible, will never reach an end, but will provide such interesting scenery *en route* that this, by itself, gives meaning to life and mind and thought.

There is nothing to Jastrow's implication that astronomers are disturbed by the prospect of a big-bang beginning because, presumably, they hate to admit that theologians were on the mountain peak before them.

This seems wrong to me.

In 1948, some astronomers (including the well-known Fred Hoyle) worked out a closely reasoned theory that made it seem that, despite the established notions of the existence of an expanding Universe, there were no big bang and no beginning; that there was an eternal Universe without beginning or end. The theory was called "continuous creation."

Did astronomers in general leap gleefully behind Hoyle, to thumb their noses at the theologians? Not at all. The majority were hostile. Evidence for the big bang was stronger than for continuous creation, and as time went on, the evidence for the big bang grew stronger still. Now the theory of continuous creation is just about dead.

Astronomers didn't hesitate to follow the trail of evidence to the big bang just because it might have led them to the theologians. I don't think many of them even dreamed of it as leading to them.

The Cyclic Universe

Then why are there anti-beginning feelings among astronomers, if not because of anti-theological bigotry? Because the results are not all in yet.

Considering the amount of matter in the Universe, as far as astronomers can now tell the Universe is going to expand forever. There was one big bang at the beginning and one infinite scattering at the end. That's it.

If, however, there should be about a hundred times as much matter in the Universe as astronomers think there is, then gravitation will be sufficiently intense to bring a halt to the expansion, force a gradual contraction, produce another cosmic egg and another big bang, over and over. This would be like a ball bouncing high, gradually brought to a halt in its upward climb by the pull of gravity, falling faster and faster, hitting the ground, bouncing upward again, and so on, over and over. In that case, alternative C, *not* alternative B, is the truth.

Frankly, I find such a cyclic Universe more emotionally satisfying (I am human; I have emotions) than a one-shot beginning and ending. My intuition, if you like, tells me that astronomers will find that missing matter and will decide that the Universe is cyclic and that there is no beginning after all and no ending, only endless repetitions, endless bouncings, endless pulsation.

That is only my intuition, I repeat, and what makes me a scientist is that if the evidence in favor of a one-shot Universe continues to increase, I will abandon my intuition without a quiver. I have done so in other cases, as for instance when I opposed the drifting of the continents and then accepted it as more evidence came in. If, on the other hand, my intuition turns out to be correct, what happens to Jastrow's mountain peak?

No doubt Jastrow will hurriedly seek and find other Biblical quotations and quickly transport his theologians to some other mountain peak with instructions to say they have occupied it for centuries.

The scientific road is painful, hard, and slow, and to some (poor souls) the delights are not worth the effort. King Ptolemy of Egypt once asked Euclid, the mathematician, to instruct him in geometry, and Euclid undertook the task. Ptolemy grew restless at the slow progress, however, and finally ordered Euclid to make his proofs simpler.

"Sir," said Euclid haughtily, "there is no royal road to geometry."

But how tempting to seek the royal road when one can't face the mental perspiration of the tedious step-by-step. That's what makes intuition seem so delightful to most people ("Oh, don't give me your arguments. I just *know* . . .").

It even tempts Fritjof Capra, a physicist at the University of California who has written *The Tao of Physics*. He thinks, apparently, that what physicists have found out with great difficulty, Eastern sages have known all along. There's the old mountain peak.

Capra cites the Chinese notion of yin and yang, where yin represents the rational mode of thought and yang the intuitional, and believes them to be the "two sides of the same reality" or "polar areas of a single whole."

All right. No argument. Every scientist uses both reason and intuition in his attack on problems—but in the end the two are *not* equal. If intuition overwhelmingly suggests a conclusion, it still must be supported by reason, or else it is only soap-bubble speculation. If, on the other hand, the conclusion of reason goes against intuition, then reason must nevertheless be supported and intuition dismissed.

Capra seems to imply that they are equal, and he points out that modern physics, in probing into the most fundamental aspects of matter and energy, has come up with a picture in which the Universe seems to be "a continuous dancing and vibrating motion whose rhythmic patterns are determined by the molecular, atomic, and nuclear structures."

He then quotes a Taoist text to the following effect: "The stillness in stillness is not the real stillness. Only when there is stillness in movement can the spiritual rhythm appear which pervades heaven and earth." This, says Capra, is "exactly the message we get from modern atomic physics."

Selected Scraps of Elliptical Obscurity

But what does the Taoist text mean? I can see that "stillness in movement" represents dynamic equilibrium and that it is the latter that is important in the Universe—but that is my *interpretation*, based on my knowledge of *science*. What did it mean to the fellow who first said it? And what other interpretations can be made of it by people who don't have my particular cast of mind?

Many Eastern sages have said many things in elliptical and obscure language, even in the original, which suffer further in the translation. Anyone as imaginative and dedicated as Fritjof Capra (or even someone more limited, such as myself) can go through the vast volume of Eastern sage-sayings and come up with remarks that can be interpreted to seem to match any scientific conclusion.

Capra says that all mystical traditions, East and West, agree with modern physics. Of course they do, if *one* person (Capra, for instance) undertakes to interpret selected scraps of statements from each of them in his own way.

And if modern physics changes its mind, as it has in the past when new evidence came in, what then? In that case, no doubt, one person (still Capra, perhaps) will find other scraps of mystical tradition and subject them to new interpretations and come up with another match.

But if intuition is as important to the world as reason, and if the Eastern sages are as knowledgeable about the Universe as physicists are, then why not take matters in reverse? Why not use the wisdom of the East as a key to some of the unanswered questions in physics? For instance:

What is the basic component making up subatomic particles that physicists call a quark? How many different quarks are there? What is the relationship between the intensity of their interactions and distance? Are the leptons—the lightest particles, such as the electron—made up of anything simpler? Are there any additional heavy leptons? How many: What is the relationship between quarks and leptons? And so on.

Physicists are attempting to find answers to these questions by using enormously expensive instruments to study cosmic rays and to promote high-energy particle interactions—that is, the smashing together of sub-atomic particles at enormous speeds to see what changes are produced. It would be much simpler to study Taoist texts for the answers. But if Taoist texts can only be properly understood after physicists reach the answers, then of what scientific use are the Taoist texts?

What nonsense all this supposed intuitional truth is, and how comic is the sight of the genuflections made to it by rational minds who lost their nerve.

No, it isn't really comic; it's tragic. There has been at least one other such occasion in history, when Greek secular and rational thought bowed to the mystical aspects of Christianity, and what followed was a Dark Age.

We can't afford another.

34

Creationist Pseudoscience

Robert Schadewald

The instructed Christian knows that the evidences for full divine inspiration of Scripture are far weightier than the evidences for any fact of science.
John C. Whitcomb and Henry M. Morris[1]

Scientific creationists consider the first chapters of Genesis the ultimate science textbook. They believe that the entire universe was created six to ten thousand years ago in six solar days. All earthly forms of life were specially created, and most of them perished in a global Deluge a few thousand years ago. The Deluge also laid down essentially all of the sedimentary rock on earth.

Many fundamentalists insist that these doctrines should be part of the public school science curriculum. In 1981, Arkansas creationists smuggled a bill through the legislature that would have forced public schools that teach evolutionary theory to give equal time to creationism, but it was thrown out by a federal court on January 5, 1982, after a challenge by the ACLU. The ACLU successfully challenged a similar law in Louisiana.

Pseudoscience

Scientific creationism is a classical pseudoscience, and it therefore differs from science in several fundamental ways. Essentially, science is an open system based on skeptical inquiry, and its ultimate appeal is to evidence. Scientists use inductive reasoning to formulate general laws from specific observations. A pseudoscience is a closed system based on belief, and its ultimate appeal is to doctrine. Pseudoscientists base their systems on deductive logic, deducing how the universe must act to conform with their doctrines. As Whitcomb and Morris indicated in the opening quotation, when the scientific facts conflict with their interpretation of the Bible, then the facts be damned!

The standard work on pseudoscience is Martin Gardner's *Fads and Fallacies in the Name of Science*, originally published in 1952. The first chapter is a profile of pseudoscience and the pseudoscientific personality and has not been improved upon in thirty years. I will therefore limit my general remarks on pseudoscience, referring readers to Gardner's work for a fuller treatment.

Pseudoscientists are of two types. One group consists of ordinary "cranks," self-proclaimed geniuses who seem motivated by contempt for conventional scientists. George Francis Gillette, discoverer of a remarkable "backscrewing theory of gravity," was an excellent example. The other type of pseudoscientist seeks to justify some sort of ideology with scientific arguments. Examples of the latter range from Nazi anthropologists to scientific creationists. Both types of pseudoscientists harbor feelings of personal greatness.

The classic scientific crank is characterized by paranoia. He (rarely she) has unshakable faith in his own ideas and considers all who fail to share his vision blind and ignorant. Conventional scientists are dismissed as hollow authority figures who maintain their positions by parroting the "party line" and preserving the status quo. The crank frequently considers them not only venal, but intellectually dishonest. Indeed, their rejection of the crank's ideas proves their venality and dishonesty and demonstrates how they plot against him.

The greatness that the classic crank claims for his own genius, the religiously motivated pseudoscientist vests in his relationship with God. For instance, Henry M. Morris, founder of the Institute for Creation Research and America's foremost scientific creationist, believes that the Second Coming of Christ may be expected momentarily. Part of the Doctrinal Position of Morris's Christian Heritage College is a belief in the Rapture of the Church. Before the upcoming "Seven-Year Tribulation Period," the saved will be bodily snatched up to Heaven, leaving the unsaved rabble to a horrible fate.[2] Such beliefs can imbue one "chosen" with a personal exaltation that no earthly force can humble. Thus Morris believes that his opinions on virtually all scientific matters are superior to those of the vast majority of scientists (i.e., all those who reject his biblical literalism).

As for plots, Morris's belief system is dominated by a supernatural plot. His universe is literally haunted by Satan and visited by various forms of supernatural intervention. He has written, for instance, that the moon's craters may be scars incidentally inflicted on that body during a cosmic battle between the forces of Satan and the armies of Michael the Archangel.[3] He has also suggested in two of his books that Satan personally met with the Babylonian king Nimrod atop the tower of Babel to plot the theory of evolution.[4,5] Therefore when Henry Morris argues that evolution is Satanic and that scientists reject creationism because Satan has blinded them,[6] he's not speaking figuratively.

Pseudoscientists are characteristically blind to what they do not want to see. Ordinary cranks come by their irrationality through ignorance or personality defect. Many creation scientists, rather than being victims of crippled intellects, are self-mutilated. To accept the contradiction-riddled Bible as inerrant, they have schooled themselves in the art of finding reasons for whatever they want to believe. They have learned to ignore or rationalize away inconvenient facts, to avoid questioning (or even thinking about) key concepts, and to avoid pursuing lines of inquiry that threaten to contradict their received dogmas. In other words, they've trained themselves to shun the unfettered inquiry and hard-nosed skepticism that are the hallmarks of science.

Creation scientists therefore tend to fall naturally into patterns of thought and activity indistinguishable from those of ordinary cranks. Like secular cranks, they base their beliefs in exaggerted self-esteem and plot theories and maintain them by mangling logic and ignoring evidence. Personally, I find little to choose between the self-proclaimed "genius" and the self-proclaimed "humble instrument of the Almighty." The latter merely covers his delusions of grandeur with a cloak of false modesty. Indeed, I consider many scientific creationists merely garden-variety scientific cranks whose religious beliefs lead them to ride the creationist hobbyhorse rather than promoting (say) pushing gravity, the nonrotation of the moon, or the astronomical fantasies of Immanuel Velikovsky. I emphasize, however, that this assumption is not necessary to explain the behavior of creationists.

Pseudoscientists frequently rely on the "if today ain't the 4th of July then it must be Christmas" mode of reasoning. That is, they throw rocks at conventional science and then pull their personal rabbit out of a hat as the *only alternative*. In keeping with this tradition, creationist books, lectures, and debate presentations consist mainly of criticisms of orthodox science, to which they add a liberal sprinkling of theology. (Before the unsaved public, creationists try to conceal the theology by removing overt references to God and the Bible.) Most of their criticisms of science are dreary reading; but when creationists try to construct alternatives, they sometimes become interesting. Their offerings, however, are a far cry from science.

Science is a method for seeking reliable knowledge about the natural world by forming and testing hypotheses. Pseudoscientists erect their hypotheses as defenses against the facts, and they rarely dare to submit them to meaningful tests. The simplest way to demonstrate that creationism is pseudoscience is to try to treat it as science by subjecting creationist hypotheses to obvious and straightforward tests. The Institute for Creation Research is creationism's foremost think-tank. In what follows, I examine ICR's best-known argument against biological evolution as well as several of the structures they have erected to replace conventional science. Presumably these are examples of the "creation science" that ICR wants taught in public schools.

Thermoapologetics

Duane Gish is vice-president of the Institute for Creation Research. A biochemist with a Ph.D. from Berkeley, he has for a decade been ICR's most skillful propagandist. On the evening of February 4, 1982, Gish headed up a panel discussion on creationism at Tulane University. At one point, he offered a favorite apologetic, arguing that the Big Bang theory of the universe violates the second law of thermodynamics. While his exact words at Tulane were not preserved for posterity, Gish has publicly made this argument numerous times, essentially as follows:

> We had that Cosmic Egg that exploded. Nothing operated on that Cosmic Egg from the outside of the cloud of hydrogen and helium. Nothing was brought from the outside. It transformed itself from that chaotic state. Certainly an explosion doesn't generate order. It generates chaos. And from that initial chaotic state, we have the highly ordered universe that we have today. Now, ladies and gentlemen, that is directly contradictory to the second law of thermodynamics.[7]

A discussion period followed the panel at Tulane. According to an account probably written by Gish:

> Most of the discussion was actually generated by two physical scientists from the Tulane faculty who were in the audience. They challenged Dr. Gish's claim that the Second Law of Thermodynamics contradicts the Big Bang theory of the origin of the universe, asserting that the universe has constantly been getting more disorderly since the Big Bang. Gish rejected this claim as an impossibility.[8]

Again, what the physicists said wasn't recorded, but they no doubt noted that, in thermodynamic terms, a relatively compact cosmic fireball at zillions of degrees is more "ordered" than our present, relatively cool universe, which is scattered over 15 billion light-years.[9] It is not surprising that Gish would reject this argument, for he has made numerous other remarkable statements about thermodynamics. On one occasion, he argued that equilibrium constants cannot vary with temperature, or else they wouldn't be called "constants."[10] (Equilibrium constants vary exponentially with temperature.) Pressed about this later, Gish called it a "slip."[11] He later published a discussion of the following reaction:[12]

$$2CH_4 + NH_3 + 2H_2O \longleftrightarrow H_2NCH_2COOH + H_2O$$

He did not explain how 15 hydrogen atoms on the left side of the reaction become 7 hydrogen atoms on the right, nor how 2 oxygen atoms on the left become 3 on the right.

Gish also seems impressed by the classic 1978 paper of creationist

David R. Boylan entitled "Process Constraints in Living Systems."[13] Boylan, dean of the College of Engineering at Iowa State University and Technical Adviser to the Institute for Creation Research, had previously written that "the Second Law has been particularly helpful in developing an apologetic against abiogenesis."[14] In "Process Constraints," he attempted to show that living systems require a certain "kind" or "quality" of energy that is "intelligence or ordered energy."[15] Such concepts are foreign to conventional thermodynamics, and another ISU scientist, John W. Patterson, challenged Boylan's thermodynamic arguments. In a blistering critique of the paper, Patterson charged that Boylan made a number of elementary errors, the worst being that perennial gaffe of beginning thermo students, trying to compute an entropy change for a system without first devising a reversible path.[16]

The editor of *Creation Research Society Quarterly* refused to print Patterson's critique,[17] but several creationists collaborated in developing an apologetic against Patterson. Essentially they argued that Patterson's criticism was misdirected, in part because Boylan was really discussing a Steady State Steady Flow (SSSF) system. Announcing this to Patterson, Gish stated categorically and repeatedly that Boylan's paper contained absolutely no errors.[18] The thermoapologists having at this point purchased new rope, Patterson broke off the private dialogue. In a subsequent paper alleging widespread creationist incompetence, he pointed out that the entropy change in an SSSF system is zero *by definition.* "Hence," Patterson wrote, "Boylan's central result—i.e., his erroneous formula for the entropy change—should have come out identically zero (!) and not the nonvanishing sum whose limiting cases he discusses at great length."[19]

Meanwhile, in late 1981, the Creation Research Society published No. 1 in its new monograph series, a volume entitled *Thermodynamics and the Development of Order.* Chapter 2 is a popularized version of the Boylan paper with most of the fatal errors (the erroneous derivations) relegated to an appendix.[20] Chapter 4 is "The Origin of Biological Order and the Second Law," in which Duane Gish discusses the ersatz reaction quoted earlier. Where this little volume leaves thermoapologetics depends on your point of view. Gish and the others apparently see themselves reaching new heights of creation science. Conventional thermodynamicists also see them as elevated—and twisting in the wind.

Flood Geology

It is no accident that the foundation work of modern scientific creationism is *The Genesis Flood.* Flood geology is absolutely crucial to scientific creationism. Not only do creationists need the Flood to float Noah's Ark, but it is their only explanation for the geologic strata. The geologic column almost screams "ancient," but the creationists claim it is young. Furthermore, if conventional geology is correct then, whether by Darwinian evolution or some other means, various life forms appeared on earth sequentially over a long period of time. Thus, of the three major elements of scientific creationism, two of them (original special creation and young earth) are founded on the third. And a shaky foundation it is.

One of the many problems with Flood geology is that it cannot explain the distribution of fossils in geologic strata. In *The Genesis Flood*, Whitcomb and Morris attribute fossil distributions to a combination of habitat, victim mobility, and hydraulic sorting.

The habitat and victim-mobility apologetics go hand in hand. The argument is that creatures of the sea and seashore would be buried first, then slow-moving lowland animals like amphibians and reptiles, then faster and more mobile land mammals, and finally wily and speedy man would be swept from the mountaintops. This explanation, poor as it is for explaining animal fossils, fails entirely for plant fossils. We can test the hypothesis by asking about angiosperms (flowering plants). The earth's shorelines are covered by a tremendous variety of flowering plants, and none of them can run very fast. Thus we'd expect to find the flowering plants overwhelmed with the amphibians. Yet angiosperms appear suddenly in the early Cretaceous, along with mammals.

The hydraulic-sorting apologetic is much more outrageous. For objects of the same composition and shape, cross-sectional area is proportional to the square of linear dimensions, while weight is proportional to the cube. Thus similar objects falling through a fluid medium are sorted according to size. If trilobites, for instance, were sorted hydraulically, we would find all large trilobites of a certain species low in the fossil record and all small ones higher. This follows directly from the hydraulic-sorting hypothesis. But it is decidedly not what we find. The hydraulic-sorting hypothesis fails the most elementary test, and it is difficult to see how Henry Morris, who flogs the public with his Ph.D. in hydraulic engineering, could fail to understand this.

Numerous other fatal flaws of Flood geology have been discussed elsewhere.[21,22] Because it is so absurd, creationists will do almost anything to avoid discussing Flood geology in public debates. In a debate at Tampa on September 19, 1981, Henry Morris refused to answer the devastating attack on Flood geology made by biologist Kenneth Miller. Morris claimed it was because he preferred to talk about the "key issue" (biological evolution), but even the creationists in the audience knew better.

The Flood Waters and Continental Zip

Creationists have a tremendous problem explaining where the Flood waters could have come from. The late George McCready Price, the

Seventh Day Adventist theologian who for thirty years was the leading anti-evolutionary writer, opted for a tidal wave caused by a near-collison with an astronomical body. Older creationists appealed to comet tails and sudden, radical shifts of the earth's axis. Most scientific creationists follow the lead of the turn-of-the century Quaker Isaac Newton Vail, who theorized that the earth was once surrounded by a vapor canopy.[23] The collapse of this canopy supposedly caused Noah's Flood, and some creationists believe that a second canopy collapsed to cause *the* (as in *one*) Ice Age.

There are numerous and insuperable difficulties with a canopy. To mention three: (*a*) there are no reasonable conditions under which a vapor canopy containing most of the ocean's waters would be stable; (*b*) the condensation of such a canopy would release as much energy (latent heat) as the earth receives from the sun in two or three centuries; and (*c*) the tremendous pressure at the base of such an atmosphere would be fatal to virtually all forms of terrestrial life. Indeed, creationist Robert E. Kofahl, science coordinator of the Creation-Science Research Center in San Diego, examined several variants of the canopy hypothesis and concluded that "without miraculous supernatural intervention into the natural order. . . . they all are doomed to failure. They cannot be made to fit with the laws of physics or with the requirements for life on earth." Kofahl, however, was not concerned about the lack of a natural mechanism, since "God is not inadequate . . . for any need or circumstance."[24]

Despite the obvious contradictions, Henry Morris adopted the vapor canopy as his source for the Flood waters, and he bases other apologetics on it as well. For instance, he suggests that Methuselah and other pre-Flood patriarchs achieved their extreme longevity because the canopy shielded them from the aging effects of harmful radiation.[25] Because the canopy hypothesis fails miserably without divine intervention, Morris and other ICR creationists avoid discussing it before the unsaved; but it is a hidden part of the "creation model" that they want taught in public schools.

One creationist not satisfied with a supernaturally supported canopy is Walter Brown, director of the Institute for Creation Research's Midwest Center in Wheaton, Illinois. A retired Air Force colonel with a Ph.D. in mechanical engineering from MIT, Brown has the kind of credentials and self-assurance that appeal to a creationist audience. As part of his ministry, he delivers daylong seminars entitled "In the Beginning . . ." In these seminars, Brown presents arguments he claims conventional scientists cannot answer. For example, he claims that coal could not have been formed in prehistoric peat bogs, since it takes at least five feet of peat to make a foot of coal and some coal seams are 100 feet thick. "Can anyone here imagine a peat bog 500 feet thick? I mean, it's ridiculous! How could the bottom layers *grow*?"[26] Frankly, a conventional geologist asked the latter question probably *would* be struck speechless. The same reaction might be evoked by Brown's solution to the problem of the Deluge waters, which might be called the Theory of Continental Zip.

Suppose the earth was originally created with an outer crust about five miles thick separated from a rocky inner core by a mile-thick layer of superheated water at 500° F. Suppose this superheated water broke through the earth's crust in the middle of the prehistoric continent conventional geologists call *Pangaea.* The crust ripped like a rotten seam, the rip extending nearly around the earth. The escaping superheated water flashed into steam, and the continuing steam explosion tore loose megatons of dust and rock and carried them high into the atmosphere. The blast set the two halves of Pangaea in motion, and they slid away from each other over the remaining water. When most of the water had escaped, the zipping continents braked to a halt. The braking action caused buckling in the areas that first began to drag, thus raising mountain ranges. Meanwhile, the escaped steam condensed and it rained for forty days and forty nights. The earth was flooded and the turbulent waters laid down most of the earth's present load of sedimentary rock, the sediments coming largely from material torn loose by escaping steam. Eventually the Flood waters ran off into the ocean basins, including the newborn Atlantic basin, which straddled the escape rift—the mid-Atlantic rift. The steam remaining in the atmosphere continued to fall as snow in the northern regions, causing a series of yearlong ice ages.

Any geologist who spends a few moments thinking about this Keystone Cops version of continental drift, which tries to compress tens of millions of years into one year, will find numerous fatal objections. The most obvious comes from high school physics. At one of Brown's seminars, someone pointed out that his theory calls for releasing an enormous amount of 500° F water onto the earth's surface and into its atmosphere within a year. The energy must be conserved. Wouldn't things get unbearably hot? Brown said he hadn't tried to calculate what the effect would be, and he attempted to slough the objection off by claiming that the expanding steam would have a cooling effect. This offers at best temporary relief, since in condensing the Flood waters would again release their latent heat. Indeed, as I subsequently pointed out to Brown, just in cooling from 500° F to 212° F, the mass of water he hypothesizes would release roughly as much energy as the earth receives from the sun in a century. Brown admitted that this is a problem, but said his model has fewer difficulties than alternatives like the canopy.[27]

Brown's marvelous theory of continental zip illustrates the perils faced by a creationist who tries to construct a scientific explanation for dogma. Brown makes a great pother about scientific methodology, offering two predictions whose failure he says would falsify his theory. First, he predicts that large amounts of primeval water remain below the earth's major mountain ranges. Second, he predicts that efforts to measure continental drift rates will be futile.

The Age of the Earth

Scientific creationists insist that the earth and universe are only 6,000 to 10,000 years old. In his first book, *That You Might Believe,* Morris argued that the conventional time-scale is "ridiculous" because it is contradicted by the growth trend for human population. Morris noted that if the original population of the world was two, doubling it 30 times would produce the 1946 population of 2,160,000,000. Since the 1600s, the population has doubled roughly every century. Morris concluded that the doubling rate was incompatible with any evolutionary scenario. "If, on the other hand, all people are descended from Noah and his wife, who according to the best biblical chronology must have lived about 4,500 years ago, then the average interval for doubling is 150 years, which is entirely reasonable."[28]

It's hard to argue with that kind of logic. If a critic pointed out that, by this reasoning, the world population when the Great Pyramid was built must have been about 16, Morris would claim he was being ridiculed. Besides, Morris now has 69 other equally valid arguments against conventional geochronology, and he published all 70 in a laundry list in his 1977 book, *The Scientific Case for Creation.*[29] (This is still 30 fewer than flat-earther William Carpenter offered in *One Hundred Proofs that the Earth Is Not a Globe.*) To get an idea of the quality of the other arguments, consider the following example, based on a discussion by Dr. G. Brent Dalrymple of the U.S. Geological Survey, Menlo Park, California.[30]

Ask a creationist for the best evidence that the earth is young and he or she will probably cite the decay of the earth's magnetic field. Thomas G. Barnes, the present head of ICR's Graduate School, developed this apologetic, and he has presented it at creationist conventions, in creationist periodicals, and in an ICR "technical monograph."[31] It is perhaps the creationists' most widely cited piece of "scientific research."

Barnes claims that the earth's magnetic moment has been measured reliably since about 1835 and that it is decreasing. He argues that the earth's magnetic field is generated by electric currents circulating in the earth's core, asserting that no mechanism (such as convection caused by radioactive decay) exists to sustain these currents. Barnes concludes that the earth's magnetic field must decay exponentially, and from data accumulated since 1835 he calculates a decay half-life of 1,400 years. He then argues that the earth cannot be older than about 10,000 years, since before that its magnetic field would have been unreasonably high. Typically, Barnes carefully avoids testing his hypothesis.

As it happens, the strength of the earth's magnetic field for the past several thousand years is well documented. Archaeologists have long used pottery styles and types to identify and date human habitation sites, and they have developed detailed and reliable pottery chronologies for many regions and cultures. Pottery is also of value to those studying paleomagnetism, for when fired pottery cools through a critical temperature (called the Curie temperature) it becomes permanently magnetized by the earth's field. By measuring the magnetic moments preserved in reliably dated pottery, specialists in paleomagnetism have reconstructed the history of the earth's magnetic field for the last several millennia. A recently developed and independent check on the pottery data is based on the moments preserved in mud bricks from historical sites.[32] Barne's hypothesis predicts that three half-lives (4,200 years) ago (when, according to Morris's "very reasonable" hypothesis, the human population was 16) the earth's magnetic field was eight times what it is now. The magnetic moments recorded in bricks and pottery from that era agree nicely: the earth's field then was slightly weaker than it is now.

Neither Barnes nor any other creationist has ever, so far as I know, acknowledged that such data exist. But Barnes's theory is not just flatly contradicted by the evidence. As Dalrymple pointed out, it is based on a hopelessly incorrect assumption.

Barnes essentially argues that the *total energy* of the earth's magnetic field is decaying with a half-life of 1,400 years, and the dynamo he hypothesizes to produce it would do just that. But the earth's magnetic field can be split into two components, the dipole and the nondipole moments. While the dipole moment is in fact decreasing, the nondipole moment is increasing at almost a matching rate, so that the total energy of the earth's magnetic field is decreasing hardly at all. This fatal fact, which demonstrates that Barnes's scheme is unequivocally wrong, is discussed at length in the very book from which Barnes drew his data![33]

Creationist Astronomy

Perhaps the most outspoken critic of conventional science among the creationists is Harold S. Slusher of the University of Texas (El Paso) and the Institute for Creation Research. Slusher is also a foremost exponent of the conspiracy theory that holds that "evolutionists" conspire to suppress creationism. He has frequently accused conventional scientists of intellectual dishonesty and/or incompetence.[34]

Slusher has long claimed the title "Doctor" based on his honorary D.Sc. from Indiana Christian University. (ICU is a Bible College that, when I last checked, had a 1/2-man graduate science department, i.e., a professor of natural science *and* mathematics.[35]) He now boasts an earned Ph.D. from Columbia Pacific University, described by creationists as "a nontraditional university with two campuses in the Bay Area. The

Graduate School is set up with a similar manner of operation to that of such European universities as the University of Vienna, the University of Bologna, and the Sorbonne (Paris)."[36] From the bologna about the Sorbonne and the universities of Vienna and Bologna, experienced creationist watchers will immediately (and correctly) infer that Columbia Pacific University is unaccredited.

Slusher is noted for championing the theory that light travels in highly curved Riemannian space and that the entire universe is actually only 15.7 (5 pi) light-years across. Even more remarkable is his discovery (published in his 1974 monograph, *A Critique of Radiometric Dating*) that the radioactive decay constant of the isotope iron-57 can be changed up to 3 percent under laboratory conditions.[37] In a recent, revised edition of this monograph, Slusher further claims that the decay constant of cesium-133 can likewise be altered.[38] Standard references unanimously agree that both isotopes are stable. (An iron-57 nucleus can exist in—and thus decay from—an excited state, but this decay has nothing to do with radiometric dating.)

Considering the foregoing, connoisseurs of pseudoscience would expect great things of a joint astronomical paper by Slusher and Barnes. In the December 1981 *Creation Research Society Quarterly*, they joined forces with G. Russell Akridge, formerly associated with Bob Jones University and now with Westminster Schools in Atlanta, Georgia. Akridge, who also teaches summer courses in astronomy at ICR's Graduate School, is author of several articles claiming that scientific speculations about decreases in the sun's diameter prove that the sun cannot be 5 billion years old. The Akridge-Barnes-Slusher paper, entitled "A Recent Creation Explanation of the 3°K Background Black Body Radiation,"[39] is a masterpiece of creation science.

The authors begin by asserting that the Big Bang theory cannot explain the 3°K background radiation. (This would surely have surprised the late George Gamow, who explicitly *predicted* the background radiation about 30 years before it was discovered.) They argue that, if such radiation were present initially, it would have escaped the universe long ago. (*Nothing* can escape a relativistic universe by any conventional means, and it's incredible that Akridge, Barnes, and Slusher, all three of whom have published articles about special relativity, do not seem to know that.) They also ridicule Big Bang theorists for basing their calculations on an expanding universe with finite volume and perfectly reflecting walls. (Cosmologists use this simple model instead of the rigorous relativistic equations because, within limits, it gives reliable answers with minimum hassle.)

But it is not the AB&S sins of omission and distortion that interest us here, but rather their pseudoscientific creation. After criticizing conventional theory, the authors offer their own explanation for the 3°K background radiation. They argue that the dust and gas of the galaxy were created at absolute zero and have been warming up ever since. According to their calculations, if the interstellar dust and gas have been warming up for about 6,800 years (roughly, since the traditional date for creation is 4004 B.C.), it should presently be at—surprise!—3°K.

Only an astronomer can properly appreciate the AB&S theory, but one contradiction is obvious to an informed layman. If the 3°K radiation came from dust and gas in the galaxy, its intensity would exhibit a strong maximum in the plane of the Milky Way, where most of the dust and gas are concentrated, and a minimum at the galactic poles. But (as AB&S state explicitly at the beginning of their paper) the intensity of the 3°K radiation is remarkably isotropic! Thus the AB&S theory fails the most obvious test imaginable.

Conclusions

The preceding examples of creation science came from the intellectual leaders of scientific creationism. Readers seeking more of the same should consult back issues of the *Creation Research Society Quarterly*, the *Bible-Science Newsletter*, *Acts & Facts/Impacts*, or any other creationist periodical. The *Creation Research Society Quarterly* especially provides a safe, slick-paper haven where creation scientists can publish their most bizarre rationalizations, confident that their colleagues will be too kind (or too blind) to insist that they subject their hypotheses to elementary tests. Because creationists refuse to test their hypotheses and accept the results, scientific creationism is a classical pseudoscience, and the gulf between it and mainstream science is unbridgeable.

Since they cannot win by the rules of science, creationists promote their doctrines by religious, political, and legal means. While Jerry Falwell promotes creationism on television, thousands of other fundamentalist ministers push it from their pulpits; and many of them pay ICR propagandists to come into their churches and make presentations. The nation's largest Protestant denomination, the Southern Baptists, has officially endorsed the teaching of "creation science" in public schools.[40] They and other fundamentalists continually pursue this goal through legislation and school board action.

Scientific creationism is the best organized movement in the history of America pseudoscience, and thus the most dangerous. It is a dubious honor.

References

1. John C. Whitcomb, and Henry M. Morris. *The Genesis Flood* (Nutley, N.J.: Presbyterian and Reformed Publishing Co., 1961), p. 118.

2. Christian Heritage College *Catalog*, 1979.

3. Henry M. Morris, *The Remarkable Birth of Planet Earth* (San Diego: Institute for Creation Research, 1972), p. 67.

4. Morris, *The Troubled Waters of Evolution* (San Diego: Creation-Life Publishers, 1974), p. 74.

5. Morris, *Remarkable Birth*, p. 52 f.

6. Morris, *The Twilight of Evolution* (Grand Rapids: Baker Book House, 1963), p. 77.

7. Debate between Duane T. Gish and John W. Patterson at Ankeny, Iowa, January 16, 1980.

8. *Acts & Facts*, May 1982, p. 3.

9. See P. C. W. Davies, *The Physics of Time Assymetry* (Berkeley: University of California Press, 1974), pp. 80-111, for a discussion of thermodynamics and cosmology.

10. Undated open letter from Sheldon Matlow to Duane Gish (circa September 1974).

11. Gish-Patterson debate, op. cit.

12. Emmett L. Williams (ed.), *Thermodynamics and the Development of Order*, Creation Research Society Monograph Series: No. 1 (Ann Arbor: Creation Research Society Books, 1981).

13. *Creation Research Society Quarterly*, December 1978, pp. 133-38.

14. *Twenty-one Scientists Who Believe in Creation*, 2nd ed. (San Diego: Creation-Life Publishers, 1977), p. 8.

15. Boylan, op. cit., p. 135.

16. John W. Patterson, "Direct Evidence of Thermodynamic Incompetence Among Creationists: Some Specific Examples with Critical Comments and Some Thoughtful Reflections" (1980, unpublished).

17. Harold W. Armstrong, letter to Patterson dated May 12, 1980.

18. Duane T. Gish, letter to Patterson dated August 13, 1980.

19. John W. Patterson, "An Engineer Looks at the Creationist Movement," *Proceedings of the Iowa Academy of Sciences*, June 1982.

20. Williams, op. cit.

21. Christopher Gregory Weber, "The Fatal Flaws of Flood Geology," *Creation/Evolution*, Summer 1980, pp. 24-37.

22. Robert J. Schadewald "Six Flood Arguments Creationists Can't Answer," *The Vector* (April 1982), pp. 16-20.

23. Isaac Newton Vail, *The Waters Above the Firmament*, 1874.

24. Robert E. Kofahl, "By Faith We Understand: Creation and the Gospel; Physics and the Canopy Models," *Proceedings of the Third Annual Creation Science Conference* (Caldwell, Idaho: Bible-Science Association, n.d.), p. 11.

25. Whitcomb and Morris, op. cit., p. 399-405.

26. Walter Brown, "In the Beginning . . ." seminar, November 21, 1981, Temple Baptist Church, St. Paul, Minn.

27. Phone conversation, December 1, 1981.

28. Morris, *That You Might Believe* (Chicago: Good Books, Inc., 1946), p. 77 f.

29. Morris, *The Scientific Case for Creation* (CLP Publishers, 1977), p. 55 ff.

30. G. Brent Dalrymple, *Radiometric Dating, Geologic Time, and the Age of*

318   Robert Schadewald

the Earth, rough draft dated April 1, 1981.

31. Thomas G. Barnes, *The Origin and Destiny of the Earth's Magnetic Field* (San Diego: Institute for Creation Research, 1973).

32. Ken Games, "The Earth's Magnetism—in Bricks," *New Scientist*, June 11, 1981, p. 678-81.

33. G. Brent Dalrymple, "Can the Earth Be Dated Using the Decay of the Magnetic Field?" draft dated January 25, 1982, p. 16.

34. Harold S. Slusher, *A Critique of Radiometric Dating* (San Diego: Institute for Creation Research, 1973), p. 2.

35. Indiana Christian University *Catalog*, 1979.

36. *Acts & Facts*, April 1982, p. 4.

37. Slusher, *Critique*, p. 38.

38. Stephen Brush, "Finding the Age of the Earth: By Physics or by Faith?" preliminary draft (dated August 1981) of a paper subsequently published in *Journal of Geological Education*, January 1982, ms. p. 45.

39. G. Russell Akridge, Thomas G. Barnes, and Harold S. Slusher, "A Recent Creation Explanation of the 3°K Background Black Body Radiation," *CRSQ*, December 1981.

40. "Baptists Back School-Prayer Amendment," *Minneapolis Star and Tribune*, June 18, 1982, p. 12A.

35

Paluxy Footprints

Steven Schafersman

Creationists have written about ten books and dozens of articles on the alleged co-occurrence of human and dinosaur tracks in the 100-million-year-old Cretaceous limestone of the Paluxy River bed near Glen Rose, Texas. All of the major books promoting "scientific creationism" mention the Paluxy River "man tracks" as evidence that the geologic time-scale is wrong and that therefore the evidence for evolution is not as good as scientists think.

Scientists currently think that the dinosaurs became extinct 65 million years before the appearance of the genus *Homo*. But creationists, in their lectures and debates before public audiences, almost invariably present the contemporarity of humans and dinosaurs as fact. Because the presence together of "human" and dinosaur footprints is so widely touted by creationists in their writings and speeches for the public, and also because a recent excavation by two Baptist ministers at a site on the Paluxy River resulted in discovery of new "human" footprints and received wide exposure in the national press, a number of scientists decided it was time for a thorough on-site examination of the elongated depressions that the creationists claimed are human footprints. The examination was both appropriate and timely: although the creationists have promoted the human tracks for twenty years, the scientific response to their claims has been either to ridicule or to ignore them, and scientists have now learned that this response is counterproductive. Therefore, anthropologists Laurie Godfrey and John Cole, geologist and paleontologist Steven Schafersman, and science teacher Ronnie Hastings journeyed to Glen Rose to examine the footprints for first-hand information to explain why the "human tracks" are not human footprints at all, and what actually caused them.

The limestone bed of the Paluxy River has an international reputation for the excellence and quantity of the dinosaur footprints found there. Vertebrate paleontologists and ichnologists (trace fossil experts) have studied the footprints for years, and Dinosaur Valley State Park was created to preserve them and make them accessible to the public. The few scientists who have professionally studied the dinosaur footprints have been the primary source of information for the press on the rare occasions when a rebuttal to the creationists' claims was desired. The scientists' explanations about the real nature of the features the creationists were confusing with legitimate human footprints were quite accurate, but they did not get the attention they deserved. They were always presented to the public in a context of disagreeing with the pseudoscientific claims of the creationists. There was no single place that one could turn to obtain the necessary information to refute the creationists about this particular problem.

To avoid this mistake in the future, Godfrey and her colleagues published a detailed and illustrated monograph on the "human footprints," including their history and place in the culture of pseudoscience, in Issue 15 of *Creation/Evolution* in 1982. Scientists are now directly answering creationists' assertions by explaining, in numerous books and popular articles, how science works, why evolution is a factual scientific concept, and how the creationists have distorted and misrepresented the scientific evidence in their campaign to deceive the public. An excellent example of this new activity among scientists is the book *Scientists Confront Creationism* (Norton, 1983), edited by Laurie Godfrey and containing papers by herself, Cole, Schafersman, and others.

The results of the Paluxy River expedition are easy to summarize: no evidence for genuine human footprints was found among all the tracks examined, but the scientific investigators were able to discover six major methods by which the elongate depressions, which the creationists claim to be of human origin, were formed. The investigators examined, measured, mapped, and photographed all the specific objects that are identified as human footprints in the creationist literature. Three sites were investigated in detail: the State Park Ledge, the McFall site (where the excavation of International Baptist College is in progress), and a site near New Braunfels, Texas.

Not a single "man track" examined had the correct anatomic features of a human foot either at rest or in stride. This observation is also true for the famous carvings of the giant man tracks from Glen Rose. The person who carved these tracks had only a primitive knowledge of human foot anatomy and produced tracks that superficially resembled giant human footprints but that are quite unmistakably artifacts to a trained physical anthropologist. These notorious carvings are not accepted as real human footprints by some of today's sophisticated creationists, but most believe they are genuine. Most older creationist books, such as Whitcomb and Morris's *The Genesis Flood*, illustrate these carved tracks and assert that they are real. None of these tracks, however, have proper human footprint dimensions or impressions of the instep, toes, or ball of the foot. The "human track trail" we examined had stride and pace lengths and directions uncharacteristic of human beings. Most of the creationist-identified human footprints are isolated from others but are adjacent to well-preserved dinosaur tracks. This fact assumes great importance in our discussion of how the "human" tracks were formed.

If we ignore these carvings, the origin of all the oblong or elongate depressions in the Cretaceous limestone that the creationists claim are human footprints can be explained by natural, nonhuman causes. Although this statement has been made by scientists for more than twenty years, the different "human tracks" have different origins, and it is now possible, after detailed investigation by the team of scientists, to ascribe the origin of each specific impression to a specific cause.

Most of the "man tracks" are simple erosion channels or gouges on the limestone river banks produced by water currents during flood conditions. Rapidly moving water is a powerful erosive force that can produce an endless variety of erosional features in hard limestone. One would expect to find a few superficially human-foot-shaped depressions in the Paluxy River banks, and most of the "man tracks" turn out to be these. Most of the State Park Ledge site "human footprints," the ones most people see, are simple erosion features. The tracklike depressions are invariably elongated parallel to the river and occur together with depressions of similar size and depth but different shape. The "insteps" of these erosional "footprints" are created by the undercutting of limestone layers of different hardness by the water current.

Since river erosion is such an easily understood and logical explanation for the oblong depressions, many persons who hear this explanation are quick to dismiss the claims of the creationists. Therefore, the creationists have gone to a great deal of effort to remove a thick layer of limestone at the McFall site to expose a limestone underneath that contains many well-preserved dinosaur tracks and, in the words of the project's director, the Reverend Carl Baugh, "twenty-nine human tracks, 27 of which are 16 inches in length." The creationists are pleased with these tracks, since there is obviously no way their existence can be ascribed to erosion by flood water. But what is one supposed to think of 16-inch-long featureless depressions that are all about 3 or 4 inches wide? The creationists believe that these are the tracks of giant men with odd-shaped feet (Genesis, after all, says only that "in those days, giants walked the earth" the book is silent about the shape of their feet). These elongate depressions have no anatomic features that suggest to a scientist that they are of human origin. Yet the question of what caused them remains. There are several reasonable, natural explanations. The depressions could be primary sedimentary structures, such as channel, gouge, or flute marks, formed during erosion of the lime mud tidal flat before it was buried by later sediment and subseqently lithified. The depressions could be *karren* features formed by leaching (dissolving) the limestone by rain water, which seeps into the rock along joints and fractures and produces elongated cavities in the soluble limestone. Both of these processes are excellent natural explanations for the origin of oblong or elongate depressions.

The best and probably correct explanation, however, is that dinosaurs produced these features. The elongate depressions that the Reverend Baugh and his creationist colleagues claim to be human footprints are almost always found associated with three-toed dinosaur tracks, are the same depth as the dinosaur tracks, are the same length as the dinosaur tracks (16 inches), and sometimes occur in trails of the same pace-length as dinosaur tracks. A genuine human footprint in the limestone would be much shorter and shallower than the observed depressions. Sometimes the two outer toes of a three-toed dinosaur track are indistinct or absent; this is caused by the dinosaur putting most pressure on the central toe, by the tidal flat mud being so firm that only the central toe produced an impression, and by water scouring the tidal flat and eroding the shallower outer toe prints before the impression hardened. Such an odd-shaped or distorted dinosaur track would look like an elongate depression of approximately the same dimensions as those seen at the McFall site. Another mechanism involves a dinosaur putting its foot onto a part of the mud flat that is extremely soft. Its foot sinks and, upon withdrawing it, the mud flows back in the hole from the sides. The resulting oblong depression would be exactly the same length and approximately the same depth as the adjacent dinosaur tracks, but would be narrower and show no toes. Since this mechanism produces features essentially identical to those observed, it is the preferred explanation.

The elongate depressions could also be caused by the dinosaur dragging its rear toe (hallux) or upper foot bones (metatarsals) in the mud. These depressions extend outward from behind a number of dinosaur tracks, and Reverend Baugh has claimed that these indicated a human had stepped on a dinosaur track or vice versa. At least one "man track" and two "sabre-toothed tiger tracks" at the McFall site were created by the trace fossil *Thalassinoides* which is extremely abundant on the limestone surface. The burrow casts of *Thalassinoides*, made by a Cretaceous crustacean in the tidal flat mud, formed ridges which Reverend Baugh and his followers misinterpreted to be ridges between five perfectly spaced "toes." Many other tracks and patterns can be seen on the trace fossil-covered limestone surface if one uses one's "imagination," as the creationists themselves admit must be done. At the New Braunfels site, creationist-identified "man tracks" are actually dinosaur tracks distorted by repeated growth of a Cretaceous algal mat in and around each footprint depression the tidal flat. After several such layers have accumulated, the dinosaur prints are indistinct, and some assume a roughly oblong shape; creationists claim these are human despite the fact that they occur in perfect spacing and orientation with the well-preserved and exposed dinosaur trails. The point is that there are many natural explanations that need to be considered before resorting to fantastic explanations for elongate depressions in Cretaceous limestone—especially explanations that conflict with much of what we know about evolutionary history and stratigraphy. It is fair to say that the investigations reported here has falsified the claims of the creationists and discredited their attempt to present themselves as legitimate scientists.

36

The Shroud of Turin: A Critical Appraisal

Marvin M. Mueller

Science is done for diverse reasons. In common with most other human enterprises, it is done by individuals having a great variety of motives, and unalloyed "truth for its own sake" is not always paramount. The main marvel of modern science is that useful, reliable, predictive knowledge *eventually* emerges from a chaos of contradictory opinions. These opinions are held by researchers who are all too obviously encumbered by common human frailties, such as fallibility. For this reason, science in its self-correcting aspect needs, and thrives on, open, uninhibited airing of divergent opinions. Such will be the spirit of this critique.

Early in the four-century history of modern science, a dominant motive for doing "natural philosophy" was usually cut from the cloth of "natural theology"—worshipping the Creator by studying and understanding his handiwork. Later, and certainly in our time, material utilitarianism has become the most conspicuous motive for carrying out the difficult, tedious, and costly experimental procedures required.

In any event, it would be surprising to find in these times a fairly large, well-funded, well-organized program of experimentation with predominantly theological motivations. (The Creation Science Research Center in San Diego does not have an experimental program.) Yet, for several years now, such an enterprise has been subjecting a most famous relic, the so-called Shroud of Turin, to an impressive panoply of scientific tests. I present here the view of an outside scientist concerning the present status and the implications of this unusual application of forensic science. I am moved to do this primarily by the half-truths and even gross distortions concerning the results, and their implications, that have recently been reported in newspapers, on television, and in several books.

Early Attitudes of the Present Investigators

My interest in the complex international shroud investigation was kindled on Good Friday 1978. My local newspaper, the *Los Alamos Monitor*, carried a long front-page story about four local scientists who were taking

324

part in the investigation. Three of the four made scientifically irresponsible statements concerning the matter. A similar story, citing the preconceptions of other investigators on the team, was given national distribution by the wire services. A few quotations by the local scientists before they had seen the shroud will suffice:

> I am forced to conclude that the image was formed by a burst of radiant energy—light, if you will. I think there is no question about that.

> What better way, if you were a deity, of regenerating faith in a skeptical age, than to leave evidence 2,000 years ago that could be defined only by the technology available in that skeptical age.

> The one possible alternative is that the images were created by a burst of radiant light, such as Christ might have produced at the moment of resurrection.

> I believe it through the eyes of faith, and as a scientist I have seen evidence that it could be His [Christ's] shroud.

> The most important thing is that any data we obtain should be interpreted through the twin communities of the eyes of the Church and of the scientific community.

In fairness, I hasten to add that these Los Alamos scientists would not make the same statements today. Nowadays they are considerably more cautious and guarded. Indeed, the Los Alamos investigators (with one exception) are now known for being, relatively speaking, the most skeptical of the whole nationwide STURP (Shroud of Turin Research Project) organization. (Nevertheless, some of the leaders among the national membership do still espouse views similar to the early ones I have quoted.) In any event, it was public statements like these that motivated me to appoint myself as devil's advocate to the shroud investigation.

In October 1978, some thirty American scientists, along with a smaller number of Europeans, spent five days in Turin, Italy, in an intensive round-the-clock experimental study of the famous shroud. The Americans brought with them more than six tons of sophisticated equipment to perform a careful, prescheduled program of observations and tests. Never in all history has a religious relic been subjected to such thorough scientific scrutiny.

About a year later, a flood of news releases and television shows indicated that the shroud's authenticity was just about established. An Associated Press report on November 20, 1979, stated: "The scientist who led the team that investigated the Shroud of Turin last year says evidence so far indicates that the linen did in fact wrap the crucified body of Jesus Christ." The report quoted the scientist: "All of us who were there, at least all of those I talked to, are convinced that the burden of proof has shifted.

The burden is now on the skeptic. ... Every one of the scientists I have talked to believes the cloth is authentic."

Then, just before Easter in 1981, the ABC-TV network show "20/20" gave a detailed presentation on the shroud investigation lasting more than twenty minutes. It left the impression that, in the opinion of the researchers, the shroud is authentic. One leading STURP scientist on the show said, "The likelihood of this being a forgery is less than one in ten million, in my opinion." And so it has been for more than four years now: the views of STURP members have nearly completely dominated the media.

The celebrated interpretation that the shroud image is a scorch formed by a burst of radiation emitted by the corpse obviously requires nothing less than a miracle. Not only would the source of the radiation be unprecedented, but the radiation emanating vertically from every element of the body surface would somehow have to be collimated to project an image and would also have to fall off to zero intensity at about two inches—properties totally unknown to science.

Science developed originally as an oasis of rational naturalism in the vast desert of superstition and supernaturalism that had existed since primitive times. It became a unique attempt to explain the observed world in its own terms—that is, without introducing supernatural forces. In all history, science has never been forced to resort to a supernatural or miraculous hypothesis to explain a phenomenon. Thus I wondered what unique observations, what total failure of naturalistic hypotheses, could have driven these men to embrace such a momentous break with scientific tradition?

From the 1350s, when the shroud first appeared on the stage of history—exhibited in a church at Lirey in north-central France—to the present flurry of scientific activity, this most revered of all relics has played the central role in controversies of many kinds. The whole story is multifaceted and fascinating, but it is much too complex to be discussed in detail here. It is, however, readily available to anyone willing to do some reading. I particularly recommend the sources cited in [1], [2], and [3]. The *Proceedings* [1] gives considerable early scientific details in general and is the best place for details of the radiation-scorch hypothesis in particular. It also gives the general tone and motivation of the organizational meeting of STURP held in Albuquerque in March 1977. The orientation of Ian Wilson's book [2] is more historical, but it also gives considerable scientific detail about the radiation-scorch hypothesis. For excellent color photographs as well as many more-recent results, see the *National Geographic* for June 1980 [3]. However, the most thorough and up-to-date summary of recent experimental results, along with numerous historical and personal anecdotes, is to be found in the November 1981 *Harper's* [4]. This article is lengthy, detailed, and generally factual. It is, however, frequently slanted in diverse ways in favor of STURP and its views, so the overall impression it leaves of the present status of the shroud investigation is misleading.

The Shroud Before 1898

By 1357 the Lirey cloth, proclaimed to be Christ's burial shroud, was being exhibited frequently for fees to large crowds of pilgrims from all over, when a skeptical French bishop named Henri de Poitiers launched an investigation into its provenance. As reported by Henri's successor, Bishop Pierre d'Arcis, the result was that the exhibitions were stopped. But three decades later they started again, prompting Pierre to write a lengthy letter [2] to Clement VII, the Avignon Pope, to counter "the contempt brought upon the Church" and the "danger to souls." The famous letter—the earliest known written reference to the shroud—begins (in translation):

> The case, Holy Father, stands thus. Some time since in this diocese of Troyes the Dean of a certain collegiate church, to wit, that of Lirey, falsely and deceitfully, being consumed with the passion of avarice, and not from any motive of devotion but only of gain, procured for his church a certain cloth cunningly painted ...

Pierre then describes the image on the cloth, which we today call the Shroud of Turin, along with the circumstances of the exhibitions, and continues:

> Eventually, after diligent inquiry and examination, he [Henri de Poitiers] discovered the fraud and how the said cloth had been cunningly painted, the truth being attested by the artist who had painted it, to wit, that it was a work of human skill and not miraculously wrought or bestowed.

Later in his letter Pierre states:

> I offer myself here as ready to supply all information sufficient to remove any doubt concerning the facts alleged.

Clement considered the matter and issued a Bull, which, although allowing exhibition of the cloth, ordered that it be advertised only as a "copy or representation." However, this directive was gradually forgotten, and the proclaimed shroud came to be the most venerated relic in Christendom. To the credit of the Church, while it did not discourage popular veneration, it never took an official stand on the question of authenticity. Within the hierarchy itself, the earlier skepticism has evolved into the guarded belief of the present day—after all, the infallibility of bishops has yet to be proclaimed.

The second important event in shroud history prior to 1898 was its near-destruction in a chapel fire in 1532. While its silver-lined reliquary casket was burning, it was doused with water. This resulted in burn holes (from drops of molten silver) and water stains that partially disfigure the

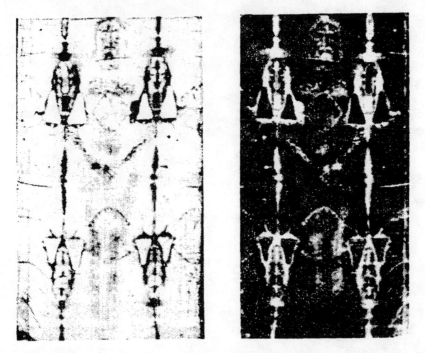

FIGURE 1. The half of the shroud containing the frontal image, normal photographic print on left, negative print on right.

body image. The patched burn holes now appear as rows of triangular spots on both sides of center. (See Figure 1.) The relic has reposed in Turin in this condition for four centuries.

Both ventral and dorsal images of an unclothed, apparently crucified man appear on the same side of a 14-foot-long piece of herringbone-twill linen cloth—as if the body had been laid onto one half of the cloth facing up and the cloth had then been folded over the head to cover the front part of the body. Needless to say, this is not one of the customary ways of wrapping a body; it is also contrary to the description given in the Gospels.

There are many other circumstantial arguments against authenticity, but these cannot be given here; the reader is referred to the writings of Joe Nickell [5]. One central problem is how fifty square feet of high-grade linen, uniquely bearing a life-size image sacred to Christendom, could have escaped explicit historical notice for thirteen centuries.

Post-1898 Investigations

Scientific interest in the Shroud of Turin dates from the first photograph of it, taken in 1898. To the unaided eye, the image on the shroud appears, at least in modern times, faint and nebulous. The image is straw-colored and only slightly darker than the old linen itself. It has the property of appearing to fade out as one approaches to examine parts of it more closely. Also, it has little visual impact, being rather tentative and unappealing.

But the first photograph changed all this. The negative (white-on-black) image showed a distinct figure with Gothic majesty of form and bearing—definitely more "Christ-like," in the canonical art sense. (Compare the negative and positive images displayed side-by-side in Figure 1.) The immediate and obvious question then presented itself: How could an artist, even an artistic genius, have been prescient enough to have contrived to create an image whose perfection would not be revealed until more than half a millennium later? Since no one could give a satisfactory answer, shroud scholars and clerics naturally postulated a miraculous origin for the image. And, as naturalistic explanations for the image—the contact-transfer and vaporgraph hypotheses, for example—were in time shown to be indefensible, skeptics, for the first time, found themselves in an uncomfortable position.

Growing Awareness of the Shroud Mystery

So it happened that different kinds of people, including a sprinkling of scientists and pathologists, began to take the problem posed by the shroud image seriously. Once clear, distinct photographs had become available, devout pathologists and surgeons like Barbet [6] had a field day showing how the image and "bloodflows" agreed in minute detail with the pathology of crucifixion. Their claim was that no artist, particularly in medieval times, could have known that much detail about the medical aspects.

The result of this was a slow but steady increase in scholarly interest in the puzzle of the shroud in the early and middle decades of this century. Nearly all of the interest, however, was restricted to Roman Catholics. Ecumenical interest developed only relatively recently.

In 1969, and more seriously in 1973, small groups of Italian scientists were appointed to make limited tests and observations. (Before word of its existence was leaked, this official commission had been kept secret for several years.) The results were mixed: While they established that the image was not the result of ordinary painting, they also failed to get a positive test for blood in the "bloodstain" areas. Overall, the tone of the 1976 report [7] of this work was slightly skeptical. Two of the experts even suggested that the photographic negativity of the image might be due to some artistic printing technique using a model or molds.

3-D or not 3-D?

The next big "break" in the shroud conundrum came in 1976. John Jackson, joined shortly later by Eric Jumper (both then at the Air Force Weapons Laboratory in Albuquerque, New Mexico), discovered that the shroud image could be interpreted as containing unique three-dimensional information about the corpse they assumed it once was wrapped around. This meant that, allowed their implicit presupposition, they could produce a statue of the "Man in the Shroud," as they called it. This unexpected revelation had a profound impact on the Church hierarchy, as well as on the organizational meeting of STURP held in Albuquerque in March 1977 [1]. The crucial role of the necessary presuppositions was poorly understood. It was swept aside in the nearly unanimous enthusiasm about having made a discovery tantamount to proof that the shroud image could neither be the work of an artist nor caused by natural processes. This discovery doubtless played a decisive role in persuading the hierarchy to allow another, much more thorough round of scientific tests after the shroud was put on public display in Turin in September 1978. (Jackson and Jumper, the founders of STURP, were also able to exert influence through the Holy Shroud Guild, where they sit on the Executive Council.)

Combined with preliminary observations indicating that the image characteristics were similar to those of lightly and superficially scorched linen, the so-called 3-D effect resulted in the hypothesis that the image was caused by a "short burst of radiant energy," probably emanating from the body of Christ at the moment of resurrection [1], [2]. Judging from published statements, it appears that nearly all of the new members joining STURP at this time accepted this hypothesis as the only viable one in the light of the evidence then available. Predictably, the media eagerly picked up this hypothesis and, sans caveats, disseminated it here and abroad as a scientific result. As commonly happens in circumstances suggesting a paranormal explanation, the old journalistic maxim that extraordinary claims demand extraordinary proof was conveniently forgotten.

Let's examine this most extraordinary hypothesis more carefully. More detail can be found in [1]; the radiation-scorch hypothesis has not been published in the scientific literature. The arguments given below were published on December 16, 1979, as part of my critical review in the *Los Alamos Monitor*.

First, an outline of the shroud image is carefully traced, full size, onto a piece of cloth simulating the actual "shroud." The cloth is then draped over the reclining figure of a suitable human male volunteer (5' 11", 170 lbs.) so body parts on the simulated image are in contact with, or vertically over, those of the volunteer. Then by optical techniques (more difficult than they appear) the vertical cloth-to-body distances are determined for a two-dimensional array of image points. Previously, the optical density (darkness) of a photograph of the shroud image had been determined by microdensitometry for a two-dimensional array also. Thus image darkness can be correlated with cloth-to-body distance over a two-dimensional field. This correlation turns out to be only fair, exhibiting a lot of data scatter, but it does indicate a roughly exponential-type fall-off to near-zero image-darkness in a little less than two inches. Jackson and Jumper then draw a smooth curve (an exponential function in the early work) through the region of the scattered data points and call this the "mapping function."

But they also have to make allowance for the sag and drape of the cloth by determining, for the same two-dimensinal array, the distances of the cloth above the table top on which the human model reposes. (By simple subtraction, one also has the shape of the upper surface of the model relative to the table top.) This done, once they have the image darkness for a specific point on the shroud image, they can mathematically invert the mapping function to give the corresponding cloth-to-body distance. From this procedure along with the drape-shape function, the height above the reference plane of a point on the presumed human shape under the shroud can be calculated. By doing this for all the points in the two-dimensional array, they obtain a three-dimensional image reconstruction that can be either viewed isometrically on a computer-generated graph or used to manufacture a statue of the "Man in the Shroud." The resultant three-dimensinal figure looks rather human but is somewhat distorted. To make it look better, the mathematical drape-shape function is iteratively changed—after all, no one knows how the alleged actual shroud might have been draped in a tomb in Palestine two millennia ago. Indeed, the mapping function itself is also modified to give a less-distorted human figure. The final three-dimensional figure then looks normally human except for one thing: If the face is adjusted to have normal relief, the body appears to be in bas-relief.

The salient point is that the three-dimensional reconstruction method depends *essentially* on circular reasoning, and the question of whether or not the shroud ever contained a human form is begged. Also, if there were no measurement errors in determining the cloth-to-body distances, and if there were no data scatter in the correlation plot of image darkness against distance, then they would wind up with just a three-dimensional relief of the human model chosen for the experiment! The irony is that the smoothing process (replacing a scatter plot with a smooth function) itself produces distortion of the relief, but it also affords the opportunity for some of the characteristics of the shroud image to be superimposed on the relief of the human model. Hence the resultant statue is some blend of the human model and the shroud image—not, as has repeatedly been asserted, a statue of the "Man in the Shroud."

What Jackson and Jumper have done is to demonstrate that they can obtain a fair correlation between the image darkness on the shroud and the corresponding cloth-to-body distance they obtain when a particular human model of the proper size is overlaid with a cloth draped in an optimized way. But, because correlation is not causality, that is all they have done. The fact that they have reconstructed a statue from the shroud image using the method above does *not* imply that the image was formed by projection across the distance from body to cloth—or by any other method, for that matter. Some degree of internal consistency in tonal gradations over *localized* regions of the shroud image is all that they have established, despite five years of brouhaha in the press.

By now, and much to their credit, many of the STURP scientists (including all of the Los Alamos branch) have changed their minds concerning the implications of the "3-D effect." They now largely agree with my views stated above. However, there is an influential faction led by John Jackson that still espouses either the radiation-scorch hypothesis or a close semantic relative, despite the compelling evidence that the image is not a *radiation* scorch. For one thing, scorches on linen, whether modern ones or relics of the 1532 fire, exhibit strong reddish fluorescence under ultraviolet irradiation, but the shroud image does not fluoresce at all [8]. For another, STURP, carrying out many trials with lasers of diverse wavelengths and pulse-widths, has not been able even to come close to duplicating the microscopic shroud-image characteristics by lightly scorch - ing linen.

For this and the other reasons given above, the supernatural radiation-scorch hypothesis should have been laid to rest long before now; but it keeps reappearing in the media, in books [2], [9], and even in the *Harper's* article [4] in more restrained language. The recent book [9] by Stevenson and Habermas (reviewed by Schafersman in this volume) reaches the conclusion that the shroud is the burial cloth of Christ largely through belief in the validity of the radiation-scorch mechanism of image formation. This despite (or, more likely, aided by) the fact that they fail to understand the first principles of the logic involved.

One by-product of the three-dimensional work was the "discovery" of strange, nonspherically shaped eyeballs, which can be made to appear to be flat objects if the mapping function is altered to reduce the normal facial relief. Much has been made of the interpretation of these flattened regions as coins over the eyes—particularly by Father Francis Filas, S.J., of the Holy Shroud Guild. Filas claims to have dated the shroud to the middle of the first century by identifying the inscriptions on ancient Roman coins covering the eyes. Press notices on this subject, as well as a half-hour television show, have appeared periodically for several years despite the fact that the claim is so patently absurd that STURP has disclaimed it. The magnified weave patterns in the image areas function somewhat as a Rorschach test—one sees what one wants to see.

Could the Image Be a Rubbing?

The discovery in 1898 that a negative photograph of the shroud image was much more visually appealing than a positive (or the shroud image itself) became a turning point in the longstanding controversy. Thenceforth, shroud scholars held that, since it was long before the invention of photography, no medieval artist would have produced a negative image. Indeed, this, along with the impressive medical evidence, constituted the main argument against forgery before the "3-D effect" was discovered in 1976. However, this situation changed in 1978 when Joe Nickell demonstrated that a rubbing technique, used since the twelfth century or perhaps much earlier [10], automatically produces an image that simulates many of the characteristics of the shroud image, including photographic negativity [5]. If some care is taken in the procedures, the resultant image is rather faint, shows no brush marks, has visually proper tonal gradations, and has a depth of color penetration limited to a few surface fibrils. The technique is to wet-mold a cloth to a bas-relief sculpture, let it dry, and then rub on the pigment medium with daubers of varying size, depending on the contours. Nickell's early work was done with myrrh and aloes, but lately he has been using either dry hematite powder or hematite in a cake matrix.

Nickell's first rubbing was done using an available bas-relief—a four-inch oblique (nonfrontal) view of the face of Bing Crosby. STURP applied its mapping function to a half-tone magazine photograph of this image and found it wanting. As expected, it exhibits the three-dimensional information it picked up from the bas-relief, but STURP's three-dimensional reconstruction from it is badly distorted in places. Whether or not this is a proper and significant test is highly debatable, but nevertheless STURP has made much of this "failure" [4]. It has used this one early result to dismiss Nickell's rubbing technique as a method a forger could have used. There is some question whether this "3-D test" (actually a test for local consistency in tonal gradations) forms a proper crucial test of a method of image formation: the particular mapping function appropriate for the shroud image could be expected to depend on many accidental circumstances attendant at image formation.

It is clear that STURP's conclusion is unwarranted at this time. First, the number of variations of the rubbing technique are enormous—who could say that one of the many could not pass STURP's procrustean "3-D test"? Second, Nickell has by now improved his technique. He has a larger, frontal-aspect bas-relief that gives rubbings bearing a distinct, but not perfect, visual resemblance to the face of the shroud image—see the comparison in Figures 2 and 3 between a negative of the shroud image and one of Nickell's images. Who could say that a few more trials, with appropriate changes in succession, might not result in an image closely simulating the shroud image in appearance and in local consistency of

FIGURE 2. Negative photograph of the face of the shroud image

FIGURE 3. Negative photograph of a rubbing image done by Joe Nickell using iron oxide as the pigment.

tonal gradations?

A Pathological Case?

Several pathologists of obviously strong convictions have long claimed that the shroud image accurately depicts anatomical details along with the wounds described in the Gospels. More recently, a pathologist associated with STURP who is a member of the Executive Council of the Holy Shroud Guild has claimed that no artist could have known enough about the pathology of crucifixion to have forged the image. This pathologist, Robert Bucklin, claims (along with several others) that the medical details of the "bloodflows" are too perfect to have been forged. Also, in the afterword to *Verdict on the Shroud* [9], Bucklin states that "there is support for the resurrection in the things we see on the Shroud of Turin."

All of these claims have recently been challenged [11] by an outside pathologist, Michael Baden, deputy chief medical examiner of New York for Queens County. Baden claims that nearly all of Bucklin's assertions concerning anatomical and pathological features of the shroud image and "bloodflows" are either dubious or incorrect, and on the basis of the image characteristics themselves it is very probable that the shroud never contained a corpse.

What Is the Image Composed Of?

So far we have considered only the macroscopic aspects of the shroud. Now we turn to the issues of microscopic and chemical constitution. Surely, one might hope, the central issue of authenticity versus forgery should be settled here if anywhere.

The details of the many microscopic and analytical tests performed by STURP cannot be given here—see [4] and [12]. In brief, STURP finds no foreign organic substance on the image fibrils. Its unanimous conclusion is that the straw-yellow color of the image fibrils was probably caused by dehydration of the cellulose. With respect to inorganics, STURP's x-ray fluorescence tests find an appreciable amount of calcium as well as detectable iron rather evenly distributed across image and clear area boundaries.

Blood on the Shroud?

In the "bloodstain" areas, STURP finds much more iron and appreciable protein, but no sodium or potassium. (All of these are present in actual blood.) Also found is intense Soret spectral absorption [13], indicating the presence of a porphyrinic material. Other tests also suggest that porphyrins are present. (The presence of porphyrins does not necessarily imply blood—many kinds of porphyrins are present in common plant and animal substances.) From these tests and other, less definitive ones, STURP concludes (apparently unanimously) that the shroud "blood" areas are indeed blood. It also states that there is nothing to suggest that anything other than blood is present in these "blood" areas.

My opinion is that, while real blood could well be present, STURP has not established that it is. More important, it has no case at all for asserting that the "bloodstains" are entirely blood. As an illustration, the reflectance spectrum of Fe_2O_3 on linen [8] fairly closely parallels that of the shroud "bloodstains." From these curves, I roughly estimate that as much as 50 percent of the bloodstain color could be due to Fe_2O_3. Some such adulteration is strongly suggested by the fact that the "bloodstain" areas appear much too red to be just blood—old blood is always brown or black. STURP cannot refute the contention that an appreciable fraction of the "bloodstains" could be due to an artist's pigment composed of iron oxide. Further support for this contention comes from the picturelike appearance of the "bloodstains"; this would seem impossible if blood on a corpse dried in contact with the linen and this was later pulled free.

Of course the putative artist would be likely to use some real blood in producing the "bloodstains" anyway, and so the presence of blood is not of great importance in deciding authenticity. What is of crucial importance is whether the "bloodstains" are at least partly composed of some pigment foreign to blood. There is strong evidence for this, as we shall see.

How Was the Image Formed?

There are only three classes of possibilities for the image formation: by human artifice, through natural processes transferring the image to the linen from a real crucified corpse, or by supernatural means. Of the third, not much can be said, because then all scientific discussion and all rational discourse must perforce cease.

But a lot can be said about natural processes. In terse summary, they can be ruled out *definitively* by the quality and beauty of the shroud image. However, since natural processes involving contact transfer and vapor transport have long been discussed as possible mechanisms, we should really lay them to rest here. In this, STURP would concur wholeheartedly.

Both contact-transfer and vapor-transport mechanisms are ruled out by the observed superficiality and lack of saturation at supposed points of contact with the corpse. Also, the image-forming process must be pressure-independent, for the maximum image darkness is the same for the dorsal image (supposedly with a body lying on it) as for the ventral. Contact transfer cannot explain image formation in the many places (like eye sockets) where the linen would not be expected to touch, and tonal gradations would be quite limited in any case. And then there is an inescapable fact of geometry: Even if the cloth were to be somehow closely molded to a full-relief (body or statue) for contact transfer to take place, then when it was taken off and flattened out the image would be grossly distorted. The fatal flaw of vapor transport is the undeniable fact that gross blurring necessarily results from a diffusive process acting across a distance.

Artistic processes are quite another matter, however. While some can be ruled out, others cannot be—quite contrary to all STURP press releases, the recent *Harper's* article [4], and all books on the subject. That the shroud image was not painted in any ordinary sense of the term is by now beyond dispute. But what about "cunningly painted"? Scorching from a full-relief, whether by contact or by radiation transport, can be ruled out for many reasons, including those given in the paragraph above. Very light scorching from a moderately hot bas-relief cannot be ruled out with certainty, but it runs into serious trouble with image superficiality as well as scorch fluorescence. In its favor is the fact that low-temperature scorches do result in dehydrated cellulose.

But there are also other methods of dehydrating cellulose, and these can be employed in a hypothetical hypothesis of human artifice, as we shall see, which STURP cannot refute at this time. Yellowing of linen due to cellulose dehydration can be caused by a large variety of common substances [8], perhaps nearly anything, in contact with linen. With the passage of time, exposure to sunlight, or elevated temperatures such as those due to gentle baking, substances from body oils to lemon juice are observed [8] to produce accelerated dehydration of cellulose fibrils. In order to obtain an image due solely to dehydrated cellulose, the medium would have to be applied very sparingly so that it would either be consumed in the reaction or would disappear (over time or temperature) by evaporation or sublimation. To limit the image coloration to superficial depths, the medium would have to be applied in the form of a dry powder or a semi-solid cake. Also, the possibility that the shroud was washed at some time after the image became fixed cannot be ruled out and is another possible mode of medium disappearance.

The only technique of medium application that appears probable at this time is the rubbing technique of Joe Nickell, discussed earlier. On October 2, 1980, I sent a Los Alamos STURP member a brief note proposing a simple rubbing scenario. STURP has not yet refuted it. All that would be necessary, at least in principle, to form the shroud image is to postulate that the rubbing was done using a semi-solid, presumably colored (so the artist could see what he was doing) medium that, over the six centuries or in the 1532 fire, has virtually disappeared, leaving behind only cellulose fibrils discolored by dehydration.

Until they have ruled out this and similar hypotheses, STURP members have no justification whatsoever for stating, as has been reported in the media for years, that their conclusion is that none of the forgery hypotheses is tenable. Lately, they've been calling the mechanism of image formation "a mystery." This is somewhat misleading; it implies that there is no viable hypothesis at this time. The one given above is not yet established experimentally, but it does deserve to be taken seriously. In private, several of the Los Alamos members will admit that they have not really ruled out human artifice. Nationwide, at least, most members still seem to regard the dehydrated-cellulose image as a probable low-temperature scorch, and the image as having been somehow "projected" across space onto the cloth. This is, of course, the old radiation-scorch hypothesis in thin semantic disguise. The main difference between 1977 and the present is that much more is known now, and STURP members are more puzzled and much less certain. Also, and this cannot be overemphasized, STURP nowadays is hardly monolithic—it is really an assembly of individuals holding rather diverse opinions.

However, ex-STURP members Stevenson and Habermas in their recent book [9] regard the radiation-scorch hypothesis as proved and as God's belated gift to believers. They fully realize that it clearly implies supernatural intervention. Their book is replete with one or more significant misunderstandings and misrepresentations—both subtle and gross—on almost every page. STURP as a whole is quite unhappy with the representation of the book as a publication of STURP results.

Enter Walter McCrone

I have purposely saved a discussion of Walter McCrone's contributions until last. Since STURP unanimously rejects McCrone's interpretations and conclusions, I wanted to demonstrate first that even without McCrone's results one can make a good case for human artifice.

But who is Walter McCrone? He is certainly the best-known forensic microanalyst in the world and is widely regarded also to be the best in his field. He runs the McCrone Research Institute in Chicago, which performs microscopic and ultramicroscopic tests of all kinds using a panoply of sophisticated instruments in addition to optical microscopes. He is best known for having produced the definitive work [14], in five volumes, on microscopic particle identification and assay techniques, as well as for showing in 1974 that the formerly celebrated "priceless" Vinland Map of Yale University (supposedly a pre-Columbian map of the American east coast) is actually a modern forgery on ancient parchment.

During the 1978 tests on the shroud, STURP used an especially pure sticky tape to lift samples from the shroud surface for later analysis. Thirty-three tape samples were obtained at carefully measured locations in different regions on the shroud. STURP gave McCrone, who was still a member at the time, thirty-two of the tape samples for analysis. A "blind" study [15] was conducted to separate the tapes by microscopic observations into two groups: those with apparent pigment particles on the fibrils, and those without. None of the ten samples from clear (no image, water stain, or "bloodstain") areas showed pigment particles; nineteen from image and "bloodstain" areas showed significant amounts of pigment. Of the nine tapes from image areas, three did not exhibit significant pigment, but these were from areas with a very weak image. McCrone subsequently identified the particles as micron-size hydrous and anhydrous iron oxide. These particles are mostly well dispersed and coat the individual fibrils in both the image and the "bloodstain" areas, with the latter having a much greater particle density. Also found, in the "bloodstain" areas predominantly, are large numbers of red clumps or aggregates of about 10- to 25-micron size; these contribute appreciably to the reddish color of the "bloodstains."

Since he finds practically no iron oxide in the clear areas, and since artists' pigments containing iron oxide (red ochre or hematite) were widely used in the Middle Ages as well as much earlier, McCrone suggests that iron earth pigment was intentionally applied by hand to form the image. STURP admits that there are some iron oxide particles on the image fibrils but claims they are of insufficient number (actually, areal density) to contribute appreciably to the visual image coloration. STURP backs this up with optical reflectance spectra [8] of the shroud image areas and claims that the spectrum of Fe_2O_3 is much too red for iron oxide to be a significant factor in the overall image color. McCrone counters that the color of iron oxide particles varies from yellow to red depending on the state of hydration and that most of the particles on the image fibrils are yellow. After reviewing the evidence on both sides, I *tentatively* conclude that ferric oxide contributes less than about 10 percent to the overall image intensity; but more experiments should be done.

My thinking is that, while the issue of visibility is important, it falls short of being crucial. What *is* crucial is whether McCrone is right in claiming that iron oxide particles are found only in the image and "bloodstain" areas and almost never in the clear areas. Because no conceivable natural process of iron oxide particle formation could so discriminate between image and clear areas, a reliable finding of iron oxide—*even in macroscopically invisible quantities*—only in the nonclear areas would be tantamount to proof that human artifice was involved in producing the image. One possible scenario would be that the image was formed originally by iron oxide pigment dispersed in some semi-solid matrix in sufficient concentration to produce deeper image coloration than now exists. Subsequently, over six centuries of exhibitions and handling (very much more frequent in the early centuries than the later) the larger particles were sloughed off, leaving a faint image colored mainly by dehydrated cellulose fibrils coated sparingly by micron-size iron oxide particles. (Micron-size particles are known to adhere *tenaciously* to any fabric.)

The red particle aggregates I mentioned are quite numerous in the "bloodstain" areas, where they contribute to the redness of the visual image. At about 30-power magnification in STURP photographs [3], they appear as numerous red specks among the threads. McCrone analyzed [15] eleven of these red specks using scanning electron microscopy, electron microprobe analysis, electron diffractometry, x-ray diffractometry, and energy-dispersive x-ray analysis, as well as optical microanalysis. He finds the specks to be agglomerates of particles containing mainly calcium and iron with frequent large quantities of mercury.

Maps of the iron distribution within the individual agglomerates usually correlate well with the observed distribution of red color, but in some cases the brightest color correlates best with separate-but-identical distribution maps of mercury and sulfur. The presence of significant amounts of HgS, a pigment called vermilion, has been demonstrated unequivocally in six of the eleven agglomerates. One agglomerate even contained enough mercury to form a tiny "mirror" when reduced onto a microscopic copper sliver in a well-known test for mercury.

Vermilion is also an artist's red pigment used widely in the Middle Ages and earlier. Because vermilion cannot come from blood or any other body substance, its presence in a "bloodstain" area strongly indicates that the "bloodstains" have been produced, at least in part, by an artist. A crucial issue here is how representative of the speck population is McCrone's sample of eleven? More work needs to be done on this.

Epilogue

So modern experimental science in its full-bore, megabuck aspect has now been used to confront the most venerated relic in Christendom, and the outcome is still far more tentative and debatable than most observers expected. That the Shroud of Turin is not a painting in the *standard* sense has been demonstrated unequivocally; that it was not produced by unaided natural processes is virtually certain. All that now remains is to decide between human involvement in "cunningly painting" it and its provenance through means totally unknown and unimaginable to science—and therefore possibly supernatural. Needless to say, the issue at stake here is one of epochal potential implication.

Because modern science has never had to introduce a supernatural hypothesis (whatever that might mean, since scientific method is absolutely and essentially based on naturalistic explanations) *scientific* investigators eschew any paranormal or supernatural explanation—even for something that has long been venerated as the actual image of Jesus Christ. In their *recent* public statements, STURP members do avoid *directly* mentioning a supernatural explanation. However, in newspapers, and particularly in books, as well as in the recent *Harper's* article, the supernatural is often directly suggested, or is, at least, logically implied by ruling out human involvement.

But what is a skeptical inquirer to make of all this? He will naturally try to formulate a human-artifice hypothesis consistent with the known principles of art and science and also consistent with the observations of STURP, McCrone, and Nickell. Such a hypothesis has been discussed in the previous section, but in a rather disjointed manner. To recapitulate briefly here: First, the image-forming method will probably have to be based on some kind of imprinting technique to impart photonegativity automatically—most likely rubbing, à la Nickell. Second, the rubbing medium must be dry or semi-solid (to limit image depth), must contain a coloring pigment (probably an iron earth), and must largely disappear (being nonproteinaceous would help, relative to STURP's tests) by reaction with cellulose over time and temperature, or perhaps by evaporation, leaving finally only dehydrated cellulose fibrils coated sparsely with iron oxide particles.

It appears certain that STURP cannot rule out this hypothesis, or similar ones, at this time. Why then do recent press releases and publications (such as the *Harper's* article, which obviously enjoyed STURP's blessing) state that human artifice has been eliminated as a possible image-forming method? First is the fact that the media do frequently distort statements to serve their needs. Media writers are insatiably hungry for important and unusual stories—particularly those with the air of a "breakthrough," as well as those involving religion or the paranormal. Just a whiff of a hint that the shroud could be authentic is sufficient to fulfill these conditions. Second is the obvious fact that STURP itself is far from blameless in this matter. Scientists, like all human beings, desire to make a difference in the world by working on something of considerable (if not epochal) importance. Concluding that the shroud is the work of an artist would dash the hope of changing the prevalent scientific world-view. Also, the strong religious inclinations of nearly all of the STURP membership doubtless play a role here.)

With its radiation-scorch hypothesis of 1977, STURP released a potent and willful genie that now refuses to get back into the bottle on command. Some STURP members admit in private that they regret the matter was handled the way it was. With the benefit of hindsight, what was their main mistake? It was, in part at least, neglecting to include card-carrying "skepticians" in their working group. Thoroughgoing, scientific skeptics would certainly have been a great help in keeping *that* genie bottled up. Of course it is easy to criticize now. In reality, very few skeptical scientists would have been willing, even if invited, to sacrifice years of family and vacation time for the dubious reward of (in their opinion) finally finding the obvious—an obvious devoid of much real scientific interest. Hence a kind of selection process among the STURP investigators was inevitable, for this and other reasons.

What then of the future? Foremost is the need to date the shroud cloth by radiocarbon methods. This could now be done with independent replications using a total amount of cloth no larger than a fingernail, and with an accuracy of about ±100 years. Any replicated date more recent than about A.D. 300 would establish that it is not the burial cloth of Christ. However, since an artist might well have bothered to obtain ancient linen, a date of even A.D. 30 ±100 could not rule out forgery. Hence, the church hierarchy has little incentive, and considerable disincentive (unless its faith in the shroud is very strong), to permit radiocarbon dating. To the credit of STURP, it has made several concerted attempts to persuade Turin and the Vatican to permit radiocarbon dating; one of these was turned down by the Vatican nearly two years ago, and a new attempt is being made now. STURP points out that there is enough charred shroud linen hanging loose under the burn patches for several hundred dating tests, so the church's argument against destroying shroud cloth is weak indeed. Ideally, of course, the shroud investigation should have *started* with radiocarbon dating, which might then have saved an enormous number of man-years—both within and without STURP.

Another point that really needs to be settled is the serious disagreement between McCrone and STURP over what is or is not on the cloth.

Replication, the tedious and costly price that science pays for objectivity, is probably the key to resolving this controversy. In the many areas of disagreement, the procedures of McCrone and STURP should be repeated (adding random sampling and extension to larger shroud areas) at a suitable independent laboratory, ideally under the close scrutiny of both McCrone and a STURP representative. It would indeed be surprising if this didn't settle the controversy once and for all.

Finally, the hypothesis I have presented, that the dehydrated-cellulose image was formed by use of a rubbing method, should be subjected to serious and careful testing. Iterative trials should be made to try to simulate the shroud image both macroscopically and microscopically. What is required is a bona fide commitment to trying to show how an artist might have forged it.

This could involve rubbing using a bas-relief (perhaps to start with, the one used by Joe Nickell in producing Figure 3) and a medium similar to the one I have discussed. Aging and the 1532 fire would be simulated by low-temperature oven-baking [8]. After-baking treatment might include washing to try to remove the last trace of rubbing medium from the cloth. The handling and folding/unfolding of the early centuries could perhaps be roughly simulated to determine the degree of iron oxide sloughing and transfer.

This is a serious challenge. STURP, having the resources and expertise required, should be eager to accept it.

References

1. *Proceedings of the 1977 United States Conference of Research on the Shroud of Turin,* Holy Shroud Guild, 294 E. 150 St., Bronx, N.Y. 10451.
2. Ian Wilson, *The Shroud of Turin* (New York: Doubleday, 1978).
3. Kenneth Weaver, "The Mystery of the Shroud," *National Geographic,* June 1980.
4. Cullen Murphy, "Shreds of Evidence: Science Confronts the Miraculous—The Shroud of Turin," *Harper's,* November 1981.
5. Joe Nickell, *The Humanist,* November/December 1978; *Popular Photography,* November 1979; *Free Inquiry,* Summer 1981.
6. Pierre Barbet, *A Doctor at Calvary* (Image Books, 1963).
7. *Report of the Turin Commission on the Holy Shroud,* Screenpro Films, 5 Meard St., London W1V 3HQ, 1976.
8. S. F. Pellicori, *Appl. Opt.* 19:1913-20 (1980).
9. K. E. Stevenson and G. R. Habermas, *Verdict on the Shroud* (Ann Arbor, Mich.: Servant Books, 1981).

10. J. J. Bodor. *Rubbings and Textures: A Graphic Technique* (New York: Reinhold, 1968).

11. R. W. Rhein, *Medical World News,* December 22, 1980, pp. 40-50.

12. L. A. Schwalbe and R. N. Rogers, "Physics and Chemistry of the Shroud of Turin" (to be published in *Anal. Chim. Acta,* February 1982).

13. J. H. Heller and A. D. Adler, *Appl. Opt.* 19:2742-44 (1980).

14. W. C. McCrone, *The Particle Atlas,* vols. 1-5 (Ann Arbor Science Publishers, 1973-79).

15. W. C. McCrone and C. Skirius, *Microscope* 28(3/4):105-14 (1980); W. C. McCrone, *Microscope* 28(3/4):115-28 (1980); 29(1/4):19-39 (1981).

37

Shroud Image Is the Work of an Artist

Walter McCrone

I am very pleased to have this opportunity to summarize my present position with respect to the Turin shroud. My position has not really changed during the past year or so since I completed the experimental work, although I have become more confident and have expressed my conclusions with a little more care and precision.

My microanalytical work on the sticky tapes from the shroud's surface has proved to my satisfaction that the entire image was produced by an artist using iron earth and vermilion pigments in a tempera medium during the middle of the fourteenth century. All of the image area tapes show varying amounts of two red pigments, which are easily identifiable as either an iron earth pigment or vermilion. The amount of these pigments is greater in the areas of greater image density, particularly the "bloodstains." Smaller amounts are present in image areas, and in some image areas only very small amounts of the pigment can be found. The dispersion of very tiny pigment particles characteristic of the image areas is totally absent in control samples from the cloth.

The image was created by an artist who was commissioned to paint a shroud, probably to be used in religious processions or to be exhibited in the newly founded church in Lirey by the de Charny family. I doubt if the artist was intending to fool anyone, and I feel that the church vergers didn't have to make any conscious effort to convince the general populace that this was the shroud of Christ. I think the vergers did allow the populace to come to that conclusion and, since that time, of course, most believers have so concluded.

Coming back to the image itself, I feel that the artist, having been commissioned to paint a shroud, did an inspired job of thinking just how one might expect an image to appear in the absence of light, hence no shadows. He had to depend on thinking of a cloth simply in contact with the human body. He naturally filled in the portions where the cloth touched the high points of the body and then artistically graded them in decreasing intensity from those high spots, thus creating a pleasing image

344

and one which, it then turns out, automatically produces a photographic negative when copied photographically and indeed develops a three-dimensional structure when interpreted in terms of cloth-body distances as Jackson and Jumper did. There is nothing unusual about this and it was in fact entirely automatic. The artist, of course, certainly knew nothing about photographic negatives, nor did he think about three-dimensional reconstructions.

Finally, I can see no possible mechanism by which the shroud image could have been produced except as the work of an artist. The faithful representation of all of the anatomical and pathological markings, so well described in the New Testament, would be difficult to produce except by an artist. They are totally without distortion and, indeed, look exactly the way we would like to have them look.

Cryptozoology

38

Sonar and Photographic Searches for the Loch Ness Monster: A Reassessment

Rikki Razdan and Alan Kielar

The existence of a Loch Ness monster has long been the subject of speculation. In recent years, mainly as a result of sonar work carried out by Robert Rines and co-workers [1-4], the phenomenon has gained sufficient scientific credibilty that a zoological classification has been ascribed to the creature [5].

Alleged scientific evidence for the existence of a Loch Ness monster dates back to as early as 1960. Copious amounts of data have been generated through expeditions conducted by Cambridge [6] and Birmingham [7] universities, the Loch Ness Investigation Bureau [8], the Loch Ness and Morar Project [9], and the Academy of Applied Science [1-4].

A thorough evaluation of this data led us to question its validity. In many instances, the findings were distorted by significant procedural and quantitative errors. We present evidence here on a case-by-case basis that conflicts with the findings of these researchers and renders inconclusive claims for the existence of the Loch Ness monster.

The questionable accuracy of this previous data coupled with our longstanding interest in the Loch Ness legend prompted us to initiate our own expedition to Loch Ness in May 1983. We designed a tracking sonar array specifically for this purpose at our laboratory in Rochester, New York. This expedition had the advantage of being able to image in three dimensions and obtain an actual tissue sample from any large underwater creature coming within range of our 25-meter × 25-meter array. The karyotype analysis of a tissue sample would provide conclusive evidence of the existence of some unique creature, in contrast to ambiguous sonar traces and photographs presented by previous researchers.

The 15-ton sonar apparatus, which can scan a depth of 33 meters, was tested during the summer of 1982 at Conesus Lake in New York State and was then shipped to Temple Pier, Drumnadrochit on Urquhart Bay at Loch Ness. The array consisted of 144 sonar transducers, arranged in a 12 × 12 grid. It imaged the area over which it floated by measuring the water depth directly beneath the individual sonar transducers. These depths were displayed on a television monitor as color-coded blocks arranged in a 12 × 12 grid pattern corresponding to the placement of the transducers. Any sonar target over 3 meters in length moving under the grid triggered an alarm, and the target was automatically tracked in three dimensions. Schools of fish were automatically recognized and ignored by the tracking electronics based on an analysis of the return echoes from each sonar transducer. Nine underwater dart guns with retrievable biopsy tips were mounted on the array and could be automatically fired upon detection of an intruder beneath. If a tissue sample were obtained from a large target, arrangements had been made for its analysis at the Zoology Department of Glasgow University. The sonar transducers [10] were encased in rubber flotation collars and were separated from one another by 2-meter-long aluminum cross-pieces.

The array, tethered to shore by a 330-meter control cable and floating in approximately 65 meters of water, was moored adjacent to the raft used by the Academy of Applied Science (AAS) expedition in 1976 [11]. The array was powered by 12 volts of direct current (DC) to avoid the possibility that alternating current (AC) might frighten away marine life. All sonar information was automatically recorded on videotape. The equipment, except for the biopsy darts, was tested with a scuba diver and air-filled calibration targets and functioned as designed. The biopsy darts were capable of reaching a depth of 33 meters. By July 25, 1983, the array was fully functional and ran continuously until September 16, 1983. During that time it recorded nothing larger than a 1-meter fish. We are aware that our stationary array could monitor only a very small area of the Loch, but this area was one that had produced the most compelling evidence to date [1-4].

Our working on the project beside the loch for four and a half months established our credibility with the local residents. They were familiar with previous investigators and their work, particularly that of the AAS. The AAS's investigations had been the most ambitious and their results the most dramatic. Our conversations with these residents uncovered background material that contrasted with the published circumstances under which the Academy's results, especially the well-known flipper photographs [1], were obtained. We continued to review previously published sonar work carried out at Loch Ness and investigated its original sources at the Inverness Library and the Loch Ness Centre in Drumnadrochit. Our findings place much of this seemingly credible data into a more proper perspective.

Sonar Searches at Loch Ness

The first organized sonar searches of Loch Ness, using echo sounding and dredging equipment, were conducted by Cambridge and Birmingham universities between 1960 and 1962. Several authors [12] have reported that this investigation produced evidence of a Loch Ness monster. However, the actual reports [6, 7] indicate a general scarcity of animal life in the loch and state that no unexplained objects were encountered in either the echo sounding or the dredging operations.

In 1968 Braithwaite and Tucker [13], from the University of Birmingham, deployed an experimental digital sonar system off Temple Pier. The system ran continuously for a two-week period and produced a 13-minute sequence showing targets rising vertically underwater, with velocities exceeding those of known fish. We found that Tucker, after continued on-site experimentation through 1970, attributed those earlier sonar contacts to gas bubbles and thermal effects encountered over the long range (1.2 kilometers) of their sonar beam [14].

In 1969, under the auspices of the Loch Ness Investigation Bureau (LNIB), Robert Love conducted extensive channel-wall and bottom-echo soundings in the loch [8]. These narrow-beam soundings gave no indication of underwater caverns or overhanging ledges in the loch walls. His transverse sonar running across the loch denoted a general wall slope of 45 degrees, leading to a fairly uniform bottom depth of 200 to 230 meters. Love also conducted midwater patrols with a sector-scanning sonar, making only one documented contact with a large and seemingly inanimate object. Although he admitted that the target's vertical movement could not be assessed, he claimed that the target exhibited horizontal movement of an unusual, hence animate, nature. However, our analysis of his published data [15] revealed that inconsistencies in the correlation of his boat's speed with the distance he covered during the fixed observation time of the phenomenon, coupled with ambiguous ship's head information, suggest he may have been tracking a fixed underwater target, possibly debris from past expeditions [16]. In addition, the 15-degree width of Love's sonar beam at the target's maximum range of 480 meters is a significant proportion of the target's calculated horizontal movement.

During the summer of 1969 the Vickers submarine *Pisces* was introduced into the loch. It has been credited with obtaining a depth reading of 320 meters off Urquhart Castle. We discovered that this reading was actually made with a depth sounder from a boat on the surface [17], presumably over the submarine. The area of the loch in question has been crisscrossed many times since 1969, and no depths greater than 250 meters have been reported [9].

The most recent sonar data to come from Loch Ness are contacts made by Shine during the summer of 1982 [9]. Although the sonar traces show apparently strong midwater targets, none of Shine's contacts conclusively demonstrate any horizontal or vertical movement. As with Love, we deduce that he probably obtained sonar echoes from stationary midwater debris.

The AAS Investigation and the Flipper Photographs

The most ambitious investigation of Loch Ness was undertaken over a period of eight years, beginning in 1970, by the Academy of Applied Science, Belmont, Massachusetts, headed by Rines. In 1970 the results of a three-day side-scan sonar test by his colleague Klein indicated that there were large moving objects in the loch and that there are large ridges or caves in the steep walls of the loch that could conceivably harbor large creatures [18]. These findings, published in 1976 [2], were regarded as important discoveries. Our study of a 1972 Academy publication revealed that Klein had interpreted the 1970 large moving sonar targets as "probably from three large fish" [19] and emphasized the necessity of probing the areas that appeared to be caves or overhangs with "a narrow conical beam sonar, to look straight in at the walls to determine the actual depth of these undercuts" [18].

In 1971 the Academy conducted "copious sonar profiles" [20] of Urquhart Bay. These profiles indicated the presence of significant underwater mounds and channels, and Rines spent considerable time correlating eyewitness surface sightings of the monster with these unusual topographical features [21]. He speculated that "it is along these deep ravines, in parts several hundred feet deep, that large creatures could frolic . . . without ever rippling the surface" [22]. In fact, we found that no sonar studies conducted prior to, or after, the Academy's survey show anything but a fairly steep slope down to a flat bottom [8, 9].

A paper entitled "Sonar Eyes Unmask Urquhart Bay" [4] contains the following statement by Rines: "In the summer of 1971, the Academy of Applied Science (AAS) team, in the company of local residents (former RAF wing commander Basil Carey and his wife Winifred), viewed and tried to photograph in the dusk, over a five minute interval, 20 feet of slowly moving black hump protruding at least five feet out of the calm waters of Urquhart Bay before submerging" [23]. On August 4, 1983, along with Ivor Newby and Dennis Vann [24], we talked to Mrs. Carey [25] about that evening in 1971. Mrs. Carey demonstrated for us how, on the evening of June 23, 1971, with Rines present, she dowsed [26] for the monster using a dowsing stick, a map, a pendulum, and a nonmagnetic pencil. She claimed that the point on the surface of the loch where Rines reported sighting the monster was the very one she had predicted to him, and she gave us a map on which she had recorded her 16 monster sightings since 1917, including the one made that evening with the AAS team— composed of only "Bob and Carol Rines" [27]. Asked if she could describe her impressions of the sighting that summer night in 1971, she said: "It was a dark evening. It could have been anything."

Also in 1971, Rines described the mysterious disappearance of a camera-strobe unit being used for underwater photography in Urquhart Bay [28]. A photograph of a loop of rope [29], taken while the camera was presumably being towed about the loch by the monster, was captioned a "rope loop" or "flipper." Rines interpreted a large, bodylike shape into the picture, even though we were told by local residents [30] that a fisherman had caught his propeller in the camera's buoy marker and that the camera had drifted out of the bay into the midwater and was later recovered by a passing LNIB member.

In 1972, Rines produced the dramatic photographs purporting to show the flipper of the Loch Ness monster, taken in Urquhart Bay off Temple Pier with a linked camera-strobe sonar system. According to an illustration published by the AAS [31], a sonar transducer was mounted on the sloping loch-bottom in about 10 meters of water overlooking the camera-strobe unit, which was also secured to the bottom in about 15 meters of water. Photographs taken by the camera-strobe unit could be confirmed by the presence of objects detected within the sonar's beam, and the linked system could provide a rough dimensional analysis of those objects. According to the Academy, on the morning of August 9, 1972, Rines and his crew (personnel from the AAS and the LNIB) waited aboard two boats in Urquhart Bay. One boat monitored the sonar equipment, the other the camera-strobe unit. At around 1:00 A.M. on August 9, the investigators detected what appeared to be large objects moving into the transducer's beam. Underwater photographs were taken and, when developed in the United States, showed only faint shapes [32]. After image enhancement by Alan Gillespie at the California Institute of Technology's Jet Propulsion Laboratory, the flipper photographs, and the photograph apparently showing two large bodies were published in books and magazines throughout the world [33].

Further discussion about the events of August 9 with Mrs. Carey uncovered previously unpublished information crucial to the integrity of Rines's findings. This information was later corroborated by Holly Arnold [34], an active participant in the events of August 9, 1972. Mrs. Carey recounted to us how, through the night of August 8 and the early morning of August 9, she dowsed for the monster from her house overlooking Urquhart Bay, while Rines remained with the boats on the water. At about 1:00 A.M. she "detected" two monsters in the vicinity of the river's mouth in the bay. Mrs. Carey then alerted Rines to the monsters' location by a prearranged signaling system, involving the flashing of automobile headlamps in a code corresponding to numbered blocks superimposed over a map of Urquhart Bay. Rines and the crews then proceeded to lower overboard the camera-strobe unit from one boat and the sonar transducer from the other.

The sonar traces [35] from the August 9, 1972, encounter indicate

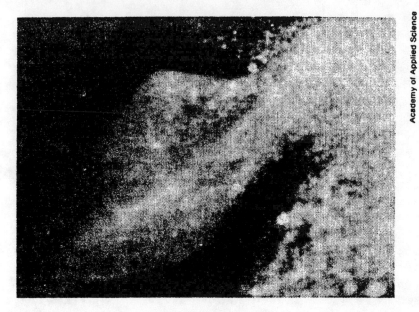

The famous flipper photo, supposedly enhanced by a computer at the Jet Propulsion Laboratory

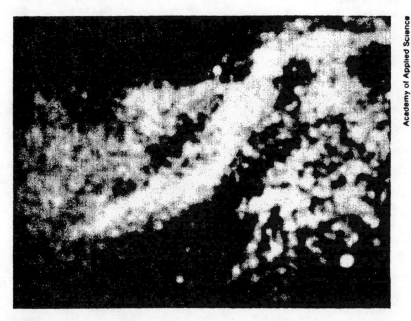

The actual JPL computer-enhanced photo

that the transducer unit was not secured to the Loch's bottom, as shown in the illustrations in most books and publications, but was free-swinging [36]. Nowhere in these sonar traces can the camera-strobe unit or its umbilical cable be seen as they would have been if the published illustrations had been accurate. With the camera and sonar units deployed in this manner, it would have been impossible for Rines to make any rational correlation between any underwater photographs and the sonar traces. Also, during the sonar contact of August 9, 1972, several investigators were rowing a large wooden fishing boat between the boat tending the sonar equipment and the boat tending the strobe cameras [37]. This could account for the sonar traces attributed to two monsters. In our opinion, the sidelobe echoes from the "flipper" sonar traces show that the transducer was moving about underwater, and the main beam traces have the obvious characteristics of boat wakes.

Apart from the circumstances in which they were obtained, the flipper photographs themselves do not stand up to scrutiny. On returning to the United States in October 1983, we contacted Alan Gillespie at the Jet Propulsion Laboratory and established that Rines had provided him with copies of his 1972 flipper photographs for image enhancement. The computer-enhanced versions were returned to Rines and were subsequently published in *Technology Review* [1]. We have copies of the two original flipper photographs before enhancement [38], the enhanced versions, and finally the published versions. Alan Gillespie sent us copies of the enhanced flipper photographs he returned to Rines, and we discovered that, prior to publication, areas of these enhanced photographs had been significantly altered to give the impression of the flipperlike objects that appear in the published versions. We feel this discrepancy is very important because it was these flipper photographs that led Sir Peter Scott to assign a zoological classification to the creature, *Nessiteras Rhombopteryx* (Ness Beast with the Diamond Shaped Fin) [5].

In 1975 Rines obtained "close up color photographs of the head, neck and body of one of the animals" [39]. They were published in 1976 [1]. The deployment of the camera-strobe unit when these pictures were taken again deserves comment. Tim Dinsdale describes how the unit was "trailed from the stern of Hunter [Rines's boat] on the end of some 30 feet of rope and allowed to swing around with the boat, as wind and chance determined" [40]. Dick Raynor, a former member of the LNIB, who was present during this expedition, recalled that the camera-strobe unit taking the pictures was retrieved from the bottom of the loch with ten feet of slack on its buoy rope because the boat had swung into shore [41]. Given these circumstances, the published photographs can be most reasonably interpreted as debris on the loch bottom or side walls. Rines claimed to have sonar traces showing the intrusion of large objects into the zone of camera coverage, but these have not been published. On the basis of

Rines's 1972 and 1975 evidence, *Nature* published "Naming the Loch Ness Monster" [5], seemingly legitimizing what the facts suggest is unscientific and careless research.*

The Academy of Applied Science returned to the loch in 1976 with the financial support of the *New York Times*. Professor Harold Edgerton (a pioneer in strobe photography) accompanied this expedition. Underwater television cameras and camera-strobe units were suspended from boats and finally attached to a raft moored off Temple Pier. Once again the equipment was not rigidly fixed underwater [11]. At one point during the summer a large sonar trace was obtained that had, said Rines, "the same dimensional extent in 1972 . . . the same type of multiple trace echo and the same type of approach to the camera from behind as in 1975" [42], but *this trace was immediately established to have been caused by the wake of a boat* [43].

A paper by Harold Edgerton in 1978 [3] outlined sonar data gathered during the summer of 1976 at Loch Ness that, while not concluding that a monster exists, suggested that further research was warranted. In this paper, he equated the width of observed sonar targets with the actual trace thickness produced on a sonar chart recorder. However, it should be recognized that there is not necessarily a linear correspondence between a target's sonar track thickness and the true thickness of an object intruding into a sonar beam. No calibration sonar traces were included in his paper for comparison.

Klein, conducting additional side-scan sonar testing in 1976, attached archaeological significance to his discovery of numerous stone circle formations, groups of which he named Kleinhenge I and Kleinhenge II, on the bed of Loch Ness [44]. Our research at the Inverness Museum (summer 1983) quickly discounted his theories, since, during the construction of the Caledonian Canal in the early 1800s, dredge steamers routinely dumped such piles of stones into Loch Ness. Formations similar to those that Klein discovered can also be found near the mouth of the canal in Loch Oich and Loch Lochy.

Conclusion

We have shown that continuous sonar monitoring for seven weeks to a depth of 33 meters in an area where many previous sonar contacts had been reported showed no evidence of anything larger than a 1-meter fish.

Note added by editor: In December 1975, *Science News,* the magazine I then edited, contacted David Davies, the then-editor of *Nature,* for elaboration. He emphasized that Rines's article and photos had been published in *Nature*'s "Comment and Opinion" section, which meant they had not undergone any peer review process. He said *Nature* did not vouch for the validity of the claims (*Science News,* 108:391, Dec. 20/27, 1975).—*K.F.*

The circumstances under which previous expeditions had obtained sonar and photographic evidence in support of the existence of the Loch Ness monster could not withstand scrutiny. The evidence itself revealed discrepancies. This is especially true of the Academy's flipper photographs, the published versions of which differ from the original computer-enhanced photographs. Careless deployment of equipment and overzealous interpretation of the data account for much of the so-called scientific evidence.

While it is not possible to prove definitely that a monster does not exist, the evidence so far advanced strongly suggests that the Loch Ness monster is nothing more than a longlived and extremely entertaining legend.

Notes and References

1. R. H. Rines, H. E. Edgerton, C. W. Wyckoff, and M. Klein, *Technology Review*, 78, no. 5 (1976):25.

2. M. Klein and C. Finkelstein, *Technology Review*, 79, no. 2, (1976):3.

3. H. E. Edgerton and C. W. Wyckoff, *IEEE Spectrum*, 15, no. 2, (1978):26.

4. M. Klein, R. H. Rines, T. Dinsdale, and L. S. Foster in *Monograph I Academy of Applied Science*, Academy of Applied Science, 1972.

5. *Nature* 258 (1975):466.

6. Report of the *Cambridge Loch Ness Expedition*, 1962.

7. Preliminary report of the *Birmingham University Loch Ness Expedition*, 1961.

8. D. James et al., *Annual Report of the Loch Ness Investigation Bureau*, 1969.

9. A. Shine, *New Scientist*, 97, no. 1345 (1983):462.

10. Radarsonics Model 231, 200 KHz with 8.5° beam width at -3db.

11. M. Klein and C. Finkelstein, op. cit., p. 9.

12. A. Shine, op. cit., p. 466; N. Witchell, *The Loch Ness Story* (Baltimore: Penguin, 1975), p. 104; T. Dinsdale, *The Story of the Loch Ness Monster* (London: W. H. Allen and Co., 1982), p. 105.

13. H. Braithwaite and D. G. Tucker, *New Scientist*, 40, no. 628 (1968):664.

14. D. G. Tucker and D. J. Creasey, *Proceedings of the Challenger Society*, 4 (1970):91.

15. D. James et al., op. cit., p. 16.

16. Our conversations during the summer of 1983 with Ivor Newby, a participant in many expeditions at Loch Ness since the sixties, revealed that monster hunters have either lost or failed to retrieve numerous bits and pieces of their underwater gear over the years. He told us that several plexiglass-enclosed cameras are still presumed floating in the midwater, moored to the lochbed, as are at least two steel gas-cylinders used for sonar calibration tests in the 1960s. This type of debris would give strong sonar returns.

17. Ivor Newby and Adrian Shine, private communication.

18. M. Klein, R. H. Rines, T. Dinsdale, and L. S. Foster, op. cit., p. 39.

19. Ibid., p. 33.

20. Ibid., p. 44.

21. Ibid., pp. 52-54.

22. Ibid., p. 49.

23. Ibid., pp. 41-42.

24. Ivor Newby and Dennis Vann are friends of Mrs. Carey and introduced us to her.

25. Mrs. Carey's sightings have been described in many books and articles about the Loch Ness monster. See W. S. Ellis, *National Geographic*, 151 (1977):763.

26. Dowsing is a method employing a divining rod to locate objects at a distance. The evidence for its success is purely anecdotal and no known controlled study for dowsing has ever produced significant results.

27. We have a copy of Mrs. Carey's map, which includes a brief description of each sighting along with the names of the witnesses present.

28. M. Klein, R. H. Rines, T. Dinsdale, and L. S. Foster, op. cit., p. 79.

29. Ibid., p. 71.

30. Ronnie Bremner and Tony Harmsworth, founders of the Loch Ness Monster Centre, Drumnadrochit, Inverness-shire Scotland.

31. R. H. Rines, H. R. Edgerton, C. W. Wyckoff, and M. Klein, op. cit., p. 31.

32. Ibid., pp. 30-31.

33. These publications include *National Geographic* [25] and *Nature* [5].

34. Ms. Arnold was a member of the LNIB for several years and told us she planned the logistics for the August 9, 1972, attempt to locate the monster by dowsing.

35. Published in *Nature* [5] and *Technology Review* [2]. We were also able to study copies of the traces on display at the Loch Ness Monster Centre.

36. In addition to evidence from the sonar traces themselves, we were told independently by Holly Arnold, Ivor Newby, Gordon Menzies, and Dick Raynor that the sonar unit was hanging off the side of the boat at the end of a rope.

37. N. Witchell, op. cit., p. 147.

38. Copies of the original photographs were given to us by Tony Harmsworth, the co-founder of the Loch Ness Monster Centre, Drumnadrochit, Inverness-shire, Scotland.

39. N. Witchell, op. cit., p. 130.

40. T. Dinsdale, *Photography Journal*, 116, no. 1 (1976):18.

41. D. Raynor, private communication.

42. D. Meredith, *Search at Loch Ness* (New York: New York Times Book Co., 1977), p. 129.

43. Ibid., p. 130.

44. See [2] and W. S. Ellis, *National Geographic*, 151 (1977):763.

Recommended Reading

Abell, George O., and Barry Singer, eds. *Science and the Paranormal.* Scribner's: New York, N.Y., 1981.

Alcock, James E. *Parapsychology: Science or Magic?* Pergamon: New York, N.Y., 1981.

Barrett, Stephen, M.D. *The Health Robbers.* George F. Stickley: Philadelphia, Penn., 1980.

Binns, Ronald. *The Loch Ness Mystery Solved.* Prometheus: Buffalo, N.Y., 1984.

Blackmore, Susan J. *Beyond the Body.* Heinemann: London, 1984.

Cazeau, Charles J., and Stuart D. Scott, Jr. *Exploring the Unknown.* Plenum: New York, N.Y., 1979.

Christopher, Milbourne. *ESP, Seers, and Psychics.* Crowell: New York, N.Y., 1970.

———. *Mediums, Mystics, and the Occult.* Crowell: New York, N.Y., 1975.

Culver, R. B., and P. A. Ianna. *The Gemini Syndrome: A Scientific Evaluation of Astrology.* Prometheus: Buffalo, N.Y., 1984.

Frazier, Kendrick, ed. *Paranormal Borderlands of Science.* Prometheus: Buffalo, N.Y., 1981.

Gardner, Martin. *Fads and Fallacies in the Name of Science.* Dover: New York, N.Y., 1957.

———. *Science: Good, Bad, and Bogus.* Prometheus: Buffalo, N.Y., 1981.

Godfrey, Laurie R. *Scientists Confront Creationism.* W. W. Norton: New York, N.Y., 1983.

Hansel, C. E. M. *ESP and Parapsychology: A Critical Evaluation.* Prometheus: Buffalo, N.Y., 1980.

Hendry, Allan. *The UFO Handbook.* Doubleday: New York, N.Y., 1979.

Kitcher, Philip. *Abusing Science.* MIT Press: Cambridge, Mass., 1983.

Klass, Philip J. *UFOs: The Public Deceived.* Prometheus: Buffalo, N.Y., 1983.

Kurtz, Paul, ed. *A Skeptic's Handbook of Parapsychology.* Prometheus: Buffalo, N.Y., 1985.

Neher, Andrew. *The Psychology of Transcendence.* Prentice-Hall: Englewood Cliffs, N.J., 1980.

Nickell, Joe. *Inquest on the Shroud of Turin.* Prometheus: Buffalo, N.Y., 1983.

Radner, Daisie, and Michael Radner. *Science and Unreason.* Wadsworth Publishing: Belmont, Ca., 1982.

Randi, James. *Flim-Flam!* Prometheus: Buffalo, N.Y., 1982.

———. *The Truth About Uri Geller.* Prometheus: Buffalo, N.Y., 1982.

Sagan, Carl. *Broca's Brain.* Random House: New York, N.Y., 1979.

Sagan, Carl, and Thornton Page, eds. *UFOs: A Scientific Debate.* W. W. Norton: New York, N.Y., 1974.

Sheaffer, Robert. *The UFO Verdict.* Prometheus: Buffalo, N.Y., 1981.

Stiebing, William H., Jr. *Ancient Astronauts, Cosmic Collisions, and Other Popular Theories About Man's Past.* Prometheus: Buffalo, N.Y., 1984.

Zusne, Leonard, and Warren H. Jones. *Anomalistic Psychology.* Lawrence Erlbaum Associates: Hillsdale, N.J., 1982

The Skeptical Inquirer (quarterly). Box 229, Buffalo, N.Y. 14215-0229

Dates of Original Publication

The articles in this anthology originally appeared in *The Skeptical Inquirer* (*SI*), a highly praised magazine devoted to the critical investigation of pseudoscience from a scientific viewpoint.

"Debunking, Neutrality, and Skepticism in Science," by Paul Kurtz. *SI,* Spring 1984, 239-246.

"The New Philosophy of Science and the 'Paranormal,'" by Stephen Toulmin. *SI,* Fall 1984, 48-55.

"Parapsychology's Past Eight Years: A Lack-of-Progress Report," by James E. Alcock. *SI,* Summer 1984, 312-321.

"Sense and Nonsense in Parapsychology," by Piet Hein Hoebens. *SI,* Winter 1983-84, 121-132.

"On Coincidences," by Ruma Falk. *SI,* Winter 1981-82, 18-31.

"Fooling Some of the People All of the Time," by Barry Singer and Victor A. Benassi. *SI,* Winter 1980-81, 17-24.

"Misperception, Folk Belief, and the Occult: A Cognitive Guide to Understanding," by John W. Connor. *SI,* Summer 1984, 344-354.

"The Great Stone Face and Other Nonmysteries," by Martin Gardner. *SI,* Fall 1985, 14-18.

"Sir Oliver Lodge and the Spiritualist," by Steven Hoffmaster. *SI,* Summer 1984, 334-343.

"Outracing the Evidence: The Muddled 'Mind Race,'" by Ray Hyman. *SI,* Winter 1984-85, 125-145.

"Remote Viewing Revisited," by David F. Marks. *SI,* Summer 1982, 18-29.

"Gerard Croiset: Investigation of the Mozart of 'Psychic Sleuths'—Part I," by Piet Hein Hoebens. *SI,* Fall 1981, 17-28.

"Gerard Croiset and Professor Tenhaeff: Discrepancies in Claims of Clairvoyance," by Piet Hein Hoebens. *SI,* Winter 1981-82, 32-40.

"The Columbus Poltergeist Case," by James Randi. *SI,* Spring 1985, 221-235.

"The Project Alpha Experiment: Part I. The First Two Years," by James Randi. *SI,* Summer 1983, 24-35.

"Lessons of a Landmark PK Hoax," by Martin Gardner. *SI*, Summer 1983, 16-19.

"Magicians in the Psi-Lab: Many Misconceptions," by Martin Gardner. *SI*, Winter 1983-84, 111-120.

"How Not to Test a Psychic: The Great SRI Die Mystery," by Martin Gardner. *SI*, Winter 1982-83, 33-41.

"Iridology: Diagnosis or Delusion?" by Russell S. Worrall. *SI*, Spring 1983, 23-35.

"Palmistry: Science or Hand-Jive?" by Michael Alan Park. *SI*, Winter 1982-83, 21-33.

"Prediction After the Fact: Lessons of the Tamara Rand Hoax," by Kendrick Frazier and James Randi. *SI*, Fall 1981, 4-7.

"Scientific Tests of Astrology Do Not Support its Claims," by Paul Kurtz and Andrew Fraknoi. *SI*, Spring 1985, 210-212.

"Alternative Explanations in Science: The Extroversion-Introversion Astrological Effect," by Ivan W. Kelly and Don H. Saklofske. *SI*, Summer 1981, 33-39.

"Book Review [The Alleged Lunar Effect]," by George O. Abell. *SI*, Spring 1979, 68-73.

"The Moon is Acquitted of Murder in Cleveland," by N. Sanduleak. *SI*, Spring 1985, 236-242.

"Intent Clear, Goal Unmet [The Claim of a Government UFO Coverup]," by Philip J. Klass. *SI*, Winter 1984-85, 171-179.

"An Eye-Opening Double Encounter," by Bruce Martin. *SI*, Fall 1984, 56-61.

"Hypnosis and UFO Abductions," by Philip J. Klass. *SI*, Spring 1981, 16-24.

"Hypnosis Gives Rise to Fantasy and Is Not a Truth Serum," by Ernest R. Hilgard. *SI*, Spring 1981, 25.

"Deciphering Ancient America," by Marshall McKusick. *SI*, Spring 1981, 44-50.

"American Disingenuous: Goodman's 'American Genesis'—A New Chapter in Cult Archaeology," by Kenneth L. Feder. *SI*, Summer 1983, 36-48.

"The Nazca Drawings Revisited: Creation of a Full-Sized Duplicate," by Joe Nickell. *SI*, Spring 1983, 36-47.

"Science and the Mountain Peak," by Isaac Asimov. *SI*, Winter 1980-81, 42-51. [Originally appeared in *Gallery* under the title "Do Scientists Believe in God?" Copyright 1979 by Montcalm Publishing.]

"Creationist Pseudoscience," by Robert Schadewald. *SI*, Fall 1983, 22-35.

"Raiders of the Lost Tracks: The Best Little Footprints in Texas [Paluxy Footprints]," by Steven Schafersman. *SI*, Spring 1983, 2-8.

"The Shroud of Turin: A Critical Appraisal," by Marvin Mueller. *SI*, Spring 1982, 15-34.

"Shroud Image Is the Work of an Artist," by Walter McCrone. *SI*, Spring 1982, 35-36.

"Sonar and Photographic Searches for the Loch Ness Monster: A Reassessment," by Rikki Razdan and Alan Kielar. *SI*, Winter 1984-85, 147-158.

Contributors

George O. Abell was a professor of astronomy at the University of California, Los Angeles, and author of several leading astronomy textbooks as well as co-editor of *Science and the Paranormal.*

James E. Alcock is an associate professor of psychology, Glendon College, York University, Toronto, and author of *Parapsychology: Science or Magic?*

Isaac Asimov is a professor of biochemistry at Boston University School of Medicine and author, at recent count, of 332 books in many fields including science, science fiction, mystery, history, and literature.

Victor A. Benassi is an associate professor of psychology at the University of New Hampshire. His research interests include perceived control and paranormal belief.

John W. Connor is a professor of anthropology at California State University, Sacramento. He has published on subjects ranging from schizophrenia to witchcraft, and he teaches courses on religion, magic, and the rise of religious cults.

Ruma Falk is with the Department of Psychology and the School of Social Work at the Hebrew University of Jerusalem. She is working on probabilistic and statistical reasoning in children and adults.

Kenneth L. Feder is an associate professor of anthropology at Central Connecticut State University. He is the founder and director of the Farmington River Archaeological Project.

Andrew Fraknoi is an astronomer, executive director of the Astronomical Society of the Pacific, and editor of its monthly magazine, *Mercury.*

Kendrick Frazier is a science writer and is editor of *The Skeptical Inquirer.*

His recent books include *Solar System* and *People of Chaco*. A former editor of *Science News*, he has been on the staff of Sandia National Laboratories since 1983.

Martin Gardner is the author of numerous books in the fields of mathematics, science, philosophy, and literary criticism.

Ernest R. Hilgard is emeritus professor of psychology at Stanford University. He is a past president of the International Society of Hypnosis.

Piet Hein Hoebens was chief editorial writer for *De Telegraaf*, Amsterdam's largest daily newspaper, and a well-known investigator and critic in the areas of parapsychology and the scientific examination of paranormal claims.

Steven Hoffmaster is associate professor of physics at Gonzaga University, Spokane, Washington. He has taught courses in scientific thought and the scientific investigation of the paranormal.

Ray Hyman is a professor of psychology at the University of Oregon.

Ivan W. Kelly is an associate professor of educational psychology at the University of Saskatchewan, Canada.

Alan Kielar is chairman of Iscan, Inc., Cambridge, Mass., a company that develops systems to track targets electronically. He is an electrical engineer and a specialist in sonar and image processing.

Philip J. Klass is a technical journalist with a leading aerospace magazine in Washington and is author of *UFOs: The Public Deceived*.

Paul Kurtz is a professor of philosophy at the State University of New York at Buffalo and founding chairman of the Committee for the Scientific Investigation of Claims of the Paranormal.

David Marks is a senior lecturer in psychology at the University of Otago, New Zealand. He is co-author of *The Psychology of the Psychic*.

Walter McCrone is director of the McCrone Research Institute, Chicago.

Marshall McKusick is an associate professor of anthropology at the University of Iowa. He is author of *The Davenport Conspiracy* and other reports exposing frauds, hoaxes, and misidentifications in American archaeology.

R. Bruce Martin is a professor of chemistry at the University of Virginia, Charlottesville.

Marvin M. Mueller is a research physicist at Los Alamos National Laboratory. He has done research in diverse areas of physics, but has concentrated on laser fusion during the last 14 years. He also has experience in image analysis and has a strong background in chemistry.

Joe Nickell is an investigator of paranormal claims and author of *Inquest on the Shroud of Turin*. He is a former professional stage magician and private detective.

Michael Alan Park is an associate professor of anthropology at Central Connecticut State University.

James Randi is a professional magician and lecturer who has devoted more than 30 years to exposing "psychic" flummery. He is author of *Flim-Flam!* and *The Truth About Uri Geller*.

Rikki Razdan is president of Iscan, Inc., Cambridge, Mass., a company that develops systems to track targets electronically. He is an electrical engineer and a specialist in sonar and image processing.

Don H. Saklofske is an associate professor of educational psychology at the University of Saskatchewan, Canada.

Nick Sanduleak is a senior research associate in the astronomy department of Case Western Reserve University.

Robert J. Schadewald is a science writer who has published dozens of articles on creation science, flat-earth cosmology, Velikovskysm, perpetual motion, and other pseudoscientific disciplines.

Steven Schafersman is a consulting petroleum geologist and college instructor in Houston, Texas. He is president of the Texas Council for Science Education.

Barry Singer is in the Department of Computer Science at the University of Oregon. His interests include human judgment and belief systems. He co-edited *Science and the Paranormal*.

Stephen Toulmin is a professor in the Committee on Social Thought at the University of Chicago. Among his many books are *Foresight and Understanding, Human Understanding, The Fabric of the Heavens*, and

The Return to Cosmology.

Russell S. Worrall is an assistant clinical professor, School of Optometry, University of California, Berkeley, and a member of the National Council Against Health Fraud.